Bridge Deck Erection Equipment

ice
Institution of Civil Engineers
publishing

Bridge Deck Erection Equipment

A best practice guide

Members of IABSE Working Group 6
(Edited by Steve Harridge and Max Meyer)

Published by ICE Publishing, One Great George Street, Westminster, London SW1P 3AA.

Full details of ICE Publishing representatives and distributors can be found at: www.icebookshop.com/bookshop_contact.asp

Other titles by ICE Publishing:

Movable Bridge Design
C. Birnstiel *et al.* ISBN 978-0-7277-5804-0
Bridge Monitoring [CSIC Series]
C. Middleton *et al.* ISBN 978-0-7277-6059-3
Steel-concrete Composite Bridges, Second edition
D. Collings. ISBN 978-0-7277-5810-1

www.icebookshop.com

A catalogue record for this book is available from the British Library

ISBN 978-0-7277-6193-4

ICE Publishing is a division of Thomas Telford Ltd, a wholly-owned subsidiary of the Institution of Civil Engineers (ICE).

Commissioning Editor: James Hobbs
Production Editor: Madhubanti Bhattacharyya
Market Development April-Asta Brodie

MIX
Paper | Supporting responsible forestry
FSC® C013604
www.fsc.org

Typeset by Academic + Technical, Bristol
Index created by Pierke Bosschieter
Printed and bound by CPI Group (UK) Ltd, Croydon, CR0 4YY

Contents

Foreword

Impressive machines and the exciting construction methods they allow for building daring and innovative bridges can be found in this book which offers very helpful and most informative guidelines to everyone involved in design and construction of bridges. *Bridge Deck Erection Equipment* illustrates the large variety of erection equipment that is available today to build bridges better, faster and more precisely and provides valuable insight into their planning and use.

Bridges are becoming more and more complex for many reasons. We want to see bridges that add to our quality of life and which contribute to 'Baukultur' (the culture of the built environment). Environmental issues fortunately have entered the public conscience and high performance materials as well as modern analysis methods allow bridges that were not possible before. All these issues, and sometimes, just sheer size, make it necessary to move away from the classic ways of using scaffolding and cranes to build bridges.

Especially for larger bridges, no successful design is possible without bearing the construction process in mind from the start. The possible options to build a bridge form part of the context which determines a given design task, and good design stems from considering all such boundary conditions. Ignoring the reality of available erection methods will lead to unrealistic designs which will later be altered beyond recognition, or get abandoned altogether for economic reasons.

In this sense, this volume closes a gap which some of us did not even know existed. This book makes us aware and grants us access to expert know-how not available to many of us so far. A rich diversity of images brings us closer to the complex contents of this book and the many innovations in bridge building techniques which are described here. Building bridges is always a team effort and here a team of renowned international experts have gathered to generously share their experience with us. Bridge-building has also become global,with workers and contractors coming from all over the world, and so have the methods and machines for erecting decks. Also, it is important to see how bridges get built in different locations such as Europe, China, India and the USA, because these examples can be translated to our local individual contexts.

International associations such as the International Association for Bridge and Structural Engineering (IABSE) are a good platform for international collaboration amongst colleagues and we can be grateful that in this case the outcome has become this book, published by ICE Publishing which has the potential to become the standard work for our community.

Mike Schlaich

About the authors

Jan de Boer

Jan de Boer was educated at the Hogeschool Utrecht. He joined the family business, a contracting firm specialising in bridge construction and civil works for the Dutch polders. He transformed the company into a European group specialising in bridge-building techniques based on jacking and skidding technologies. He has been involved in more than 2500 projects in this field in many European countries.

He is an active member in different working groups of civil engineering societies, such as IABSE, the Dutch Concrete Society and other bridge associations. One of his major contributions is the initiative to create broadly accepted guidelines for this field of activities.

Michael Däbritz

Michael Däbritz received his degree in civil engineering and welding engineering in 2005 from the Technical University of Dresden, Germany. He started his professional career at PeinigerRöRo, the former name of thyssenkrupp Infrastructure, with site management of heavy scaffolding projects for tunnels and bridges in Europe, mainly focusing on movable scaffolding systems.

In 2008 and 2009, he worked as project manager for form travellers and movable scaffolding projects for Doka Austria. He returned to thyssenkrupp in 2010, where he took over the management of the movable scaffolding department in 2013. He is now engaged with many international heavy-scaffolding projects, not exclusively in bridge construction. One of his major fields of activity is the deconstruction of viaducts within existing infrastructure.

Steve Harridge

After six years working for an international steel contractor, Steve Harridge joined Tony Gee and Partners in the UK, working on design-build bridge projects and developing erection schemes for many types of complex structures in over 40 countries. These were mostly on behalf of contractor clients.

He opened TG's Hong Kong office in 1982 and later took responsibility for the firm's international work. He became a partner of the firm in 1984 and an executive director in 2006, retiring from full-time work in 2013. He now does occasional Expert Witness and other advisory tasks. He is a fellow of the Institution of Civil Engineers and the Institution of Structural Engineers in the UK, and was elected a fellow of the Royal Academy of Engineering in 2015.

Elie Homsi

Elie Homsi holds degrees from the University of Texas at Austin and from the Lebanese American University. He is a senior vice president and fellow at Parsons, with 35 years of experience in complex construction technologies, supervising the construction engineering efforts on many large-scale, high-tech transport projects in the Americas, the Middle East, Asia and Africa. His main contributions in construction engineering include the development of means and methods, temporary works, material and equipment selection, constructability reviews, planning, phasing and scheduling.

Elie is an active member of many other committees dealing with bridge and transport technologies, such as the American Segmental Bridge Institute (ASBI) and the Post Tensioning Institute (PTI). He is also a former Board of Directors Chair of, and a volunteer with, Bridges to Prosperity. Elie has been elected to the US National Academy of Construction.

Mohan Jatkar

Mohan Jatkar gained his bachelor's and master's degrees in structural engineering from the College of Engineering, Pune, India. He joined the design section of Gammon India in 1980 and over the years worked on various assignments for the design of bridges, power stations, temporary works and so on. Since 2011, as executive director (technical), he was responsible for design and construction planning functions. He is now technical advisor at Gammon Engineers and Contractors.

His main areas of interest are front-end engineering, resolution of interface technical issues and training. He is a member of various professional bodies and member-secretary for the Indian Roads Congress Committee on Temporary Structures.

Tristan Lecomte

Tristan Lecomte received his master's degree in civil engineering from the National Institute for Applied Sciences (INSA) in Lyon. He then gained a specialised degree in steel structures at the Centre des Hautes Etudes de la Construction, in Paris. He first joined SYSTRA in 2003 as a structural engineer and worked on the design of railway bridges. In 2009, Tristan joined the technical head office of VINCI Construction Grands Projets, becoming involved in the design, specification and fabrication of permanent steel structures. In regards to bridge erection, he supervises the design, outsourcing and commissioning of bridge deck erection equipment and large temporary works.

Max Meyer

Max Meyer has a master's degree in civil engineering and an honorary doctorate of science from ETH in Zurich, Switzerland. He has been working in Asia for more than 30 years and is presently VSL International's Group technical officer. His expertise is in the development of alternative designs, construction concepts and in the detailing of prestressed and stay-suspended structures. He has been involved in the conceptual and detail design of bridge deck erection equipment and temporary works and in the construction engineering for many state-of-the-art bridge projects all over the world.

Alessandro Rovera

Alessandro Rovera gained his master's degree in civil engineer at the Politecnico di Milano. After two years working as an independent professional engineer, he has been working for more than 30 years with Gruppo de Eccher, Italy. As project manager, he worked on various bridge projects in Italy and internationally, developing innovative construction methods, including design and fabrication of the relevant bridge deck construction equipment.

He was part of the team that opened DEAL, a Rizzani de Eccher subsidiary, in 1994, becoming chief executive officer in 2004. He has been involved in many high-tech projects around the world, developing and implementing state-of-the-art equipment for many kinds of bridge deck construction methods and helping the company to advance in the field of permanent and temporary structures design, as well as in the fields of post-tensioning, bearings, expansion joint and anti-seismic devices.

Giulio Maria Scotto

Giulio Maria Scotto gained his bachelor's degree in structural engineering at the University of Firenze, Italy, with his master's thesis prepared at the Ecole Nationale des Ponts et Chaussés, France. After work experience in Freyssinet, Italy, and Italferr, he spent more than 10 years in the technical department of Condotte, being involved in design and construction of both high-speed railway lines and highways, with most of his time spent on site assistance.

Since 2012 he has worked for Freyssinet. He was appointed director of the construction methods department in 2014. The main activities of the department are launched bridges, steel structures and erection methods.

He became a member of the Italian Order of Engineers in 1999.

Philippe Vion

Philippe Vion is a graduate of the Conservatoire National des Arts et Métiers, Paris, where he received a bachelor's degree in civil engineering.

He worked for 33 years for SETRA, the technical arm of the French Road Directorate, and 8 years for SYSTRA, an engineering company, as chief engineer in the civil and structural engineering department. Since January 2017, he has worked for VINCI Construction Grands Projets as chief structural engineer in the design and engineering department.

Philippe's main contribution is in the design and project management of large road bridges, but he also designs various rail bridges for metro, mass transit and high-speed rail projects.

He is a member of associations dealing with bridge and transport technologies, such as the Fédération Internationale du Béton (\,fib), the Association Française de Génie Civil (AFGC) and the International Association of Bridge and Structural Engineers (IABSE).

Introduction

Over the last 30 years, the methods for building medium- to long-span bridges have changed dramatically. Pressure to shorten construction time has demanded an increase in the productivity of bridge deck construction, which has led to the use of more sophisticated methods and equipment. The machines that have been developed are commonly called Bridge Deck Erection Equipment (BDE).

New methods of deck construction have allowed bridges to be built in urban areas with much less disruption to the public than would have been possible a generation ago. With the burgeoning demand for high-level urban roads and light rail systems, this progression has not come a moment too soon. Medium- to long-span bridges are now built using equipment that was rarely used 30 years ago, thus requiring permanent works designers to adapt their designs to cater for the techniques now available to them. The benefit to the public is a saving in materials, better value for money and a reduction in environmental disruption while new projects are built.

The purpose of this book is to provide a Best Practice Guide for the bridge construction industry, ranging across the whole spectrum of stakeholders, from the prime client, the main contractor, the permanent works designer, the deck construction contractor, the temporary works designer and the supplier of the equipment, through to the operators who work on the site. It is not a 'recipe book' for the design of BDE. The emphasis here is the need for an integrated approach to bridge design and construction that respects the needs of all the stakeholders and manages the engineering interfaces between temporary and permanent works.

The book is written in four sections. Section 1 gives a classification of different bridge types and covers project management and safety aspects of bridge construction. Section 2 discusses in detail the various types of BDE. In Section 3 there is important information about conceptual bridge design and detailing which should help bridge engineers to make their designs easier to build. It also covers BDE design, the mechanical and electrical components of BDE, precasting and lifting attachments. Section 4 gives recommendations for the industry that will hopefully contribute to an improvement in bridge design and construction.

It covers all aspects of the methods associated with building many types of bridge, principally concrete ones although many of the topics discussed are equally applicable to the construction of steel or steel/concrete composite bridges.

There has been a notable trend among main contractors and even some of the established specialist deck construction contractors to reduce their in-house engineering teams. With this dispersal of expertise, there is a risk that the next generation of bridge builders will find it harder to acquire the level of understanding of bridge construction techniques that currently exists. This book should help to maintain the current levels of knowledge within the industry. It should also be regarded as important reading for engineering students as it is essentially a book of practical engineering guidance.

The intention of the book is to fill the gap in current published information with advice for all the stakeholders listed above. It does not purport to be a manual on permanent works bridge design. There are already many of those available, although some detailing aspects are discussed that can improve the 'buildability' of bridges.

The authors are all industry leaders in their field of expertise. Each has written one or more of the chapters but the core aim of the book is to produce a single source of information, avoiding duplication – a distillation of the knowledge of us all.

Section I

Definitions, Management and Safety

Bridge Deck Erection Equipment
ISBN 978-0-7277-6193-4

https://doi.org/10.1680/bdee.61934.003

Chapter 1
Glossary of terms and abbreviations

Balanced Cantilever Method (BCM): a method of building bridge decks starting at a pier table and proceeding symmetrically on each side

BOT: a build-operate-transfer contract

Bridge Deck Erection Equipment (BDE): Equipment specifically designed and manufactured to build bridge decks

cantilever method: a method of building bridge decks in cantilever

closure pour: the cast-in-situ portion of a concrete bridge deck between adjacent cantilevers [see also midspan stitch]

dry joint: a match-cast joint without epoxy [see also match-cast joint]

element: part of the bridge deck [see also girder and segment]

Form Traveller (FT): a self-launching BDE used to cast in situ segments by cantilever method

Full Span Method (FSM): a method of building bridge decks where the entire span is prefabricated and erected as one unit

gauge: distance between the pair of rails of a rail track

girder: longitudinal element of a bridge deck

glued joints: a match-cast joint with epoxy

HT bars: bars made from high tensile steel

Incremental Launching Method (ILM): a method of building bridge decks where the deck is progressively built segment by segment behind one of the bridge deck's abutments and is pushed or pulled into its final position

jack: device used for moving a load by tension or compression or to apply tension or compression forces

kinematics: a series of construction steps that defines the way in which a BDE is operated

Launching Gantry (LG): a self-launching BDE bridging one span or more and used to erect bridge deck elements

lifting beam: a component used to connect the lifting point device to the element to be lifted [see also spreader beam]

Lifting Frame (LF): a self-launching BDE used for the erection of segments by cantilever method

Main Girder (MG): the main longitudinal elements of a BDE

Main Truss (MT): the main longitudinal elements of a BDE with truss section

mast: the vertical element used to transfer the load of the temporary stays to the substructure used in the PCM

match-cast joint: a joint between segments which are cast against each other

match-casting: a precasting technique whereby adjacent segments are cast against each other

midspan stitch: the cast-in-situ portion of a concrete bridge deck between adjacent cantilevers [see also closure pour]

Movable Scaffolding System (MSS): a self-launching BDE used to cast in situ one complete span of a bridge deck

nailing: the temporary connection of a bridge deck to the structure below, usually a pier or a cross head, to provide a rigid restraint to the deck during construction; this is usually either by stressed HT bars or PT strands

overhead gantry/MSS/traveller: a BDE whose main structural elements are placed above the deck

pier crosshead: the transverse element located on top of the pier that transfers deck reactions to the pier

pier head: the top part of the pier

pier segment: the segment located on top of the pier

pier table: the section of the bridge deck located over the pier and that can consist of more than one segment

Post-Tensioning (PT): the action of stressing a steel tensile element (bars or strands) used to prestress a concrete member after concrete is poured and cured

pre-camber: a geometry imposed to the deck to compensate for a defined set of deflections

Prefinished Bridge Installation method (PBI): a method of building bridges where the entire deck or bridge structure (which may include piers and foundations) is built near its final location and moved, as a rigid body, into its final position

prestressing: the action of precompressing a member

pre-tensioning: the action of stressing a steel tensile element (bars, strands or wires) used to prestress a concrete member before concrete is poured

Professional Engineer (PE): an individual who complies with requirements regarding education and experience defined by a country's legislation, or by a nationally-recognised engineering institution, and is therefore permitted to offer engineering services directly to the public in this country

Progressive Cantilever Method (PCM): a method of building bridge decks whereby precast segments are installed in cantilever from one pier to the following one, using temporary mast and stay cables

PT bars: a stressed bar

pylon: the vertical element used to transfer the load of the stays to the deck or the pier

segment: part of the bridge deck usually a full cross section and of variable length

segment loader: a lifting device used to load segments on an underslung gantry

skewing forces: horizontal forces caused by skewing of the equipment in relation to its movement along the runway

skid system: a system that consists of tracks and moving devices to move elements horizontally

Span-by-Span method (SBS): a deck construction method that addresses the construction of one full span at a time

Self-Propelled Modular Transporter (SPMT): used for moving elements

spreader beam: a component used to connect the lifting device to the element to be lifted [see also lifting beam]

Stressing Platform (SPF): a platform used to access anchorages of PT tendons for installation and stressing operations

stays: stressed tension elements that support the main girder of a BDE or the bridge deck from a pylon

straddle carrier: a rubber-tyred transporter that consists of a portal frame and a lifting device

stitch: a cast-in-situ portion of the bridge deck; it can be reinforced or not reinforced

strand: twisted cable, made up of 2 to 19 high-strength steel wires

Strand Moving Unit (SMU): a hydraulic jack which can move a load in steps and uses one or more strands as tension elements between the load and the jack

swivel crane: a type of crane consisting of an L-shaped base, a tower and a horizontal rotating jib that supports a movable hoist

underslung gantry/MSS/traveller: a BDE whose main structural elements are placed below the deck

wet joint: an unreinforced concrete stitch

Bridge Deck Erection Equipment
ISBN 978-0-7277-6193-4

https://doi.org/10.1680/bdee.61934.007

Chapter 2
Classification of bridge decks

2.1. Introduction

This book deals with bridge decks that are built using Bridge Deck Erection Equipment (BDE). It does not cover bridge decks erected by cranes, derricks or on stationary falsework.

Bridges are classified according to the way their decks are built and the type of material used for their decks. These two aspects have a big impact on the design and detailing of the bridges and the interface between permanent work and BDE.

This chapter gives an overview of various types of bridges and of the specialist bridge deck erection equipment that can be used for the construction of their decks.

2.2. Deck segmentation

Bridge decks usually need to be broken down for construction into elements to

- enable the use of highly mechanised erection equipment, which eliminates the need for labour-intensive falsework systems and so reduces construction time
- produce as many bridge elements as possible off site in factories, where productivity is higher and it is easier to achieve quality
- minimise disturbance to people living near or commuting through the project sites.

The objective to simplify construction should, however, not divert the designer's attention from the bigger picture, which requires engineers to build structures with improved sustainability and a lower carbon footprint. The designer should therefore always try to make decks continuous over as many spans and with as many monolithic connections between deck and piers as is technically feasible and compatible with the local seasonal temperature range and foundation restraints. These types of decks allow more slender structures and cut down on maintenance-intensive bearings and expansion joints, therefore reducing the carbon footprint of the bridge. Such structural systems can indeed be designed and detailed to be compatible with bridge decks sliced into elements, even when built with precast elements. For railway bridges, the length of deck continuity is, however, also controlled by the objective to avoid installation of expansion joints as much as possible. If bearings are required, the designer needs to detail the structure so that there is enough space for inspection and replacement.

Last but not least, bridges are generally huge structures with a big visual impact on the surrounding area and a lifetime of 100 years or more. Hence it is the engineer's duty to design bridges that harmonise with the environment and are aesthetically pleasing.

Decks can be broken down into elements longitudinally or transversely.

Figure 2.1 A concrete bridge deck broken down longitudinally into elements (girders) (Meyer 2009)

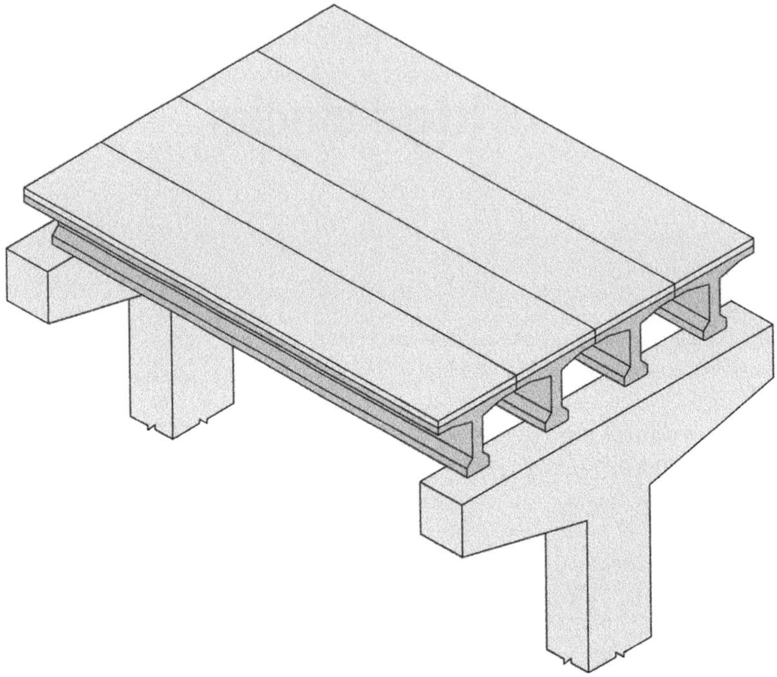

Bridge decks cut longitudinally (Figure 2.1) are made up of two or more elements, called girders, which bridge the entire span simply supported, at least initially, for the deck's self-weight. Such decks may or may not be made continuous through subsequent in-situ stitches. The girders sit on cross heads, which are as wide as the deck, and support a deck slab, which is generally made from concrete, cast in situ. It is possible to precast the part of the deck slab above the girder together with the girder and achieve transverse continuity in the deck by longitudinal in-situ stitches between the girders and transverse Post-Tensioning (PT).

Bridge decks broken down transversely into elements (Figure 2.2) are called segmental bridges.

Segmental bridge decks made from concrete have either cast-in-situ or precast segments. Depending on the erection method selected, decks are broken down into segments that vary in length from a few metres to as long as the span. Unlike decks made from precast girders, the decks of precast segmental bridges can be directly supported on the piers. This removes the need for cross heads and the entire deck can be precast fully off the erection site.

2.3. Construction of the bridge decks

2.3.1 Bridge decks cut longitudinally

The elements of longitudinally cut bridge decks – the girders – are almost always placed Span by Span (SBS) with a pair of ground-based cranes. Use of BDE for placing of the girders is therefore viable only under special circumstances. Additional construction steps are needed to complete the bridge deck's cross section after the placing of the girders.

Figure 2.2 A concrete bridge deck broken down transversely into elements (segments) (Meyer 2009)

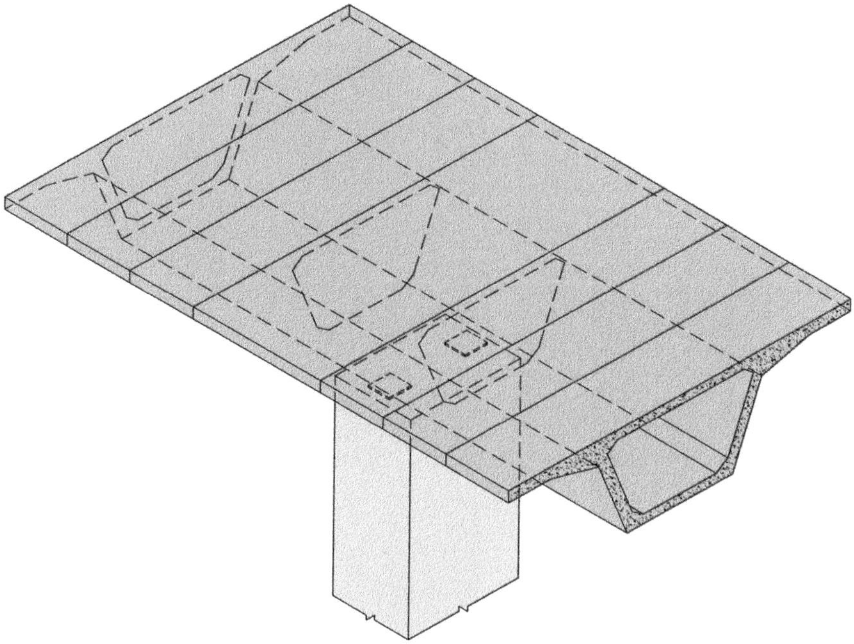

2.3.2 Bridge decks cut transversely

Bridge decks cut transversely into elements – the segments – are built by any of the following methods.

- Balanced Cantilever Method (BCM): The deck is progressively built in cantilever mode, segment by segment, starting from a pier table and extended until the centre of the spans on either side of the pier table has been reached (or in the case of an end span, until the out-of-balance part of the deck in the end span has been reached). In general, the deck is cantilevered symmetrically about the pier to minimise out-of-balance moment at the pier. Upon completion of two adjacent symmetrical cantilevers, tips of facing cantilevers are connected to achieve deck continuity. Decks built by cantilever method may or may not be suspended from stay cables.
- Progressive Cantilever Method (PCM): A special case of the cantilever method is the progressive cantilever method, whereby, with the help of a temporary mast and temporary stays, an entire span is erected by cantilever method using the previously erected spans as back spans.
- Span-by-Span method (SBS): The decks are either built simply supported or as temporary end spans of a continuous deck.
- Full Span Method (FSM): The entire span is precast in a single unit. This method is a subclass of the SBS method.
- Incremental Launching Method (ILM): The deck is progressively built segment by segment behind one of the bridge deck's abutments and is pushed or pulled into its final position. ILM is for this classification considered to be a subclass of SBS.

2.3.3 Prefinished Bridge Installation method

Another method to build bridges is to build the entire deck, the bridge superstructure including the piers or even the entire bridge with pad footings near to its final location and to move it as a rigid body

Table 2.1 Deck construction methods and means

Deck construction method		Material	Deck construction means (BDE, crane/derrick, scaffolding/temporary towers)		
Type	Deck elements	Cast in situ	Precast	Steel	Composite
BCM	Multiple segments per span	Traveller (II/11)	Lifting frame (II/9) or LG (II/5) or Crane/Derrick	–	–
	Multiple segments per span with permanent stays	Traveller (II/11)	Lifting frame (II/9) or LG (II/5) or Crane/Derrick	Lifting frame (II/9) or Crane/Derrick	Lifting frame (II/9) or Crane/Derrick
PCM	Multiple segments per span with temporary stays	–	Temporary mast, stays and swivel crane (II/10)	–	–
SBS	Multiple girders per span	–	Cranes or LG (II/7)	Cranes	Cranes
	Multiple segments per span	On scaffolding	LG (II/6) or Scaffolding + Crane	Temporary towers/ Crane {or LG (II/6)}	Temporary towers/ Crane {or LG (II/6)}
	Full span (FSM)	MSS (II/12) or On scaffolding	LG (II/8) or Cranes	Cranes {or LG (II/8)}	Cranes {or LG (II/8)}
ILM	Length of segments = N, 1/3 or $\frac{1}{2}$ of span	–	ILM (II/13)	ILM (II/13)	ILM (II/13)
PBI	Full deck, superstructure or bridge	–	(II/14)	(II/14)	(II/14)

into its final position. In this book, this method is referred to as the Prefinished Bridge Installation method (PBI).

2.4. Material

Bridge decks can be made of cast-in-situ concrete, precast concrete, steel or a combination of steel and concrete (called composite construction). Use of BDE is very common for the construction of segmental concrete bridges due to the weight of their decks and the difficulties involved in splicing segments made from concrete. For steel and composite bridges, deck erection by cranes is the natural choice, except when building bridge decks by the Incremental Launching Method (ILM) and for stay-suspended decks.

2.5. Classification of bridge decks and overview of their construction means

Table 2.1 indicates which type of BDE can be used, according to the deck construction method and deck material. The deck construction is either by the Balanced Cantilever Method (BCM), the Progressive Cantilever Method (PCM), the Span-by-Span method (SBS), the Incremental Launching Method (ILM) or the Prefinished Bridge Installation method (PBI). The deck material is cast-in-situ concrete, precast concrete, steel or composite (steel/concrete).

Subclasses for the BCM are decks without and with stays. Subclasses for the SBS method are decks with multiple girders per span, multiple segments per span and with one segment per span (Full Span Method/FSM).

The type of bridge deck using longitudinal elements is shown in yellow, whereas the type of bridge decks using transverse elements is shown in green.

The abbreviation LG stands for Launching Gantry and MSS for Movable Scaffolding System.

BDE is shown in red and the relevant chapters where it is discussed in this book are given in brackets. If shown in curly brackets, BDE can be used, but its use is not covered for this application in this book.

REFERENCES

Figure 2.1 and 2.2: Meyer M (2009) 'State of the Art Bridge Construction – The Challenge to choose the right method', IABSE Symposium BKK

Table 2.1: Adapted from VSL International

Bridge Deck Erection Equipment
ISBN 978-0-7277-6193-4

https://doi.org/10.1680/bdee.61934.013

Chapter 3
Project management

3.1. Introduction

There are no 'rules of thumb' for the selection of BDE. Every project is unique and will require a thorough analysis of all relevant factors to make sure that each project is matched with the safest, most efficient BDE. A high level of diligence should be maintained throughout the project phases. Properly experienced and qualified team members are essential to the efficient deployment of BDE on the project. The authors recommend the creation of an accreditation system to test and verify the competence of various key personnel engaged on the design and operation of a BDE.

BDE has the following special characteristics.

- It is usually a 'tailor-made' erection device specifically designed and fabricated for a particular project.
- It is often a prototype, which cannot be completely tested out before being dispatched to a project.
- BDE can be reused unchanged or modified and reconfigured for a new project. Nevertheless, it should always be treated with the same level of care and diligence on every project.
- The design basis for such equipment always has project-specific aspects.
- BDE will often impose additional temporary loads on the structure being built that will require additional analysis of the interface between BDE and the bridge structure.
- BDE is not a crawler, mobile, tower or derrick crane. It is not an 'off-the-shelf' piece of equipment.

The procurement, commissioning and operation of BDE are not straightforward tasks. In addition to the typical process of constructing a bridge using the conventional methods, the use of BDE adds another layer of complexity. It is a unique and complex method that requires expertise that is not necessarily available in a typical construction team. There is a distinct difference between opening a catalogue, studying the charts and ordering a crawler or tower crane and developing a site-specific/prototype piece of equipment that is 'tailor-made' for the project. The team ordering BDE has to

- consider all the project constraints
- study all the site-specific requirements and conditions
- analyse the interface with the structure
- understand all the applicable solutions
- review all the available BDE solutions
- investigate possible new solutions
- select the most efficient and economical BDE method
- develop the specifications

- issue the BDE tender documents
- analyse all the quotes received
- understand the subtleties between the different proposals
- equalise the offers
- select the best overall value for the project (often it is not the lowest apparent quote)
- issue a proper purchase order
- monitor the fabrication, testing and quality assurance/quality control (QA/QC)
- develop an assembly plan
- take delivery of the equipment
- assemble the BDE
- commission, test and certify the BDE
- train the crews
- operate the BDE.

Many of these tasks are very specialised and cannot be accomplished successfully without the required expertise and collaboration between all the stakeholders, including the BDE supplier in particular.

The use of BDE will often require modifications to the bridge structure. These modifications could result in an increase or decrease in the structure's permanent material quantities, depending on the selected erection method and corresponding BDE.

Depending on the delivery method and contractual model used for a project (refer to Table 3.1), the fine-tuning of the permanent work design and detailing will vary in difficulty. Contract delivery models where the contractor and equipment supplier are already on board when the permanent work is being designed allow more flexibility in modifying and optimising the structure to permit an efficient use of the appropriate BDE. Optimum solutions with regard to permanent work design and methods can only be achieved with the early involvement of all stakeholders. Traditional design–bid–build type projects, whereby the detail design is carried out by the owner prior to the tender, are harder to optimise.

The typical project stakeholders forming the team that will be involved with BDE include the

- project owner
- regulatory agencies, including environmental and permitting authorities
- neighbouring property owners or operators
- main contractor
- bridge design engineer
- specialised subcontractor
- construction engineer who will be responsible for the design of the temporary works and definition of means and methods
- BDE supplier
- BDE designer
- independent third-party checker/reviewer for BDE and permanent work design
- BDE fabricator
- BDE certifier at commissioning.

Depending on the project requirements and the project team, many of these roles can be filled by the same entity. For various options, refer to Table 3.2 later in this chapter.

One of the most important duties of management is to ensure the safe operation of BDE. Accidents in general, and BDE-related accidents in particular, can put at risk the workers as well as the public and cause serious harm to life and damage to property. In addition, these accidents will often result in costly delays to the projects.

A detailed discussion of the safe operation of BDE is addressed in Section I, Chapter 4.

Table 3.1 Contract delivery models

	Owner	Engineer	Contractor	Operator	Evaluation
Design–Bid–Build	A	B	C	A	Open bid Lowest bid
Design and build	A	C with B		A	Open bid Two envelopes
Build–Operate–Transfer (BOT)	A	Concessionaire			Restricted bid Negotiated
Alliance	Global alliance				Restricted bid Negotiated (TCE)

A, B and C are independent contractual parties
TCE stands for Target Cost Estimate

3.2. Project delivery

A main contractor tendering for a project requiring the use of BDE needs either to have in-house expertise in the procurement and operation of such equipment or should team up early on, at the design/tender stage, with a partner, subcontractor or supplier with expertise in BDE and complex bridge erection means and methods. This expertise, whether in-house or acquired, is essential to the safe and successful deployment of BDE on the project.

A BDE project requires a specially trained and qualified individual to lead the erection team. That qualified individual needs to have a complete understanding of BDE characteristics, operations, limitations, the interface with the structure and the project means and methods, as well as being able to communicate with, train and lead the construction team.

The main steps in managing BDE on a project are shown in Figure 3.1.

3.3. Preparation

3.3.1 Selection and organisation

The proper selection of BDE is one of the key decisions in planning a project or estimating a tender. The decision is not simple and is driven by a large number of variables which are always unique for any given project. The main project-specific criteria and constraints that influence the selection of erection methods and means are

- safety
- project schedule and construction duration
- superstructure design

Figure 3.1 Flowchart for project delivery

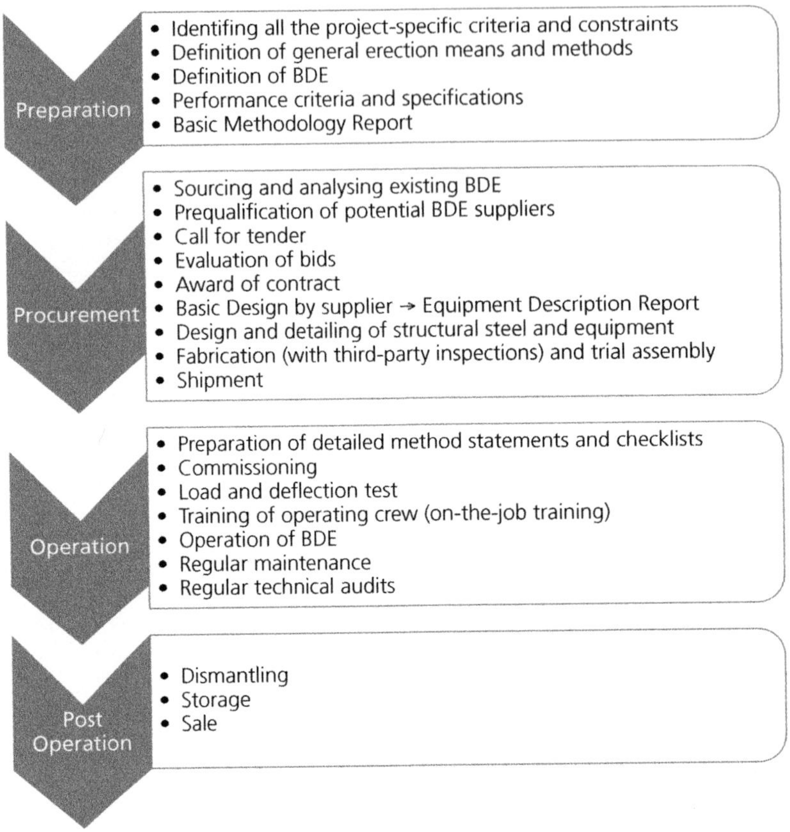

- access and site conditions
- project team
- equipment availability
- cost.

3.3.2 Identification of project-specific criteria and constraints

3.3.2.1 Safety

Safety of the public and workforce is obviously the main concern for all stakeholders. A detailed discussion of BDE-related safety issues can be found in Section I, Chapter 4.

3.3.2.2 Project schedule and construction duration

The project completion date is typically specified in the contract documents and is often the key driver for the choice of the construction method. It also drives the decision regarding the number of sets of BDE that will be required. A generous schedule, for example, may allow one set of form travellers to complete one balanced cantilever and then be reused to construct several other cantilevers. A tight schedule on the same project might dictate the need for several construction fronts working simultaneously on multiple cantilevers. Each front could require its own set of support logistics, such as tower cranes, concrete pumps, supervision and labour. Project schedules are normally fixed and the contractor must adopt the appropriate means and methods to deliver the project on time and per the contractual obligations. The target cycle time for a BDE is an important factor but it needs to be balanced with other factors, such as access, costs and resource availability that also influence the selection of an appropriate BDE.

3.3.2.3 Superstructure design

The geometric characteristics of the superstructure – including span length, pier height, horizontal curvature, maximum gradient and cross slope – and the degree of consistency and repeatability of these characteristics throughout the project have a major impact on the optimisation of the erection methods. It should be noted that for a conventional contract delivery model, the owner's design engineer has the responsibility to develop a workable erection method that is in compliance with the tender design.

It is important to verify that the superstructure is capable of carrying the additional construction loads introduced by BDE. Ground or barge-mounted cranes introduce minimal construction load to the structure. BDE may introduce loads onto the bridge, the impact of which on the permanent work needs to be verified. A good BDE concept will in general avoid the need to strengthen the permanent work.

3.3.2.4 Access and site conditions

Access for delivery of plant, material and precast elements is often the most significant factor in determining the erection method. Other considerations are

- crane access to all areas of the works
- clearance restrictions from power lines, buildings, rail catenaries, or close proximity to an airport or other facilities
- proximity to radio towers and mobile phone towers (risk of possible stray current and arcing).

When working over land, the factors that should be considered range from environmental issues such as wetlands and protected habitats to interference with local traffic, businesses and railway crossings.

When working over water, the factors that affect the selection process include

- water conditions (depth, flow, ice, etc.)
- required permits
- the feasibility and cost of constructing access trestles
- tidal variation, wave action and barge bounce (to be taken into consideration when connecting the lifting beams to the segments)
- saltwater splash on the match-cast faces and inside the Post-Tensioning (PT) ducts of precast segments
- proper design, analysis and certification of transport barges
- loading and unloading facilities, including the need for permits.

Weather can also be a major factor in the selection of the erection method. For project sites in cold climates, there may be a need to enclose and heat the segments if they are being epoxied. BDE might have to be designed for extreme cold temperature. Provisions for hurricanes and tornadoes will vary depending on the frequency, severity of the events, reliability of the warning system and on the type of BDE selected. Snow and ice loading should be taken into consideration where applicable.

For precast solutions the availability of suitable land in the vicinity of the project site for a casting yard or the presence of an existing precast plant will influence the erection and segment delivery methods, but also the maximum permissible size and weight of the precast elements. The distance from the precast yard to the bridge site will determine the number of transport vehicles, the need to provide staging areas and double handling or barge docking facilities at the erection site. In case the limited storage area requires double stacking, the segments must be checked for that loading condition, ground improvement must be performed and proper support provided in order to avoid differential settlement.

Potential locations for the precast yard will be dictated by the size and type of segment that can be transported, especially if the route between the casting yard and the erection site includes low overhead clearance sections due to power lines or overpasses, or weight restrictions due to existing structures, buried utilities or culverts. Local regulations and permit requirements and costs also must be considered when selecting the casting yard location.

Heavy-load travel restrictions during the thawing season are a major consideration in certain geographical locations. These restrictions can prohibit the transport of segments during early spring.

Transport permit requirements vary widely from one location to another and seasonal restrictions may apply.

3.3.2.5 Project team

The experience of the whole project team is another major factor that may affect the selection of the erection method. The more experienced the team, especially the contractor, the more they will be open to considering a sophisticated erection method if it is deemed to be the most efficient way to construct the project.

The availability of skilled labour and operators in the project location is another factor that should be considered during the selection of complex erection methods.

3.3.2.6 Equipment availability

The availability of erection equipment (cranes or BDE), new or used, and the competition among the suppliers during the selection process will affect the cost and delivery schedule of the erection equipment, and will influence the selection process. Similarly, if a contractor has retained BDE from previous projects, such as a launching nose, it may be more efficient to reuse that equipment rather than use a different technique.

The additional costs associated with the mobilisation, assembly and commissioning of new BDE compared to the cost of refurbishing and modifying BDE should not be underestimated as it is a major component of the cost and time analysis in the selection process.

3.3.2.7 Cost

The prime objective of the selection process of the construction method and means is to get best value for money. Safety and quality are considered to be non-negotiable criteria. The contractor must balance all of the key issues outlined above with the requirement to produce the most economic solution. This does not necessarily mean minimum cost for BDE.

In general, precast solutions will produce time savings on a project but will require larger or more specialised equipment and have the disadvantage of needing a complete off-site infrastructure established for the precast yard (unless standard precast elements, such as standard bridge beams or boxes, can be procured from established local suppliers). The more expensive specialised equipment will benefit from the economy of scale of larger projects where the initial investment can be depreciated over a larger project.

For the in-situ traveller or falsework solutions, the investments in equipment are smaller, but these need to be set against the longer project durations and the costs of supervision, ancillary equipment and labour.

3.3.3 Definition of general erection means and methods

After analysis of all the project-specific criteria and constraints, or based on the erection method identified on the tender documents for a conventional contract delivery model, the main contractor has to develop the general construction means and methods. These methods and sequences should be in line with the erection sequences as assumed by the bridge designer. The construction schedule and an initial risk assessment will also be developed at this stage.

For conventional delivery contract models, the owner's bridge designer is expected to develop and identify at least one constructible erection method for the main contractor to follow. It will be up to the main contractor to follow, refine or modify that method, as long as the identified design criteria are met. Subsequently the type of BDE can be selected and its basic working sequences (kinematics) can be defined. The number of BDE needed for the project can also be established.

It is important to note that this is an iterative process where the contractor needs to be prepared to revisit assumptions and willing to adjust the approach accordingly in order to come up with the best overall solution for the project. This solution might encompass several different construction methods, including use of some conventional equipment.

Subsequently the BDE's performance criteria and specifications are developed. They need to refer to the construction means and methods and sequences selected. They have to address safety, the permanent work interfaces, the assembly, commissioning and relocation durations, the typical cycle time, the crew size, the degree of mechanisation, the control systems and the access requirements, as well as other project-specific requirements. It is recommended that the contractor clearly identifies the 'base model' functionality as well as the desired 'upgraded options'.

If a main contractor wants to increase the residual value of a BDE, they may want to broaden the performance specifications, so that the BDE can more easily be reused on future projects.

The BDE's performance criteria and specifications are best summarised in the Basic Methodology report. This report forms the technical part of the tender documentation when calling tender for design and supply of a BDE.

3.4. Procurement

When procuring a BDE there are different models. Table 3.2 shows the most common models.

Model I requires the main contractor to be well experienced in complex bridge deck construction means and methods, including the use of BDE, and to have an experienced staff to manage and coordinate the process. At the other end of the spectrum is model III, where the main contractor entrusts the bridge deck construction with BDE to an experienced specialist deck construction subcontractor (the specialist) with extensive complex bridge means and methods experience utilising BDE. This model allows an inexperienced main contractor to get involved in bridge construction projects requiring the use of BDE, even if they are not very familiar with such equipment. The specialist subcontractor might even own some used BDE which could reduce the cost of the construction. Model II, halfway between model I and III, allows a moderately experienced main contractor to get involved

Table 3.2 Procurement models

Model	General methods	Performance specs	Design and detailing		Steel fabrication	Equipment procurement	Commissioning	Operation	Audit
			Steel	Equip					
I	MC	MC	CS	CS	MC → SC	MC	MC (→SC)	MC (→SC)	MC
IIA	MC	MC (CS)	supplier/specialist				MC (→SC)	MC (→SC)	MC
IIB	MC	MC (CS)	supplier/specialist					MC (→SC)	MC
IIC	MC	MC (CS)	supplier/specialist						
III	MC	specialist							

MC = main contractor; CS = consultant; supplier = supplier of the BDE; specialist = specialist deck construction contractor; SC = subcontractor

with a complex bridge project. Under this model, the main contractor would engage a team of experts with relevant BDE experience in construction engineering, means and methods selection, specification development, procurement and operations. The main contractor will then acquire BDE from the supplier. In this model, the main contractor may self-perform the commissioning and operations, may sublet the commissioning and/or operations to a specialist subcontractor, or may include this scope of work in the BDE supply contract.

Model II is probably the most common scenario and in this case the main contractor should consider the following when comparing BDE options.

- Number of sets of BDE required to meet the project schedule: the cost of BDE will increase as more mechanisation and options are added to the basic design. However these additional costs should be analysed as they are often offset by savings in direct and indirect costs, schedule duration savings and safety and quality risk reduction. The upgraded BDE might result in a reduction in the number of sets of BDE required to build the project.
- Weight of BDE and interface with the permanent structure: these two factors have a significant impact on the overall cost of the project. The impact of BDE on the structure should be analysed prior to the final selection. Even similar types of BDE might have significant differences in the way they interface with and impact the structure. Weight, length, anchoring details, required bracing, temporary shoring, support types and locations could all result in added post-tensioning, rebar, hardware and concrete quantities.
- Operational costs: cycles, crew sizes, consumables, additional support.
- Assembly, commissioning, relocation and dismantling costs (size of cranes/SPMTs and crews required, time, etc.).
- Shipping: costs, special load permits and FOB (free on board) location.
- Reputation, experience and past performance of supplier.
- Liquidated damages in case of late delivery.
- Performance bonds.
- Supplier's engineering support and location.
- Supplier's fabrication facilities.

- Risk of equipment quarantine due to country of fabrication.
- Risk of political upheaval in country of fabrication.
- Design and fabrication QA/QC programme.
- Request of an independent stamped/sealed peer review of the design of the BDE.
- Preassembly and testing in factory (full or partial).
- Source of material.
- Cost, availability and lead time of spare parts and consumables.
- The residual value of the BDE and/or its reusability on other projects.
- Warrantees.
- Buyback offer guarantees.

The selected supplier should be given enough time to develop the basic design of the BDE in order to confirm its basic methodology and operational characteristics. The supplier will summarise the results of the study in a draft form of an Equipment Description Report that will be submitted to the other stakeholders for concurrence and to confirm that the current concept meets the overall project needs.

This preliminary Equipment Description Report shall include as a minimum

- basic description of the equipment design
- design basis reflecting specific characteristics of the BDE
  - loading, load combinations and load factors
  - applicable wind load
  - material properties
  - list of design, safety and quality codes referred to
- dimensions and weights of the BDE and its elements
- performance characteristics
- energy sources and consumption rates
- detailed typical launching sequences/kinematics, including reactions on the structure (location, contact area, magnitude)
- equipment reactions on the structures
- lifting and anchoring points details and insert requirements
- special case studies
- commissioning and load-testing studies
- anticipated details of cycles and crew sizes for different operations.

Enough time should be allocated for all the stakeholders to properly review and comment on the draft report. An interactive process should follow to optimise the BDE concept and fine-tune the details. At the conclusion of this period, the BDE Equipment Description Report is accepted, the details are frozen, and the scope of works and supply are defined. A final version of the Equipment Description Report is issued before the BDE can proceed to the final design and detailing stages.

The final version of the Equipment Description Report shall include as a minimum the following information, either further developed from the draft report or in addition:

- basic description of the equipment design
- design basis reflecting specific characteristics of the BDE
  - loading, load combinations and load factors
  - applicable wind load

  - material properties
  - list of design, quality and safety codes referred to
- dimensions and weights of the BDE and its elements
- performance characteristics
- detailed equipment list with performance data
- components' operational characteristics (capacities, speed, etc.)
- consumables and spare parts list, usage and lead time
- energy sources and consumption rates
- detailed typical launching sequences/kinematics (covering all controlling cases) including reactions on the structure (location, contact area, magnitude)
- crane support requirement for all the operations
- general risk assessment study
- lifting and anchoring points details and insert requirements
- special case studies
- assembly and disassembly studies, including cranes, temporary structures and layouts
- commissioning and load-testing studies
- ancillary operation studies (PT, pier brackets, closures pours, material deliveries, etc.)
- equipment relocation studies
- anticipated crane support for all the operations
- access provisions on BDE (shown on drawings)
- anticipated details of cycles and crew sizes for different operations
- available options and upgrades
- deliverables
  - design calculations
  - general assembly drawings (overall, individual elements)
  - special rigging and tools required for the assembly and operation
  - operation manual (with kinematics)
  - equipment manual
  - detailed launching sequences

  [Note: fabrication drawings are not necessarily part of the deliverables]
- fabrication
  - country-of-origin requirements
  - material and component source traceability
  - QA/QC programme and certifications
  - third-party inspection/audit/certification
  - trial assembly and testing expectations
  - witnessing and approval of trial assemblies
  - release for shipping documents
- Shipping: FOB, duration, shipping constraints (no-travel periods, load size, weight restrictions), permits.

Detailed performance specifications are a key factor for the successful procurement of the BDE that will meet the project requirement and the stakeholders' expectations. In addition to the Equipment Description Report, the following points need to be addressed in the Supply Contract:

- expected reporting and auditing during the procurement phases
- anti-corruption regulations
- approval of BDE designer and fabricator if seconded from supplier

- insurance and bond
- payment schedule and start-up delays
- third-party design review cost
- certifications of BDE
- warrantee duration
- latent defects
- on-site spare parts: ownership and buyback
- BDE buy-back clauses.

It is critical that the procurement documents include a clause requiring the confirmation of the equipment weight, reactions on the structure and anticipated structural behaviour of the BDE. These verifications can be performed during the load test and commissioning phases of the project. Any discrepancies with the design documents should be addressed and resolved prior to the commencement of the erection operations.

Upon completion of the fabrication drawings and the Operation and Maintenance Manual, the design and detailing of the BDE should always be subjected to an independent design check and review by an expert. Such design checks and peer review can be under the supplier's scope of work, as long as they are performed by experts not directly involved in the design and detailing of the BDE. More details about this subject are given in Chapter 16.

3.4.1 Fabrication, corrosion protection, trial assembly and delivery

The fabrication specification for a BDE should be at least as high as for permanent works. The loads for which it will be designed are real and stresses within the structure are deliberately kept high so as to achieve the power-to-weight ratio demanded by good BDE design. It is possible that the BDE may not be subject to fatigue effects. That is a matter for the designer to check. The BDE drawings must specify the required fabrication quality that is appropriate for the product and the fabricator must comply with those requirements.

Even though BDEs are generally temporary structures, they need to be protected against corrosion. It may be possible to accept a lower degree of paint treatment for an item of BDE than would be demanded for permanent works. However, experience shows that the cost of reasonable paint treatment is worth spending for several reasons. Reuse of BDE on subsequent projects (a very attractive way of spreading the cost of what is usually an expensive item of construction plant) may be impossible if corrosion becomes significant. Some items of BDE have had to be scrapped simply because corrosion levels could not be established and thus the safety of the machine could not be proved.

It is important that tubular members are fully sealed. Sealing also prevents cracking of tubular members, which are stored outdoors in cold countries, since rainwater cannot get trapped and subsequently freeze when the temperature drops below zero.

Other light forms of bracing can be equally problematic. Angle or channel sections placed back-to-back with gusset plates between them are good for making easy connections, but the gap between them can never be painted. ('Starred' angles are better than back-to-back ones.)

Paint reduces the tendency of the machine to rust (although practical operation of BDE almost always causes paint damage, which can be mitigated by 'touching up' the damage as part of the BDE's ongoing maintenance regime). Rust will stain the structure on which it sits and those stains are very hard to remove. They will almost certainly be a source of criticism from the client. The rust may also present a safety hazard to the workforce, as well as being a potential source of damage to the mechanical components of the machine. Refer to Section III, Chapter 16, regarding the ponding of water and sealing of sections for which access is not easily available.

Machined components will usually not be painted but grease should be specified to protect those parts, particularly if transport to site is by sea.

Trial assembly of mating components at the fabrication stage is essential. It is not essential to trial-assemble the whole BDE, although, if reasonably possible, it is an excellent check and gives great confidence to all the stakeholders. Experience shows that the involvement of operating staff during trial assembly or even earlier creates a climate of ownership among the site staff and increases the level of knowledge about the BDE.

Care taken in the packaging, handling and delivery of the elements of BDE is well worth the time and cost involved. Some parts of BDE are light and fragile (for example, secondary brace members and cleats for mechanical equipment) and repairs to damaged components will take time and money on site. Temporary local strengthening, or the removal of vulnerable items from otherwise bulky and heavy components, will be worthwhile. The supply of specially made lift brackets will avoid the site operatives placing lift slings around components that will be damaged during lifting into the assembly location.

It is usually necessary to drain fuel tanks and machine oil sumps during transport to site and care must be taken to ensure that the correct fuel and lubricants are available at site when the BDE arrives.

In addition, it is important to have an independent third party confirm that the BDE was fabricated in accordance with the specifications. This can be accomplished by performing independent inspection of the fabrication and overseeing the trial assembly.

3.5. Operations

The equipment supplier has to provide the contractor with a comprehensive Operation and Maintenance Manual, which covers general operational aspects of the BDE and project-specific ones. For any future reuse of the same equipment on another project, the BDE supplier or someone of equal competence should get involved again to rewrite the project-specific features of the Operation Manual.

The party in charge of the operation of the BDE – that is the main contractor or specialist subcontractor – needs to have a qualified team familiar with detailed planning of the erection operations and capable of producing all the work-package documentation needed for the safe and efficient operation of the BDE.

3.5.1 Preparation of detailed method statements and checklists

The site team will develop, in compliance with the Operation Manual, the following documents:

- work plans
- method statements
- erection sequences and kinematics
- step-by-step operational sequences
- schedules/programmes
- training material

- checklists
- documentation.

The work packages need to be developed for all the different phases of the project, including:

- assembly
- required special rigging
- commissioning
- load test and deflection tests
- typical launching
- special launching cases
- typical structure erection
- erection of special cases
- equipment relocation
- dismantling.

Each method statement needs to include a risk assessment/job hazard analysis.

3.5.2 Commissioning

A proper layout of the assembly area needs to be identified. It is usually preferable to assemble the equipment on the ground at or near the abutment area. However, sometimes site conditions might dictate assembling the equipment away from the erection location and moving it to the final position or even assembling the equipment in an elevated position on top of the piers and/or temporary supports.

A detailed list showing the weight of all the components needs to be developed by the supplier, who should also supply the exact erection sequence for all the assemblies and sub-assemblies.

An exact layout of the assembly area should be developed, showing:

- boundaries
- the access and exit routes
- the lay-down areas
- crane positions and movements
- lifting radii
- preassembly area
- temporary foundations locations
- temporary supports locations
- existing above- and below-ground obstructions.

Detailed analysis of the crane lifts plans should be engineered to ensure that all the lifts can be performed within the safe operating limits of the equipment. Below-ground utilities and structures need to be analysed and clearly surveyed and marked on site. Off-limit areas should be physically blocked with barriers or fences to prevent accidental equipment travel. Similar precautions should be taken to protect overhead utilities.

The equipment needs to be properly anchored during the assembly and launching operations. Additional counterweights or similar measures might be required.

At the completion of the assembly and prior to the start of the commissioning activities, the equipment shall be certified by the supplier and by the equipment designer or his representative as stipulated in the contract. This certification shall state that the equipment has been assembled and tested in accordance with the equipment design and the fabrication and assembly drawings. The certification shall also state that the equipment has been operated within the expected parameters during the commissioning and that any required modifications were performed according to the approved procedures.

Occasionally field modifications might be required due to errors in design, detailing or fabrication. Sometimes field modifications are required due to system improvement, optimisation and upgrades. These modifications should be performed only after the preparation of a field modification request, also known as a temporary work modification request. These requests have to be reviewed by all the relevant stakeholders.

It is advisable to have representatives from the equipment supplier, the controls supplier, the electrical supplier and the mechanical supplier on site to troubleshoot any problems and coordinate the repairs and adjustment to the components. It is also advisable to identify various local electric and hydraulic repair shops and have them assist with the assembly, testing and commissioning in order to familiarise themselves with the BDE's components and allow them to perform the required maintenance and minor repairs in accordance with the equipment Operation and Maintenance Manuals. Getting these repair shops involved early with the equipment will also prepare them to work with the equipment supplier (remotely) to troubleshoot and repair any major breakdown during the lifetime of the project. This step will greatly reduce the downtime during the operations stage.

3.5.3 Load and deflection test

The required checks to be performed on the BDE after assembly include the following.

- Measurement of the BDE's self-weight and position of its centre of gravity as a whole or by measuring self-weights and centre of gravities of individual components. The objective of this check is to confirm the BDE weight and its reactions on the structure.
- Load tests are performed on lifting devices, including lifting beams, spreader beams and rigging. These items are typically statically tested at 1.1 to 1.25 and dynamically at 1.1 times the rated Safe Working Load (SWL) of the BDE's lifting device. However this does not mean that the equipment should ever be used for more than the rated SWL. It is acknowledged that in some countries factors higher than these are used to prove some cranes. In the case of BDE, these values are justified, because the loads are reasonably accurately known.
- Deflection tests during operation. The primary function of these tests is to verify the anticipated service deflections and subsequent recovery of the main girders due to segment or span weight and during launching. The loads applied to the girders are as specified in the contract and are usually between the rated safe working load and no more than 10% over the rated safe working load. The cantilever deflection recovery of the main girder can be checked by measuring the front/rear support leg's stroke versus jacking force. The objective of the deflection tests is to verify the assumed elastic deformations and rebounds used in the design of the BDE. The deflection tests also allow verification of the functionality of the front/rear support and its capability to reverse the cantilever deflection of the main girder. It is important to review the BDE's stability during this kind of test.

3.5.4 Training of operating crew

After the successful commissioning and certification of the BDE, it is critical to implement a project-specific training and certification programme that will involve all the team members who will be associated with the operations of the BDE. The following team members should be included in the training as a minimum the BDE supervisor/superintendent, the BDE field engineer, the BDE operators, the BDE crew, the support crews, the field supervision and the site inspectors. This training should be repeated and refreshed as new team members are added to the operations. The initial training should preferably be provided by the equipment supplier. If not practical, the training can be provided by another qualified person familiar with the operations and limitations of the BDE as well as the interface with the structure. It is very beneficial to set up a 'train the trainer' programme to be able to provide multiple and frequent training sessions and refreshers.

The training should include both classroom and hands-on training. At the completion of the training, the trainer should make recommendations to the project management identifying the trainees that have demonstrated proper understanding of the BDE and its limitations and shown practical operational competencies in their respective areas of expertise. The project management will review these recommendations and issue the proper certifications in BDE operations. There should be multiple categories of certifications, such as BDE operation supervisors/superintendents, BDE field engineers, BDE operators, launching crew, general operations crew, support crew, supervision, inspection, etc. It is recommended to have multiple people certified in each category.

3.5.5 Operation of the BDE

The BDE should be operated only by trained and certified crews. It is helpful to have a board at the BDE identifying the certified crew members by name and pictures. Their roles and responsibilities should also be stated on this board. It is important to clearly communicate who the primary qualified individual is on any given shift/operation. There should be no confusion about who is in charge.

Proper communications mechanisms should be provided to allow safe and efficient communication between the operator and the other team members. Hands-free communication devices are preferred.

The use of the checklists that were developed, reviewed and approved by the equipment supplier, the construction engineer and the project management is mandatory. Total adherence to this procedure is essential to the safe operation of the BDE. The checklists should include hold points that will require additional signatures and approval before proceeding to the next step. In order to provide traceability and accountability, the checklist points should be initialled with indication of date and time and not just checked off. It is important to recognise that the checklists are live documents that will be evolving throughout the project. Modifications and improvements to the checklists are encouraged. However they need to be reviewed and approved by the authors and reviewers of the original checklists, or their authorised replacements/delegates.

A pilot/co-pilot scenario is a proven method to manage the checklist. An example of this scenario is when the BDE supervisor/superintendent (pilot) is directing the operation and the BDE field engineer (co-pilot), working closely with the supervisor/superintendent, is responsible for filling in the checklists and making sure that the operations are proceeding per plan. Hold points can be identified on the checklist and, depending on the criticality, will require joint signatures of the pilot and co-pilot as well as the signature of an inspector. The required signatories should be clearly defined on the checklists.

3.5.6 Regular maintenance

A BDE is a complex machine that requires regular inspections and maintenance. This work needs to be performed as recommended in the Operation and Maintenance Manual by qualified and trained mechanics. In addition to the safety-related advantages, regular inspections and maintenance will drastically reduce the severity and duration of work stoppage due to equipment breakdowns and accidents. The entity responsible for the inspection and maintenance activities needs to make sure that it has the proper resources allocated and that it performs these activities daily, weekly, monthly and/or as defined in the Operation and Maintenance Manual of the BDE.

3.5.7 Regular technical audits

It is beneficial to perform occasional independent audits by qualified individuals who are familiar with the BDE operations to ensure proper adherence to operational procedures and documentation of the BDE operations. These individuals should be independent of the daily operations of the BDE and could be internal or external resources.

No modifications of any kind to any component of the BDE are allowed without explicit written approval of the BDE's supplier or a qualified engineer if the BDE's original supplier is unavailable.

Strict adherence to the BDE Operation and Maintenance Manual, launching sequences, work plan and other approved documents is essential to the safe operation of the BDE. Any deviations from these documents will require approval by the supplier or a qualified engineer if the BDE's original supplier is unavailable, as well as by the construction engineer and the independent checker.

3.6. Post operation

Preferably prior to completion of the project, a decision should be made regarding the disposal of the BDE. Scrapping it, selling it or keeping it have different financial implications that need to be analysed. Keeping the BDE makes sense only if it can be professionally stored; significant financial commitment is required for:

- proper dismantling,
- storage area cost
- containers cost
- storage on mats to prevent water pooling
- removal of mechanical, hydraulics and electrical components
- storage of electrical and controls components in climate-controlled facilities
- periodic inspections and movement of mechanical components to prevent them from seizing up.

At the completion of the project, the equipment should be dismantled per the approved work plan and stored properly. Care should be taken to protect the electrical, mechanical and hydraulic components from exposure to the elements. Electrical and electronic components can suffer severe deterioration and damage if not stored in climate-controlled facilities. All the available documentation, including fabrication records, QA/QC paperwork, load testing, shop drawings, calculations, checks, certifications, modification, upgrades, accident records, inspections, maintenance, etc., should be kept in a safe place.

3.7. Conclusion

The management of the selection, procurement and operation of a BDE can be a complex and daunting experience. However, the retention of qualified contractors, engineers, suppliers and field staff who follow the processes described in this chapter, and proper collaboration among all the stakeholders, can make the experience a lot simpler and will result in a safer, more efficient and profitable operation.

Bridge Deck Erection Equipment
ISBN 978-0-7277-6193-4

https://doi.org/10.1680/bdee.61934.029

Chapter 4
Safety

4.1. Introduction

To conceive, design, build and commission a BDE safely requires a complete understanding of bridge behaviour during construction, as well as of the interaction between BDE and the structures under construction. A BDE provides safe and reliable construction when proper procedures and processes during the design, fabrication and construction phases are implemented.

The most important factor in mitigating accidents on the project is the selection of the proper team. This team should include experienced bridge designers, construction engineers and temporary work designers, equipment fabricators, QA/QC technicians and, of course, competent contractors.

It is important to remember that accidents are often the result of a combination of events, errors and omissions. It is sometimes difficult to pinpoint the root cause, especially if the accident resulted in catastrophic failure.

For the purpose of this chapter we will classify the main apparent cause of construction accidents under four categories:

- equipment design, fabrication and commissioning
- bridge structure design and construction, including the interface between the BDE and the permanent work
- temporary structures design
- BDE field operations.

Risk-management measures to eliminate, mitigate and/or reduce the stakeholders' exposure will be discussed as well.

4.2. BDE design, fabrication and commissioning

The first category of accidents is caused by an error or defect in the equipment design, fabrication or commissioning.

One of the surest methods to mitigate equipment design and fabrication problems is to start with the selection of a competent equipment supplier who is experienced in the type of equipment being considered for the project. The following items need to be considered during the design, fabrication and commissioning of the BDE

- design codes
- modelling
- loading and load combinations

- access for personnel
- stability and longitudinal fixity
- equipment history
- raw materials and fabrication facilities
- supplier role during commissioning.

For more explanations about design codes, modelling, loading and load case combinations, access for personnel, stability and longitudinal fixity, refer to Section III, Chapter 16.

4.2.1 Design and detailing of BDE

Nobody is perfect. Independent design checks and a peer review by a BDE expert are highly recommended to minimise risks of design and detailing errors. For details, refer to Section III, Chapter 16.

4.2.2 BDE history

When considering a used BDE, all the stakeholders must know the full history of this BDE. It is critical to know if the BDE was involved in an accident and to study the circumstances and causes of the accident, as well as the repairs that were performed.

It is also as critical to know if the BDE has undergone any modifications since its original fabrication. The modifications history should be reviewed and analysed to assess its impact on the BDE design, suitability and operations.

When considering the use of existing BDE, it is important to check if the codes that were used to design and fabricate the equipment meet the requirements of the current project's codes and specifications.

The age of the BDE, number of projects, storage conditions, operating environment, level of corrosion, modifications and accident history are all factors that might contribute to a need for modifications or redesign of the BDE prior to certifying and rerating it and developing its operating parameters.

4.2.2 Raw materials and fabrication facilities

The whole is as good as the sum of its parts. The construction team must be able to verify the quality of the BDE's raw materials and be able to trace its origins 'back to the mills' and to get proper inspection and testing certifications that meet the specifications. Traceability is also important for any material used in modifications or field repairs.

The designer needs to define testing requirements for structural welds, which have to be carried out during the fabrication by an entity which is independent from the fabricator. Trial assemblies of part or the entire BDE are highly recommended, since they allow verification of whether geometrical tolerances – specified on fabrication drawings and all-important for obstruction-free movements of BDE elements – have been achieved.

It is therefore critical to select a reputable and experienced fabrication shop and to make sure that the proper QA/QC procedures are followed during the fabrication process.

4.2.3 Supplier role during commissioning

In general a BDE is assembled by the main contractor or the specialist subcontractor. In addition to the need to produce a detailed method statement with risk assessment and checklists for the assembly of a BDE, the BDE supplier should always be requested to check and certify the commissioned BDE to be fit for use after a thorough inspection and testing. Testing of the BDE as part of its commissioning is probably one of the most important steps in ensuring the proper operation and safety of the equipment and it should not be omitted. For details, refer to Section I, Chapter 3: 3.5, 'Operations'. Accidents have happened during assembly of BDEs due to lack of proper planning; BDEs have been damaged because they were not correctly assembled and commissioned as intended by the BDE supplier.

4.3. Bridge structure design and construction

The second classification for construction accidents is those caused by a failure of the permanent structure, either due to design or construction errors. All members of the team must recognise the importance of understanding the magnitude and locations of the loads being imposed by the BDE on the permanent work and the consequences of cutting corners during the construction process.

4.3.1 Designing to proper standards

As with any design, the permanent structure must meet all pertinent design codes and standards. When using a BDE which is sitting on the structure it builds, the permanent structure also has to be checked for load case 'construction with BDE', where loading may be even higher than during the completed structure's service life. This load case is not defined in the design codes and standards but has to be defined by the BDE designer. The proper implementation of a design QA/QC process and the practice of engaging an independent peer-review process are essential steps towards eliminating accidents caused by design errors.

4.3.2 Understanding loads imposed by BDE and making appropriate provisions

As mentioned above, the BDE designer has to define the load case 'construction with BDE' (defined in the Basic Methodology Report and the Equipment Description Report). The permanent work designer or the engineer in charge of the construction engineering (the design team) and the construction team must have a full understanding of the loads imposed on the permanent structure by the BDE. Well before construction of the permanent work starts, the design team must verify by calculation that the permanent work is able to take the loading imposed by the BDE. If it is not, the design team should propose and initiate either that the BDE is modified to reduce the magnitude of its reactions or that the permanent work is temporarily or permanently strengthened. During operation, measures have to be taken in the field to ensure that construction is proceeding according to the approved erection procedures to avoid overloading of the permanent and temporary work.

4.3.3 Geometrical requirements to support the BDE

Apart from properly assessing the loading imposed by the BDE, it is also crucial to ensure that the design and detailing of the permanent work gives enough space for proper and safe support and fixity of the BDE.

4.3.4 Wind, cold weather and seismic loads

The design and construction team must also be aware of and understand the effects of the environmental loads on the BDE and the structure. The BDE supplier and the designer have to properly evaluate these effects and need to define provisions (e.g. wind speed limits for operation; tie-down arrangements in case of a typhoon, etc.). These provisions have to be communicated to the rest of the team early on in the tender process, to allow for proper pricing. The construction team will be more reluctant to implement such provisions and more likely to cut corners if the provisions were not properly understood and priced during the tender stage.

4.3.5 Construction QA/QC

Another leading cause of accidents involving BDE is the failure of the permanent structure due to substandard construction practices, such as missing reinforcing steel, inadequate concrete, improperly stressed Post-Tensioning (PT), missing embeds, inferior material, improper curing, and not using the latest shop drawings revisions. Any of these irregularities could lead to a collapse of the structure and result in major injuries to people and damage to property.

Field inspection and quality control and assurance are an essential element in ensuring a safe working environment. Inspectors should be regarded as a second set of eyes whose main duties are to make sure that the work is being completed in accordance with the approved design and shop drawings.

A trusting relationship between the inspectors, QA/QC personnel and the production crews should be nurtured. The production crew should accept the inspectors and the QA/QC personnel as members of the team whose main role is to provide timely checks and feedback to the crews. These timely comments will minimise the amount of rework and will ensure that the final structure will be safely completed in accordance with the design.

Management should recognise that inspectors and QA/QC personnel are the first independent line of defence against shoddy work that could result in expensive rework, or worse, in unsafe structures that could collapse. Accordingly, management should encourage collaboration between all the team members.

4.4. Temporary structures design

The third classification for construction accidents is those caused by a failure of the temporary structures used in association with the BDE. These temporary structures include scaffolding, shoring towers, access ramps, work platforms, trestles, props, temporary foundations and other construction structures which are not part of the permanent installation.

Just as in the permanent structures, engineers and designers must adhere to the appropriate standards and specifications for the design of the temporary structures. In addition, they need to fully understand the interaction between the BDE, the permanent and temporary structures, the anticipated construction sequences and the magnitude and position of loadings imposed by the BDE, including the space needed to allow safe and proper support and fixity. Thus it is important that these temporary structures are designed by a competent and experienced design engineer who is familiar with the type of construction under way. This specialist design engineer should be capable and responsible for the proper coordination between the designer of the permanent structure, the BDE supplier and the contractor's design and construction team.

As in all phases of the project, proper QA/QC and independent design checks and peer reviews are needed to reduce the risk of errors and omissions. Although these structures are temporary in nature, it is critical to implement the same level of QA/QC as with the permanent structures.

4.5. BDE field operations

The fourth classification for construction accidents is those caused by BDE field operations. Even with the implementation of the best practices for BDE and permanent and temporary structure designs, a major component of BDE safety lies in the hands of the field personnel operating that equipment.

It is the responsibility of the BDE supplier to provide the construction team with manuals addressing the proper and safe operation of the equipment, as well as inspection and maintenance requirements and schedules. The supplier should also provide guidance regarding the reuse of certain components such as rigging, lifting bars, tie-downs, splice hardware, bolts and more.

The most common cause of accidents is probably the failure to follow established and approved procedures. One of the drivers behind this behaviour is a desire on the part of the crew to speed up the operations by finding a 'better', 'faster', 'easier' or 'more economical' way of doing things. Although most significant innovations do originate in the field, it is dangerous to implement modifications without going through the proper approval channels, with sign-offs from the designers and engineers who developed and approved the original procedures. Unchecked 'on the spot' modifications could be devised and implemented by unqualified personnel making decisions without having a full understanding of the complex interaction between the structures and the BDE. These decisions often endanger the lives of their co-workers and can result in significant losses.

Another driver behind the failure to follow established procedures is complacency. The operation of BDE is often repetitive. The crews perform the same delicate tasks over and over for long periods of time and become desensitised to the risks involved with launching gantries or erecting segments. Complacency is particularly dangerous when the crew, accustomed to a certain routine, does not recognise a special-case situation and fails to apply the appropriate special procedures. It is this overconfidence that leads to complacency and accidents.

Inexperience, lack of training and an inability to recognise hazards are also major contributors to field operation accidents. Although the operation of BDE can be relatively simple when the proper procedures are followed, it does require a higher level of skills from the construction crews than is demanded by other bridge-construction operations. In addition to the standard skills that are required for bridge construction – such as rebar tying, carpentry, forming, shoring and concrete pouring – operation of a BDE requires the skills to operate and work around complex and sophisticated machinery.

To reduce safety accidents, the entire crew, not only the operators, need to be trained in every aspect of the construction project and need to understand the criticality of the operations. The crew should also gain an appreciation of the complexity of the interaction between the structure and the BDE and the importance of following the approved procedures, as well as the consequences of failing to do so.

Several steps can be taken to reduce the risks of field operation accidents. One of the most important steps is getting supervision staff who are experienced and competent in the type of work and BDE being used on the project. They should also be able to train the workforce assigned to the project and should be able to command respect and adherence to approved procedures.

Daily 'toolbox safety meetings' and 'play of the day' discussions at the beginning of the shift are very useful. In these meetings the supervisory staff and the crew will discuss the day's upcoming activities, which will help raise the awareness of the crew and will identify any special cases or procedures, as well as safety concerns and hazard recognition. This process is especially helpful when new members are added to the team or when multiple crews are being shifted around the project.

Detailed training sessions should be held before the start of any new activity on the project. The commitment for training should be enforced by the supervision staff. Skilled and trained labour is a major contributor to the safe and successful completion of the project.

At the completion of the appropriate training sessions, management will designate competent staff with differing levels of responsibility and authority. Checklists (for an example, refer to Figure 4.1)

Figure 4.1 Extract from a checklist for segment erection with an overhead launching gantry (VSL International)

VSL	OVERHEAD LAUNCHING GANTRY KLANG VALLEY MRT					Doc. No:VSL/MRT/LG/006
	SEGMENT ERECTION					Rev.03
From Pier P(N)		To Pier P(N-1)		Pick Up	Below	Date
Span Length		Radius		Track	Single / Double	

Item	Activity	Sign off / Time
7	Ensure all stressbars,bearing plates,washers,nuts are properly connected. Stress the bars to 300 kN/bar if sement lifting holes have no conical inserts.	
8	Beam assembly is tied to segment with four number Ø32mm tension bars.	
9	Check tension bars have not been bent,cut or heat damaged.	
10	Stressing platform is installed at FSF.	
	Segment Erection	
11	Hang pier segment P(N-1) [ ] in correct longitudinal position. Transverse position and level should match the previous span.	
12	Hang adjacent segment at the same level and 50mm away from the pier segment; make sure the shear keys are not engaged. The 50mm gap will allow pier segment P(N-1) [ ] to be adjusted longitudinally, transversely and vertically.	
13	Continue preloading the LG, ensure minimum gap of 20mm between segments. Place the second last segment on the ground below its final position.	
14	Align P(N-1) [ ] pier segment. The vertical alignment of this pier segment should be adjusted so that P(N) [ ] pier segment soffit level is +30mm above the theoretical final level; this will ensure we do not clash with the bearing on pier P(N) [ ]	
15	Raise the span support cylinders until they touch P(N-1) [ ] pier segment soffit, continue raising span support cylinders until presure of 10bar is reach; tighten locking ring and release hydraulic pressure. Check sledge is longitudinally guided to allow 20mm rebound of pier during load transfer. Note: all sledges on pier P(N-1) [ ] to be guided to allow 20mm pier rebound.(Refer above drawing of pier top view)	
16	Match fit remaining segments until second last segment.Lift second last segment and match fit. The last segment to be positioned is the pier segment on pier P(N). [ ] Max level difference between segments to be +/-5mm. Max span hogging level during segment erection to be no more than -20mm.	
17	Raise / lower FULRS cylinders until pier segment P(N) [ ] at correct level.	
18	Raise the span support cylinders on P(N) [ ] until they touch the pier segment soffit, continue raising span support cylinders until presure of 10bar is reach; tighten locking ring and release hydraulic pressure. Check sledge is longitudinally guided to allow 10mm rebound of pier during load transfer.	
	Note : Below step to be performed only in PRESENCE of ______ and to be checked and signed below before progressing further Name ______ Signature/Date ______ HOLD POINT	

with hold points should be used and should be initialled by the designated competent personnel during the actual field work. It is important that the checklists, which have been derived from the BDE's Operation Manual, reflect the correct sequences, are practical and are written in a language which is understood by the crew A checklist should be a tool for the crew to operate the BDE safely and not be a bureaucratic burden.

As with the operation of aeroplanes, operation of a BDE should be entrusted to a pilot and co-pilot – the BDE supervisor and the BDE engineer – with the supervisor managing the crew and the engineer in charge of the checklist. Hold points have to be initialled by both. For each operation, the supervisor and engineer in charge of the BDE have to be appointed and this appointment has to be communicated to the crew. When working two shifts per day, proper handover procedures between the people in charge of the operation of a BDE are important, so the new crew knows exactly what stage the BDE is at when taking over.

Daily and periodic inspection and maintenance schedules should be followed and properly documented.

Management should also recognise the role inspectors play on the project and should welcome their involvement and provide them with all the tools, procedures and sequences to be able to verify that the work is executed according to plans.

Management should plan regular audits of the BDE and its operation to ensure that operating standards are maintained through the entire project.

It is also advised to establish an 'authorised person' training programme with hard-hat stickers to train and identify the personnel authorised to access the BDE. That training should cover the BDE's operations, hazards recognition and the operations 'Dos and Don'ts'. Mini training sessions can be developed for escorted visitors.

4.6. Risk mitigation measures

A common tool used to manage risk is the Risk Assessment Code (RAC). The RAC is determined by combining the severity with the likelihood of an event in a risk matrix. The resulting analysis will yield a risk score (e.g. extreme, high, medium, low). The higher the risk, the more efforts are needed to reduce the exposure. All the stakeholders should collaborate to reduce the risk to the lowest possible level.

Developing effective risk control measures requires the project team to convene subject matter experts on the activities in order to determine and assess the risk and to develop the most effective risk elimination and reduction measures.

The resulting risks can be typically classified as:

- avoid: the activity should not proceed
- transfer: the risk should be transferred to another stakeholder who is better equipped to manage the risk or maybe insure the risk
- reduce: manage the severity and/or probability of the risk to reduce the likelihood of an incident
- accept: after all other measures have been analysed and implemented, the residual risk level could be deemed acceptable.

The risk/hazard identification and assessment process involves:

- detailed information about where the work is being conducted and the associated construction activities (including plant and vehicles, materials and substances in use, construction methodologies, etc.)

Figure 4.2 Example of a risk assessment (VSL International)

RISK ASSESSMENT FORM

BACH DANG FORM TRAVELLER OPERATIONS

Risk Matrix

N.B. For more details regarding use of this matrix/definitions refer to final page of this document

Consequence \ Likelihood	Rare	Unlikely	Possible	Likely	Almost Certain
Severe *Eg. Potential fatality or injury or illness with permanent disability*	MEDIUM	MEDIUM	HIGH	EXTREME	EXTREME
Major *Eg. Potential lost time injury (but non-permanent disability)*	LOW	MEDIUM	MEDIUM	HIGH	EXTREME
Moderate *Eg. Potential medical treatment injury or illness (but no lost time)*	LOW	LOW	MEDIUM	MEDIUM	HIGH
Minor *Eg. Potential first aid injury*	LOW	LOW	LOW	MEDIUM	MEDIUM
Minimal *Eg. Hazard or near miss requiring reporting and follow up action*	LOW	LOW	LOW	LOW	LOW

Hazard Identification				Risk Evaluation			Mitigation measures
S/N	**Work activity**	**Hazard**	**Possible Accident/injury**	**Consequence**	**Likelihood**	**Risk Level**	
1	Coupling of suspension stress bars	Insufficient threaded length engaged on both sides of the coupler	Collapse of the FT	Severe	Possible	HIGH	• Use modified coupler with mid length pin to physically indicate full engagement to the operator • Paint end of stressbars to indicate the required thread engagement • Use backup bars for main suspension front support • Stressing of the bars to their full required load, ie active tie down
2	Site welding/Flame cutting	Accidential damage to stress bars in tension	Collapse of the FT	Severe	Unlikely	MEDIUM	• Use protection sleeve around exposed part of the stressbars • Use backup bars for main suspension front support • Redondant tie down system for tie down of precase block
3	Placing of precast block on FT by crane	Accidental impact load on the FT	Collapse of the FT	Severe	Unlikely	MEDIUM	• Design allows for an accidental loadcase with impact factor of 2.0 on the weight of the block • Operation to be carefully coordinated on site and unloading from the crane to be very slow and progressive
4	Stressing of tie down bars	Fall from height	Fatalty/Lost time injury	Severe	Possible	HIGH	• All stressing activities done at deck level • Stressing on top of precase block to be carefully considered, with adequate fall restraint measures provided by site
5	Propping of the precast block	Hydraulic jack failure during loading	Collapse of the FT	Severe	Unlikely	MEDIUM	• Ensure all ringnuts are engaged on the prop jacks before tensioning the stay cable
6	Launching of the FT	Accidential differential movement of the launching jacks	Collapse of the FT	Severe	Possible	HIGH	• Design allows for an accidental loadcase with differential movement of 1 jack stroke (800 mm) • Operators to coordinate both sides and check position on rail beam marking every 200 mm
7	Lowering of front support	Bending and damage of suspension bars	Collapse of the FT	Severe	Possible	HIGH	• Back up bars present at all time during lowering • Lowering to happen in small steps (100 mm), with 1 operator below the deck to check the bar contact against the concrete • Design allows for a sliding longitudinal movement of the bar at the FT level
8	Extension of rear leg	Hydraulic jack failure	Collapse of the FT	Severe	Unlikely	MEDIUM	• Operator to make sure rear leg ring nut is always engaged at the end of the extension/retracting operation
9	Operate the form work	Hydraulic jack failure	Lost time injury	Major	Unlikely	MEDIUM	• Proper access provided below formwork for jack resetting • Safety pin/turnbuckle provided to lock the formwork in position at the end of jack extension/retraction
10	Placing of reinforcement	Fall from height	Lost time injury	Major	Possible	MEDIUM	• Guardrails provided all around the formwork at deck level • Proper access to soffit level into the segment shall be provided by the operator
11	Concreting	Local failure of formwork	Medical injury	Moderate	Unlikely	LOW	• Ensure all the tie bars are correctly installed
12	Stressing of post tensioning	Fall of heavy jack	Lost time injury	Major	Possible	MEDIUM	• Proper access provided around stressing areas • Operator to design support system for jack weight

- proximity and location of adjoining hazards and how they might interact with the work being conducted
- duration and frequency of tasks associated with the work
- potential for failure and associated failures or loss of risk-control measures
- human factors (behavioural, training, experience, physical, etc.) of the people involved in the work being conducted
- proximity of the public, visitors, delivery drivers or other contractors to the work being conducted
- environmental considerations of the workplace and their impacts
- emergency planning and support considerations.

The risk/hazard identification process and the risk assessment process should be conducted by individuals knowledgeable and experienced in the various risk methodologies being used. For each identified risk/hazard, it is vital to identify who or what might be harmed or damaged in order to determine the best way to manage the risk.

The risk/hazard identification and the subsequent risk assessment are conducted routinely. When new activities are anticipated or when new hazards or risks are introduced, the risk register must be updated. In addition, if new information indicates that a previous risk assessment was incomplete or inadequate, that risk should be reassessed.

For an example of a risk assessment, refer to Figure 4.2.

4.7. Conclusion

By applying standard engineering and construction best practices, bridge construction using BDE is a safe, efficient and profitable solution to building complex bridge decks. These best practices include selecting the best-qualified team, employing independent design checks and peer reviews, empowering QA/QC and inspection teams, conducting training and certification, and implementing a checklist process.

Safety and quality are matters of corporate culture that need to trickle down from the team leadership to the ranks. Educating our workforce at recognising that safety and quality are not an obstacle but rather a tool to achieving a successful project is essential. The team management must lead by example.

REFERENCES

Figures 4.1 and 4.2: VSL International

Section II

Bridge Deck Erection Equipment

Bridge Deck Erection Equipment
ISBN 978-0-7277-6193-4

https://doi.org/10.1680/bdee.61934.041

Chapter 5
Precast balanced cantilever erection by launching gantry

5.1. Introduction

This chapter covers the construction of precast segmental concrete bridge decks with Launching Gantries (LG) using the Balanced Cantilever Method (BCM).

Balanced cantilever construction by LG is a widely applied method for building continuous bridge decks by assembling precast concrete segments in a series of approximately symmetrical pairs of cantilevers, centred about piers. The pairs of cantilevers are erected in sequence. They are connected against each other by cast-in-situ 'midspan' closure pours. The LG is self-launching. It is supported by the bridge deck and the piers. Generally no ground support is required.

LGs are suitable for BCM erection for medium to large projects. Deck erection on smaller projects and bridges with large spans are generally done by crane and lifting frames.

A precondition for the use of LGs for BCM erection of precast segmental decks is that the pier table is either placed by the LG or by other means (cast in situ or precast and placed by crane).

The LG is used to lift the segments whenever possible directly from the ground level. If it is not feasible to get segments to the erection front at ground level, segments are delivered over the already completed deck to the rear of the LG.

5.2. Application range

Due to the considerable mobilisation cost of an LG, LGs are generally not used for BCM on small projects. They are suitable for use when spans are from 40 m to about 120 m and on projects where a minimum length of about 1 km of single decks or 500 m of double decks has to be erected (that is about 350 to 400 segments).

Weights of segments depend firstly on the given deck width and secondly on the length. The upper limit of the segment length is controlled by traffic restrictions and, when segments are delivered over the deck, by the available clearance along the LG. The lower limit of the segment length is between 1.5 and 2 m. Typical segment weights vary from 400 to 2000 kN. Curvature radius in plan can be as tight as 150 m (or even 120 m for LG-type III).

5.3. General arrangements of the BDE

BCM launching gantries can be classified according to their Main Girders (MG) as follows:

- LG-type I: LG with a pair of main trusses
- LG-type II: LG with a single main truss
- LG-type III: LG with a single main truss and stays.

This section of the chapter describes the main components of these three types of LGs. The kinematics are presented later in the chapter (refer to 5.4, 'Kinematics'). The selected LG configurations are typical, but may be modified for specific projects.

5.3.1 Launching Gantry with a pair of main trusses (LG-type I)

A typical LG-type I is made up of the following main components (refer to Figure 5.1).

A. **Main support of main girders:** During erection of field segments about pier N, the LG-type I is supported on two main supports. The front main support is centred over pier N and the rear main support over pier N – 1 (or in some cases a distance away from the centre line on the downstream cantilever at pier N – 1). The supports are always placed on top of the deck and their legs are normally centred over the deck's webs. The main element of these supports is the transverse beam, called the Lower Cross Beam (**LCB**). The LCB is orientated perpendicular to the bridge axis and is made up of a pair of I-beams. Between the LC and the main girder there are mechanical devices, called Universal Lower Roller Support (**ULRS**). The ULRS allows vertical, transverse, longitudinal translations and rotation about horizontal and vertical axes between the main girder and the LCB. In the case of twin-deck erection, the LCB can extend over both decks, which allows it to shift the main girders from the left to the right deck and vice versa to install the segments of the two decks during the same launching cycle.

Figure 5.1 General arrangement and main components of LG-type I: erection of field segments about pier N (for clarity, access and working platforms are shown only partially) (VSL International)

A. Main support of main girders
B. Main girders
C. Trolley with lifting device and segment lifting beam
D. Temporary supports
E. Stressing platform

B. **Main girder**: For LG-type 1, the main girder is a pair of trusses, called main trusses (**MTs**). During segment erection and when the LG is moved longitudinally during launching the MTs are sitting on the main supports. Each MT has at both main supports its own set of ULRS. Main girders of BCM launching gantries are always trusses in order to minimise the MG's self-weight. They have either a triangular or rectangular cross section. The MT's chords generally have an I-section with a single web or a pair of webs. The chords not only carry the stringer forces from the MT's global bending but also serve as rail beams during placing of segments and launching of the BDE. Chord splices have therefore to be detailed carefully so as not to obstruct passage of rollers during segment placing and launching.

C. **Trolley with lifting device**: The trolley consists of an Upper Cross Beam (**UCB**), which spans as a simply supported beam or a frame between the pair of MTs. It supports the segment-lifting device. The lifting device for the lifting and lowering of the segments is mostly a winch but can also be a Strand Moving Unit (SMU) or even a long-stroke jack when segments are delivered over the deck. The UCB is mounted on wheel bogies, which permit the longitudinal movement of the trolley along the MTs. A segment-lifting beam connected to the lifting device's lower end allows for adjustment of the segments in space, needed when joining segments to each other.

   Usually an LG-type I is equipped with a pair of segment lifting devices. This allows it to minimise the out-of-balance moment introduced into pier N during segment erection. A pair of corresponding upstream and downstream field segments can be kept suspended symmetrically from the LG at the deck's cantilever ends until both segments are attached to the deck. Once the segments are attached, the lifting devices are released slowly and simultaneously.

D. **Temporary supports**: The MTs are equipped with temporary supports at their front end, called Front Support Leg (**FSL**), and at their rear ends, called Rear Support Leg (**RSL**). They are utilised during launching for the relocation of the LG's main supports. These legs are equipped with hydraulic jacks to compensate for the MTs' deflection, when the MTs are in front or rear cantilever mode.

E. **Stressing platform**: For the installation and stressing of the cantilever tendons, access platforms – called Stressing Platforms (**SPF**) – are needed. They are handled by the LG but are generally fixed to the tips of the cantilever under construction when in use. A pair of SPFs is required.

F. **Bracket** (not required for the LG shown in Figure 5.1): Brackets, which are attached to the downstream side of the pier or cross head N, may be required if pier segments are placed by the LG and there is not enough space available to activate the FSL.

This type of gantry can receive segments from behind or below and can be used for single- or twin-deck erection.

5.3.2 Launching Gantry with a single main truss (LG-type II)

A typical LG-type II is made up of the following main components (refer to Figure 5.2).

A. **Main support of main girder:** During erection of the field segments about pier N and pier segment N + 1, the LG-type II is supported on two main supports and the front support. The front main support is centred over pier N and the rear main support over pier N – 1. The supports are always placed on top of the deck and their legs are normally centred over the deck's webs. The main element of these supports is the transverse beam, called the Lower Cross Beam (**LCB**). The LCB is orientated perpendicular to the bridge axis and is made up of a pair of I-beams. Between the LCB and the main girder there are mechanical devices, called Universal

Figure 5.2 General arrangement and main components of LG-type II: erection of field segments about pier N and pier segment at pier N + 1 (for clarity, access and working platforms are shown only partially) (VSL International)

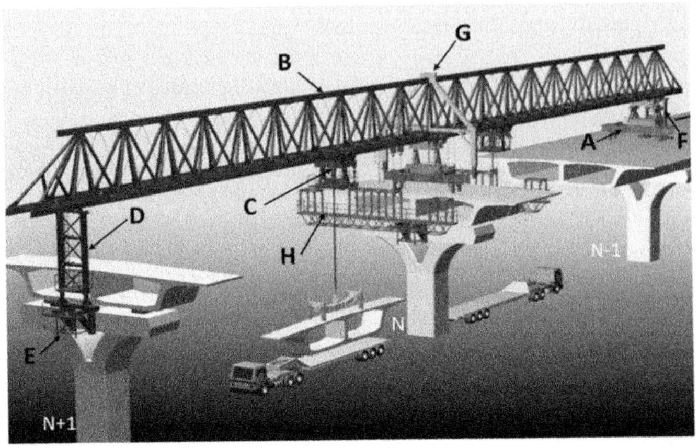

A. Main support
B. Main girder
C. Trolley with lifting device and segment-lifting beam
D. Front support
E. Bracket
F. Temporary support
G. Centre support
H. Stressing platform

Lower Roller Support (**ULRS**). The ULRS allows vertical, transverse, longitudinal translations and rotation about horizontal and vertical axes between the main girder and the LCB.

B. **Main girder**: For LG-type II the main girder is a single truss, called the Main Truss (**MT**). During segment erection the MT is sitting on the two main supports (at pier N − 1 and N) and the front support (at pier N + 1). When the LG is moved longitudinally during launching from piers N − 1/N/N + 1 to pier N/N + 1/N + 2 the MT is rolling over a ULRS at pier N and pier N + 1.

C. **Trolleys with lifting device**: The trolley is basically the upper part of the URLS, but equipped with the segment-lifting device. The lifting device for the lifting and lowering of the segments is mostly a winch but can also be an SMU. The trolley is suspended from and can be moved along the MT's bottom chords. The LG-type II has a front and rear trolley. During longitudinal movement of the MT, the front trolley's ULRS is swapped with the ULRS of the front main support.
A segment-lifting beam connected to the lifting device allows for adjustment of the segments in space, needed when joining segments to each other.

D. **Front support**: The Front Support Leg (**FSL**) provides the front support of the MT during placing of segments and during relocation of the front main support. It is equipped with hydraulic jacks to compensate for the MT's deflection, when the MT is in front cantilever mode.

E. **Bracket**: A bracket, which is attached to the downstream side of pier N + 1, is required to provide space for activation of the FSL.

F. **Temporary support**: The MT has at its rear end a temporary support, called the Rear Support Leg (**RSL**). During launching, this leg is utilised to relocate the LG's main supports. It is equipped with hydraulic jacks to compensate for the MT's deflection, when the MT is in rear cantilever mode.

G. **Centre support**: The centre support fixed to the MT at its centre is used during launching for swapping the ULRS of the front trolley with the main support's ULRS and for relocation of the main supports. It is equipped with hydraulic jacks to enable its activation.

H. Stressing platform: For the installation and stressing of the cantilever tendons, access platforms – called Stressing Platforms (**SPF**) – are needed. They are handled by the LG but for operation they are generally fixed to the tips of the cantilever under construction. A pair of SPFs is required.

In the configuration shown in Figure 5.2, the LG also erects the pier segments. It can only receive segments from below and is erecting a single deck. The gantry can be launched forward immediately after completion of a balanced cantilever, that is before the midspan closure pour has been temporarily or permanently established, since during launching the supports are always centred over the pier segments. Of course it is possible to configure LG-type II so it can also pick up segments from behind. This is achieved by raising the MT higher above deck. The rear main support has to be replaced by a rear portal and the middle support of the MT during segment erection is on the centre support and not the front main support. With these changes, the trolleys can pass with suspended segments through these two supports.

5.3.3 Launching Gantry with a single main truss and stays (LG-type III)

A typical LG-type III is made up of the following main components (refer to Figure 5.3).

A. Pylon: The pylon, together with the rear portal, is the main support of the LG-type III during erection of segments about pier N. For erection of the field segments, the pylon is centred on

Figure 5.3 General arrangement and main components of LG-Type III: erection of field segments about pier N (for clarity, access and working platforms are not shown) (VSL International)

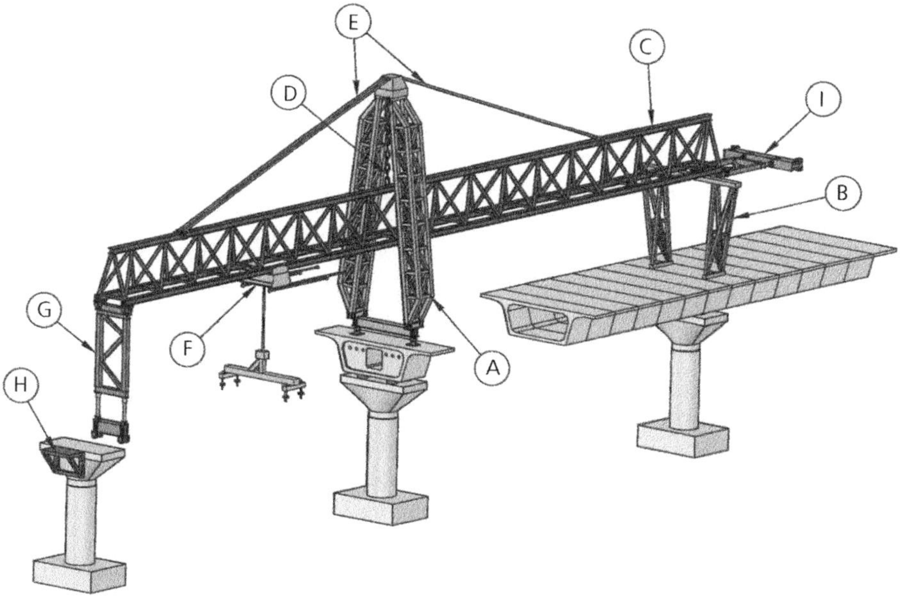

A. Pylon
B. Rear portal
C. Main girder
D. Vertical stay
E. Stays (inclined)
F. Trolley with winch
G. Front support
H. Bracket
I. Temporary support

pier segment N. For placing of pier segment N + 1, it is standing on the completed downstream cantilever of pier N. The pylon is an A-frame, which supports the LG's MG by vertical and inclined stays. It sits on a Lower Cross Beam (LCB), so that the support reactions can be introduced into the deck's webs. The LCB also provides the horizontal tie for the A-frame. The LCB's supports are hydraulic jacks to enable the pylon's activation.

B. **Rear portal:** The Rear Portal (**RP**) supports the LG during placing of segments and during longitudinal movements of the LG. For placing of the field segments, it is located on the deck over pier N – 1, and for placing of the pier segment N + 1, some distance away from the centre line at pier N – 1. The main element of the RP is a two-legged frame, whose legs are centred over the deck's webs. Between this frame and the main girder there is a mechanical device, called the Universal Lower Roller Support (**ULRS**). The ULRS allows vertical, transverse, longitudinal translations and rotation about horizontal and vertical axes between the main girder and the RP.

C. **Main girder**: For LG-type III, the main girder is a single truss, called a Main Truss (**MT**). During segment erection the MT is supported from the pylon at pier N and the RP at pier N – 1. When the LG is moved longitudinally during launching, the MT is rolling at the front over the roller frame (converted trolley) and at the rear over the ULRS of the RP.

D. **Vertical stay**: The MT's vertical support at the pylon is provided by a vertical stay. The MT can therefore be rotated about the pylon's vertical axis by transverse translation of the MT on top of the RP, which enables the LG-type III to cope with very tight radii in plan (about 120 m).

E. **Stays (inclined)**: Bending in the MT is reduced during placing of the field segments by one or two sets of symmetrically arranged stays, normally transversely in pairs spanning between the top of the pylon and the MT's top chord. These stays are stressed prior to and de-tensioned after the placing of the field segments.

F. **Trolleys with lifting device**: The trolley carries the lifting device. The lifting device for the lifting and lowering of the segments is mostly a winch but can also be an SMU when the segments are delivered over the deck. The trolley is suspended from and can be moved along the MT's bottom chords. The trolley can be converted into a roller frame (**RF**) to provide the front support during longitudinal movement of the LG. A segment-lifting beam connected to the lower winch block allows for adjustment of the segments in space, needed when joining segments to each other.

G. **Front support**: The Front Support Leg (**FSL**) provides the front support of the MT during placing of the pier segment and during relocation of the main supports. It is equipped with hydraulic jacks to compensate for the MT's deflection, when the MT is in front cantilever mode.

H. **Bracket**: A bracket, which is attached to the downstream side of pier N + 1, is required to provide space for activation of the FSL.

I. **Temporary support:** The MT is at its rear end equipped with a temporary support, called a Rear Support Leg (**RSL**). The RSL is utilised during launching for the relocation of the LG's main supports. This leg is equipped with hydraulic jacks to compensate for the MT's deflection, when the MT is in rear cantilever mode.

J. **Stressing platform**: For the installation and stressing of the cantilever tendons, access platforms – called Stressing Platforms (**SPF**) – are needed. They are handled by the LG but for operation they are generally fixed to the tips of the cantilever under construction. A pair of SPFs is required.

5.3.4 Minimum vertical clearance between the MT and the deck

Minimum vertical clearance between MTs and the deck depends on the type of LG and whether segments are delivered from the ground below or from behind over the deck. The vertical clearance for the MT of an LG-type II is the lowest, since segments are always delivered from below, whereas it is highest for an LG-type III if segments are delivered over the deck.

Whenever segments are delivered over the deck, the Rear Portal (RP), the Centre Support (CS) and the Rear Support Leg (RSL) have to be detailed as portals wide enough to allow the passage of the segments.

5.3.5 Vertical tie-downs

It is a good practice to provide vertical tie-downs between the deck and the LCBs, where the MT's reaction isn't introduced vertically but perpendicularly to the MT's gradient (the MT is sitting on a ULRS, which has articulated rollers). The rear portal, the centre support, the roller frame and (if used for placing of segments) also the front support leg should always be actively stressed down with HT bars to the deck. Transfer of horizontal forces from the LG to the deck, which is almost always by friction, should be guaranteed independently from size and orientation of the MT's vertical support reactions. During heavy wind, for example, a downwards reaction in an MT support can quickly be reduced to almost zero.

5.3.6 Longitudinal fixity

Provision of longitudinal fixity of the LG is, of course, key to ensuring equilibrium at any point of operation. The stressing of the inclined stays for the placing of the field segments with an LG-type III, for example, creates a flexural connection between the MT and the pylon. The resultant force of the longitudinal forces acting on the MT is transferred to the deck at the base of the pylon, whereas the overturning moment is stabilised by relatively small oppositely acting vertical reactions at the rear portal and the pylon. The MT can remain on rollers at the rear portal. The situation is very different for an LG-type I, where the MTs have to be fixed longitudinally to either of the two main supports. The LCBs themselves need to sit longitudinally on pairs of supports spaced sufficiently apart to be able to cope with the overturning moment due to longitudinal resultant forces.

5.4. Kinematics

5.4.1 LG-type I

Typical kinematics for an LG-type I are shown in Figure 5.4. This example presents the erection of the pier segments by the LG and segment delivery over the deck. LG-type I is very suitable to be used for twin-deck erection. For such an arrangement, the LCB is, as already mentioned above, lengthened to extend over both decks and is equipped typically with four pairs of legs positioned over the webs of the two decks. Such a set-up allows erection of decks with very tight radius in plan.

Different options for how pier segments can be erected are shown in Figure 5.5.

5.4.2 LG-type II

Typical kinematics for an LG-type II are shown in Figure 5.6. The placing of the pier segments is by the LG and segments are delivered from below. The LG is in a two-span configuration for placing of all segments. The LG-type II is very suitable for erection of single decks with delivery of segments from below. For decks which are curved in plan, this LG can also be used in a single-span configuration.

Figure 5.4 Kinematics for LG-type I: placing of field segments about pier N; launching from piers N – 1/N to pier N/N + 1 including erection of pier segment N + 1 (VSL International)

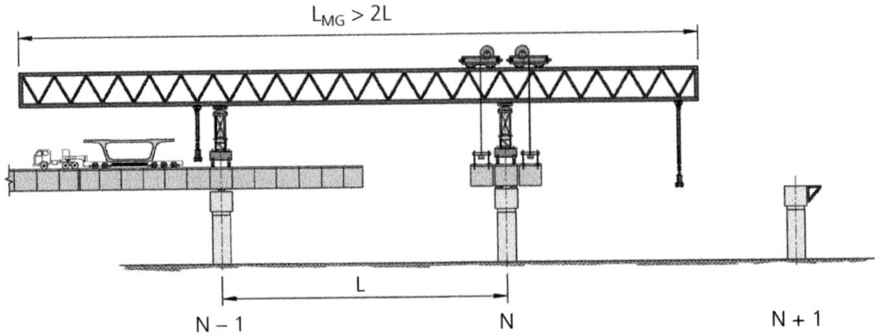

LG over span N – 1/N for placing of field segments about pier N

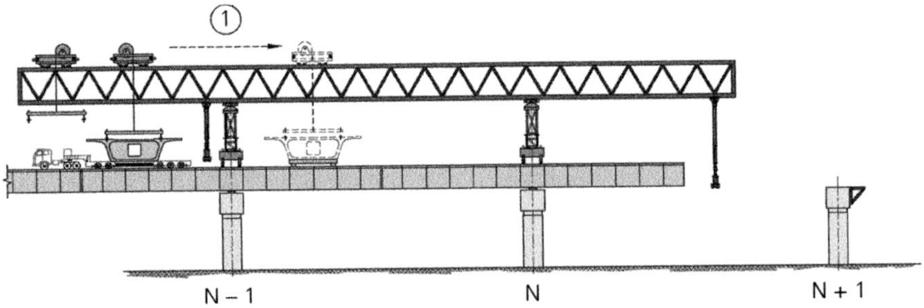

1. Pick up the pier segment N + 1 with the front trolley and park it on the deck in front of the rear main support.

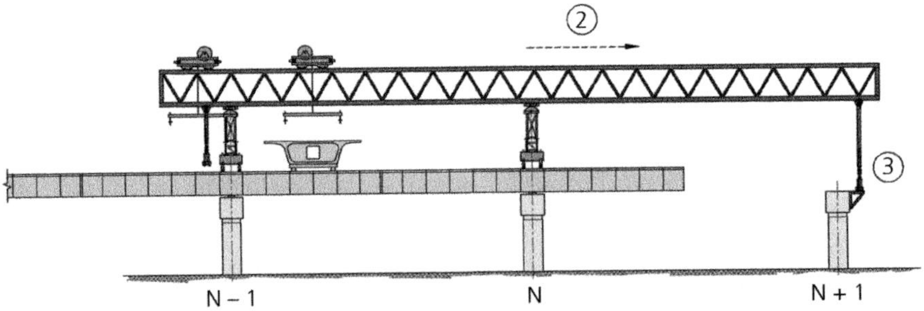

2. Launch the Main Girder (MG) forward until its tip is over the bracket at pier N + 1.
3. Activate the Front Support Leg (FSL) on the bracket.

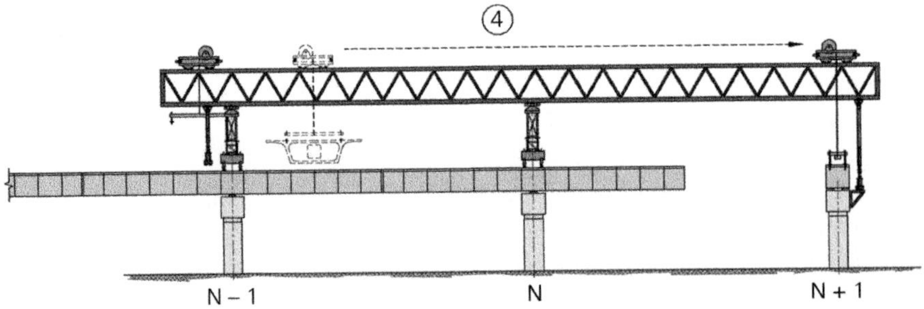

4. Place the pier segment N + 1 with the front trolley on pier N + 1 and nail it to the pier.

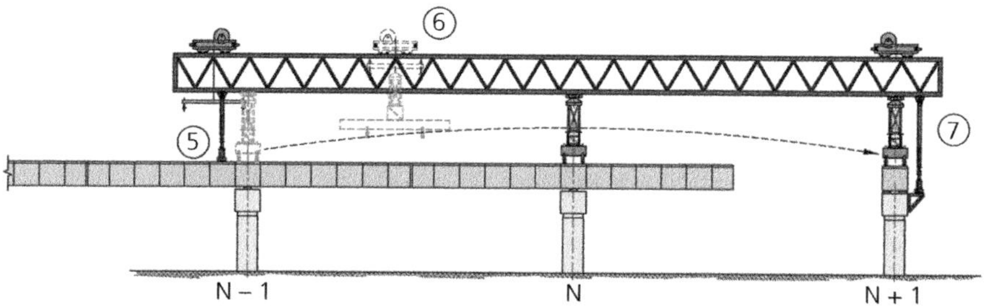

5. Activate the Rear Support Leg (RSL).
6. Move the main support from pier N − 1 to pier N + 1.
7. Transfer load from the FSL to the main support.

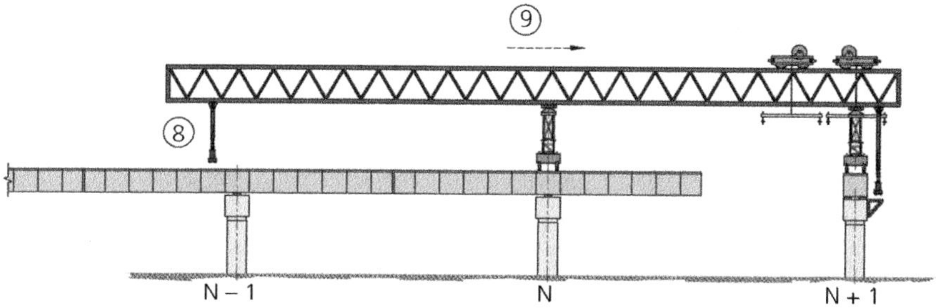

8. Deactivate the RSL.
9. Launch the MG forward until it is centred over span N/N + 1.

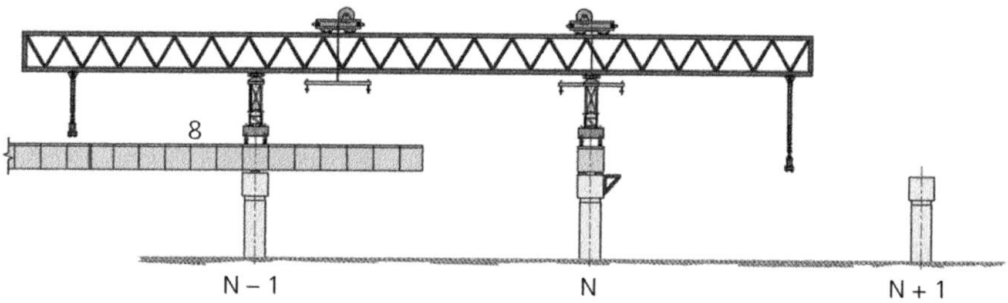

LG over span N/N + 1 for placing of field segments about pier N + 1.

5.4.3 LG-type III

Typical kinematics for an LG-type III are shown in Figure 5.7. The example presents the placing of the pier segments by the gantry and segment delivery over the deck. The LG is in a two-line support configuration for placing of the field segments. The LG-type III is very suitable for erection of single decks, for decks with a tight radius in plan and for decks built through narrow corridors.

5.5. Permanent work interfaces

5.5.1 Pier table

When the pier table is cast in situ there is a need for wet joints between the pier table and the first pair of field segments held in place by the LG or by hanger beams during casting and curing of the wet joints. The pier segments are often split in two to keep the weight to a level acceptable to the BDE. Sometimes the pier segments are cast without diaphragms to reduce weight and/or to allow monolithic connection between the deck and the pier by vertical reinforcement. For more explanations about the detailing of pier tables for concrete bridge decks, refer to Section III, Chapter 15.

Figure 5.5 Options for placing of pier segments by LG-type I (VSL International)

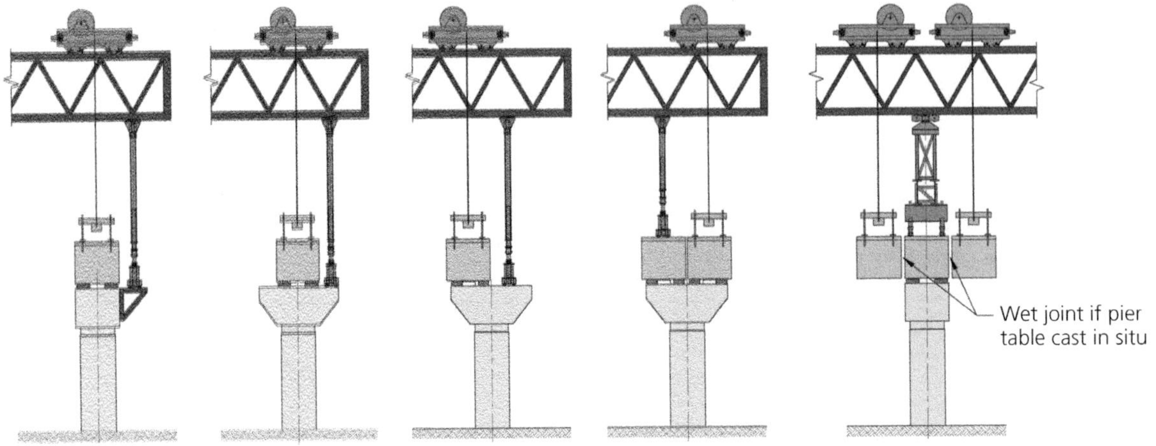

5.5.2 Support reactions

BCM gantries are heavy pieces of equipment with weights ranging from 2000 kN to more than 10 000 kN depending on span length and segment weight. During launching, the reaction at the front support reaches almost the total weight of the gantry. Therefore the supports during launching and segment placing always need to be directly over webs, diaphragms or on piers.

If a support that sits on the deck cannot be centred over a pier, particularly when located on the cantilevering part of the deck, the deck has to be checked to demonstrate that it is able to cope with the temporary loading. If it cannot, it may be possible to strengthen the deck, for example by additional temporary cantilever prestressing tendons.

5.5.3 Deck stability

As explained in Section III, Chapter 15, decks that are line-supported on piers need to be stabilised during cantilever construction against overturning. This can be done either by nailing the pier table to the pier or by provision of steel props in front of and behind the pier. These props sit directly on the pier's pile cap. As explained earlier, out-of-balance moments at piers can be significantly reduced when gantries equipped with two trolleys are used, as long as the self-weight of a pair of segments is transferred only after both segments have been temporary stressed to the deck. For single-deck erection it is possible to stabilise a line-supported deck after its construction has progressed some distance away from the pier axis by the LG itself. A special leg, acting as a pendulum between the MT and the deck behind or in front of the LG's main support, is provided for this purpose. Since the leg is progressively moved outwards as the cantilevers advance, there is need for two legs, so that stability can be maintained during the resetting of a leg.

On completion of a pair of cantilevers, the structure is stabilised by casting a closure pour against the already built deck on the upstream side. This closure pour is usually 20 to 30 cm in length. The length

Figure 5.6 (Part 1 of 2) Kinematics for LG-type II: placing of field segments about pier N and pier segment at pier N + 1; launching of LG from piers N – 1/N/N + 1 to piers N/N + 1/N + 2 (VSL International)

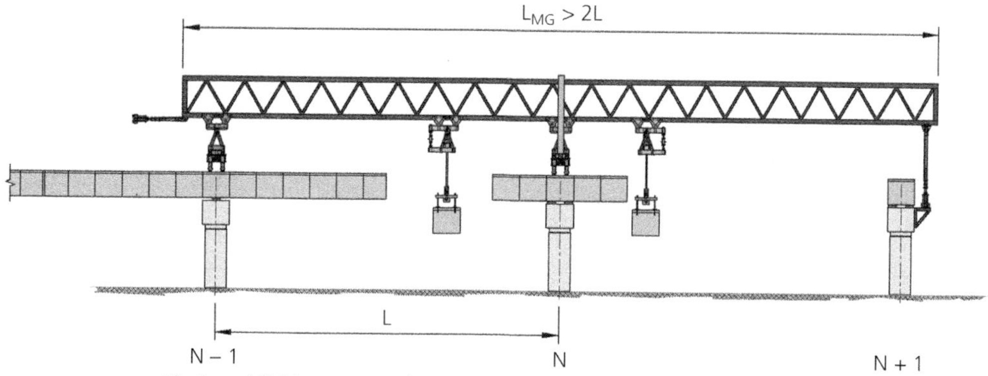

Placing of field segments about pier N and placing of pier segment N + 1

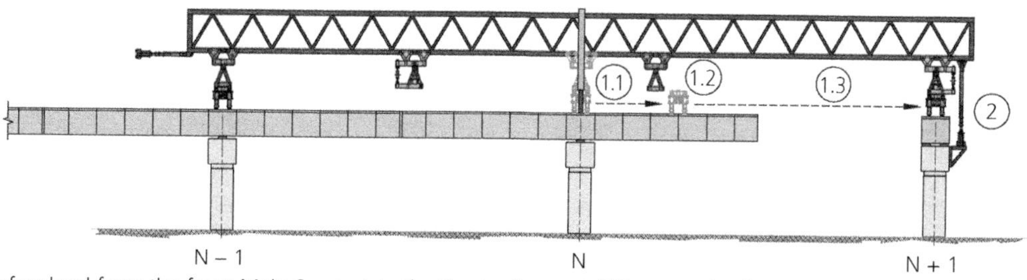

1.1 Tranfser load from the front Main Support to the Centre Support (CS); move the front Main Support forward
1.2 Park front Lower Cross Beam (LCB) on deck; disconnect from front Universal Lower Roller Support (ULRS); move and keep front ULRS suspended behind the front LCB
1.3 Connect front Trolley as ULRS to front LCB and move both to and place on pier N + 1
2 Transfer load from Front Support Leg (FSL) to front Trolley/LCB

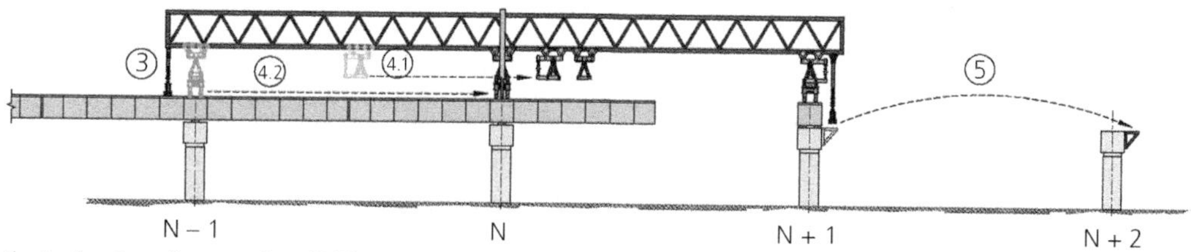

3 Activate the Rear Support Leg (RSL)
4.1 Move the rear Trolley through the CS behind the suspended front ULRS
4.2 Relocate rear Main Support from pier N – 1 to pier N; transfer load from the CS to the rear Main Support
5 Relocate the bracket from pier N + 1 to pier N + 2 (by crane)

is chosen to allow for geometrical construction tolerance, while keeping it short enough that it does not have to be reinforced with reinforcement bars. Sometimes the stitch concrete is not poured immediately. Cantilever stability is achieved instead by the installation of some temporary steel beams (stitching beams) across the joint. When the stitch is fixed (by whichever method), the main bearings under the new cantilever must be activated so thermal and shrinkage movements can start. If stitching beams are used, they have to be designed not only to provide cantilever stability but also to overcome friction at the bearings at the new pier.

Figure 5.6 (Part 2 of 2)

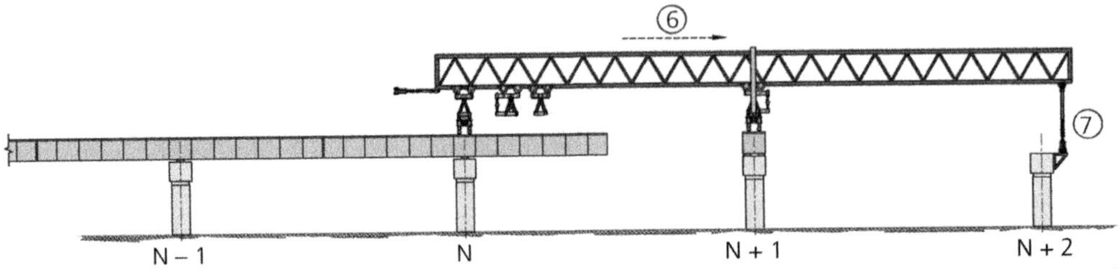

6. After deactivation of the RSL, launch the Main Girder (MG) forward until the FSL can be activated on the bracket at pier N + 2.
7. Activate FSL.

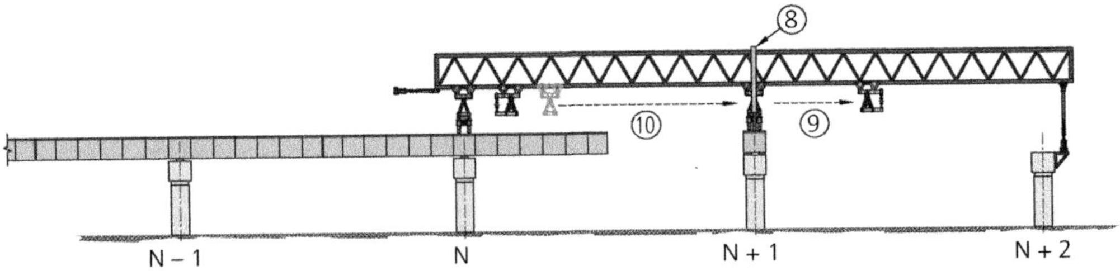

8. Activate CS.
9. Disconnect front trolley from the front LCB and move forward.
10. Move the front ULRS over pier N + 1 and connect it to the front LCB.

For details on how to erect the out-of-balance part of an end span and how to erect decks over internal expansion joints, refer to Section III, Chapter 15.

5.5.4 Arrangement of temporary and permanent PT

A segment is initially attached to the already erected part of a cantilever by temporary PT bars, which are also used to precompress the entire segment joint to ensure effective distribution of the epoxy resin applied to one of the matching segment faces.

Temporary PT bars are typically anchored in temporary concrete blisters or (as an option mainly only for top bars) in temporary steel brackets. This arrangement allows stressing from the already erected deck.

Locating the temporary PT bars inside the segment's slabs is not encouraged, since such bars can only be stressed from the front of the segment. This layout therefore requires stressing access from the stressing platform when the segment is still suspended from the LG. Working from a suspended stressing platform should, however, be permitted only if there is redundancy in the suspension system.

Sometimes the permanent cantilever tendons are anchored in alternate pairs of segments, requiring bigger tendons and a significant increase of the temporary PT rarely justifying the nominal increase in the speed of construction due to reduced number of stressing operations for the permanent tendons.

5.6. Cycle times

Table 5.1 shows a cycle for the erection of a 50 m span. The symmetrical cantilever is made up of a 2.07-m-long pier segment and eight pairs of 2.98-m-long field segments. The cycles shown in the table

Figure 5.7 Kinematics for LG-type III: placing of field segments about pier N; launching of LG from piers N − 1/N to piers N/N + 1, including placing of pier segment N + 1 (VSL International)

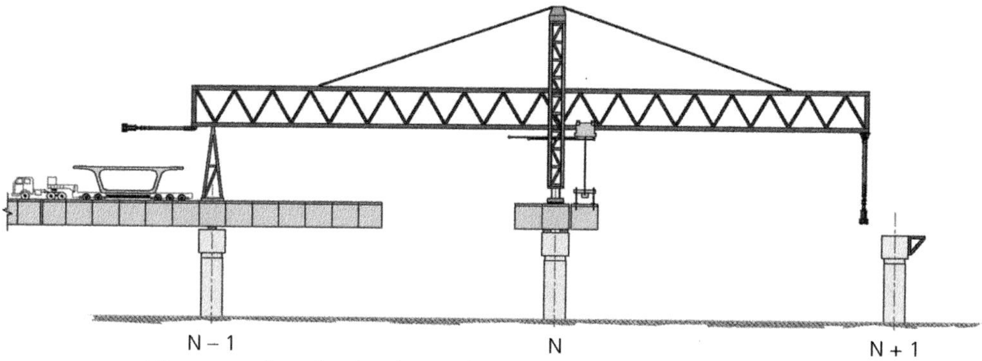

LG supported on pier N and N − 1 for placing of field segments about pier N

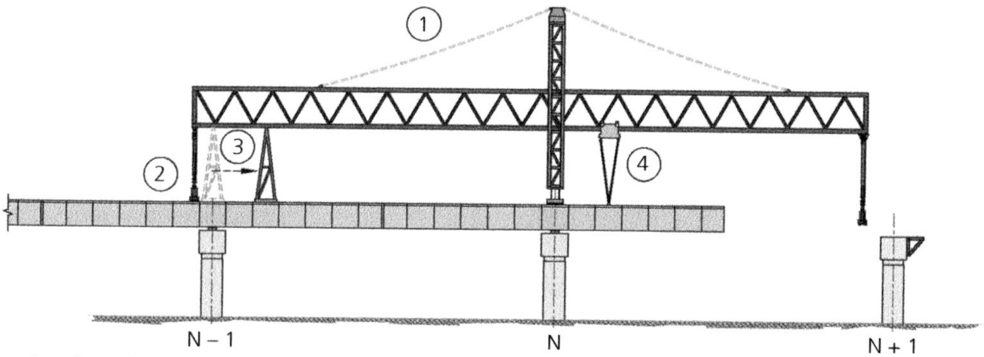

1. De-tension the inclined stays.
2. Activate the Rear Support Leg (RSL).
3. Move the Rear Portal (RP) forward and park it on the deck; transfer load from the RSL to the RP.
4. Move the trolley in front of the Pylon and convert it into the Roller Frame (RF), fixed to the deck; transfer load from the Pylon to the RF.

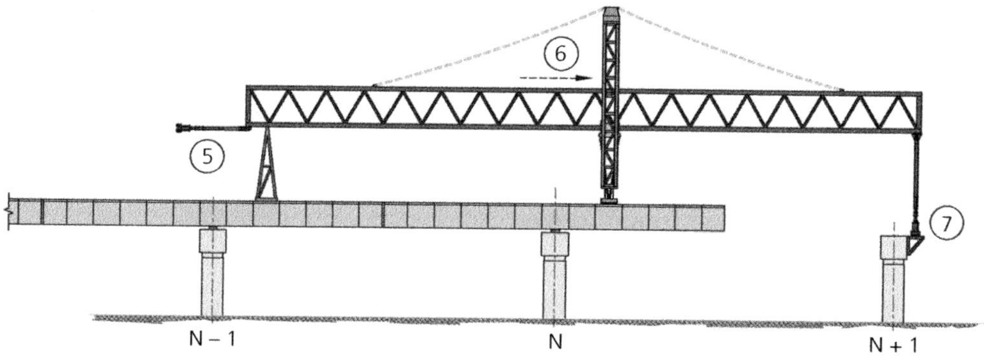

5. Rotate the RSL upwards.
6. Launch the Main Girder (MG) forward until the Front Support Leg (FSL) can be activated on the bracket at pier N + 1 and the Pylon is over the RF, respectively.
7. Activate the FSL and transfer load from the RF to the Pylon.

Figure 5.7 Continued

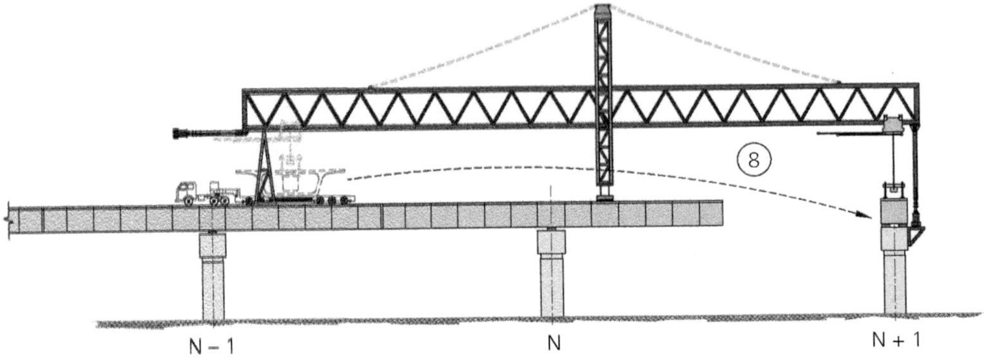

8. Pick up the pier segment N + 1 with the trolley and place it on pier N + 1; nail down the pier segment N + 1.

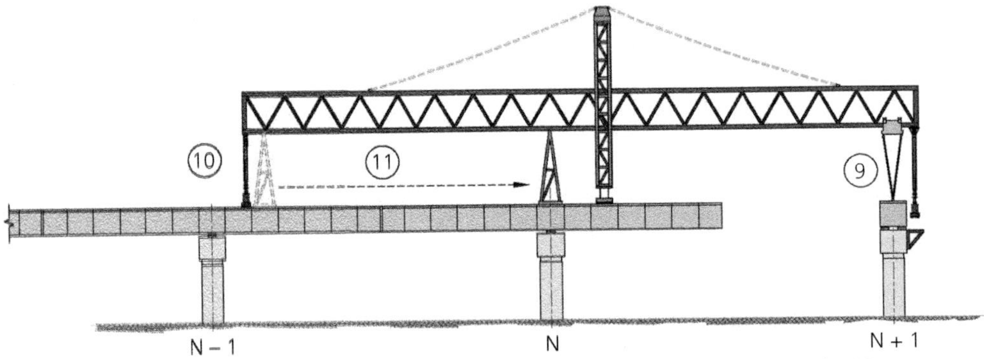

9. Convert the trolley into the RF, fixed to the pier segment N + 1; transfer load from the FSL to the RF.
10. Rotate the RSL downwards and activate.
11. Move the RP to and place on the pier segment N; transfer load from the Pylon to the RP; deactivate the RSL and rotate upwards.

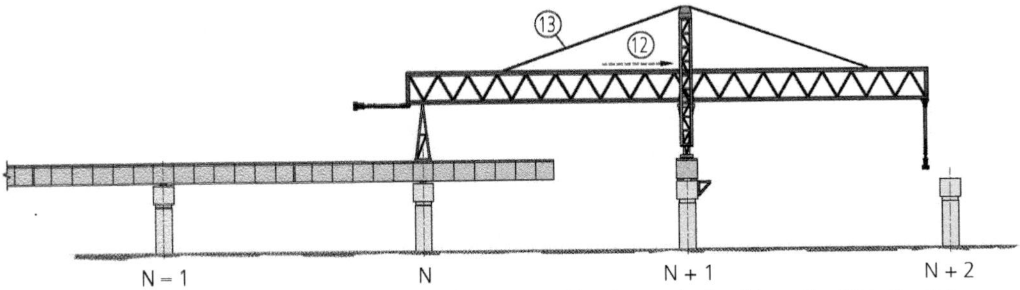

12. Launch the MG forward until the Pylon is over the pier segment N + 1; transfer load from the RF to the Pylon.
13. Tension the inclined stays.

are for illustration only and the overall cycle times should not be used for actual projects unless verified against project-specific site constraints.

Assumptions are as follows:

- starting point: balanced cantilever over pier N – 1 and upstream midspan closure completed and continuity PT in span N – 2/N – 1 stressed; LG ready for launching

Table 5.1 Cycle for an LG-type I

Activity	Duration In hours	Day 1				Day 2				Day 3				Day 4				Day 5			
First-stage launching of LG and placing of pier segment N	5		X																		
Alignment and nailing of pier segment	5			X																	
Second-stage launching of LG	5				X																
Placing of first pair of field segments: application of epoxy glue, joining, temporary PT, permanent PT	5					X															
Placing of remaining seven pairs of field segments	7 × 5						X	X	X	X	X	X	X								
Alignment of cantilever tips, preparation and casting of midspan closure pour and bearing plinths	10														X	X					
Curing of midspan closure and bearing plinths	12																X	X			
Installation of continuity PT	5															X					
Stressing of continuity PT	5																		X		

- one pier segment and eight pairs of field segments
- epoxy-glued joints between segments and unreinforced midspan closure (200 mm)
- use of early-strength concrete for the midspan closure and the bearing plinths (12 hours to reach minimum strength required for load transfer)
- pier segment support during cantilever construction: on two lines of hydraulic jacks with two lines of vertical stressed HT bars between pier head and pier segment
- load transfer to permanent bearings
- two shifts per day (10 hours per shift)
- segment delivery over deck; no delay with segment delivery.

5.7. Risks

The list of risks given in Table 5.2 is by no means comprehensive, but addresses risks that are particular to LGs used for balanced cantilever deck erection with precast segments. For a specific project, a main contractor needs to carry out a full risk assessment of the operations performed with the launching gantries, considering the actual features of the LG, the erection sequence and the site constraints.

Table 5.2 List of main risks and mitigating measures

Activity	Risk	Mitigating measures
Working on top of and inside of segment Attachment of segment-lifting beam (SLB) to segment delivered on trailer	Falling down from segment	Provision of temporary handrails around all exposed edges on top and inside of segment, installed prior to lifting of segment
Segment erection over live traffic	Drop of segment (unlikely), falling of debris, tools (possible)	No segment handling over live traffic
Segment suspension over live traffic	Drop of segment (unlikely)	Permitted after stressing of permanent cantilever PT
Epoxy gluing of segment joints	Dropping epoxy	No traffic if over road, plastic sheets or similar below joints to catch epoxy
	Closing of gap between segments due to swinging of segments	Use of timber shims, which are supported on top of the deck slab
Support of LG	Failure at support's contact area	Adequate edge distances, so that support's contact area is within concrete confined by reinforcement; shimming with similar material over entire footprint of support's contact area
Longitudinal movement of main girders	Over-launching or loss of control of longitudinal movement → loss of equilibrium → overturning of LG's supports and loss of entire LG	Back-up fixity, nominal tie-down of supports, presence of LG engineer and LG supervisor
Temporary fixity of segments to deck by temporary PT bars	Failure of temporary blisters/ anchorage blocks and/or its fixity to the deck	Either permanent blisters or, if temporary, with fixity to deck not relying on friction
Casting of wet joints between pier segment and first pair of field segments	Field segments move during concreting	Provision of 'rigid' fixity scheme to minimise movement of field segments during concreting
Casting and curing of midspan closure pour	Cracking of cast-in-situ concrete	Provision of stitching beams able to transfer global bending moment about transverse and vertical axes and vertical and horizontal shear (alignment forces, self-weight of stitch concrete, wind, temperature variations {global and gradient}); casting of closure pour in the evening with start of stressing with the continuity PT early next morning (before sun starts to warm up the top surface of the deck → work with and not against temperature gradient's daily fluctuation)

5.8. Performance criteria

The following items need to be addressed when drafting a performance specification for an LG.

- Segment
  - segment characteristics (per segment type): weight, dimensions, centre of gravity
  - delivery: from below, from behind, both
- Pier table: cast in situ (need for wet joint between pier and first pair of field segments), precast placed ahead by crane or precast placed by LG
- Alignment: gradient, curvature in plan and cross-fall variations
- Permanent work interfaces
  - permissible deck loading and locations (various support positions of LG on deck during erection and launching and loading of deck during segment delivery if over deck): load intensity, position in longitudinal and transverse direction, contact area
  - stability of deck during BCM erection
    - magnitude of out-of-balance moment that can be taken by the foundation, the pier and can be transferred from pier segment to the pier head
    - nailing:
      - location of temporary support jacks/shims (on pier head) or props on pile cap if required
      - location for temporary vertical tie-down bars/strand tendons if required
      - provision to fix cantilever longitudinally, transversely and against rotation about vertical axis
  - permissible geometrical alignment of cantilever tips prior to casting of midspan closure pour
  - permissible eccentric loading of piers during placing of pier segments
  - stability of twin decks supported by single piers with T-type cross heads: magnitude of permissible transverse out-of-balance moment (due to LG plus pair of segments)
  - space constraints for supports of LG on pier segment and, if applicable, on pier head
  - permissible recesses for support of LG's pier bracket
  - deck erection of out-of-balance part in end spans: suspension of segments in LG or support on falsework
  - deck erection over internal expansion joints: suspension of segment in LG or false cantilever construction
  - deck geometry at bifurcations and ramps
  - anchoring point for cantilever PT: does every pair of segments have its set of permanent cantilever PT? Traffic should never be allowed below segments that are attached to the cantilever only by temporary PT bars
  - access requirements for stressing of temporary high-tensile bars and permanent PT tendons (anchoring temporary PT bars at front face of segment being installed should be avoided)
  - starter bar location and dimensions (for parapets/central dividers, rail support plinths, etc.): obstructions to relocation of LCBs, obstruction for segment delivery over deck
  - stitching requirements
    - wet joints between pier segment and first pair of field segments: minimum concrete strength to permit stressing of cantilever PT and continuation with field segment erection
    - closure pour at midspan: minimum concrete strength and minimum number of stressed continuity tendons to permit the release of the nailing at the leading pier, then the launching of LG and delivery of segments over deck (if required). Note: if stitching beams are left behind to remove the casting and curing of the closure pour from the critical path, there must be enough vertical clearance to move the LCBs over the stitching beams during the launching of the LG

- Follow-up work (closing of lifting holes, removal of swift lift studs, etc.)
- Clearances: vertical below and above deck (overhead power lines, bridges, etc.); horizontal (houses, power lines, etc.)
- Cycles (double or single shift, restrictions for segment delivery, restriction for noisy type of work {strand pushing}, etc.)
- Mechanisation (labour costs, skill level, etc.)
- Longitudinal fixity (redundancies)
- Access: into LG (from deck, front pier) and inside LG during different stages of operation; details of temporary safety barriers for segments being erected (on top, inside); access from ground to piers (ahead of LG's arrival, during deck erection, for follow-up work) and onto and into deck
- Shipment costs (initial; if reused)
- Reusability
- Method, schedule and crane definition for assembly, relocation, dismantling, bracket relocation if applicable.

In addition, the buyer needs to give a time frame for fine-tuning of the LG's concept, support reactions/inserts and for delivery and commissioning of the LG.

5.9. Photo gallery

Figure 5.8 LG-type I: twin-deck road viaduct (Rogerville Viaduct, France) (Bouygues-Construction)

Figure 5.9 LG-type I: twin-deck road viaduct (Telok Blangah, Singapore) (VSL International)

Figure 5.10 LG-type I: twin-deck road viaduct (Gateway Upgrade Project, Brisbane, Australia) (VSL International)

Figure 5.11 LG-type II: single-deck road viaduct (Bayan Baru Viaduct, Malaysia) (VSL International)

Figure 5.12 LG-type III: single-deck rail viaduct (LRT Star project), Kuala Lumpur, Malaysia (Freyssinet International)

Figure 5.13 LG-type III: single-deck road viaduct (Pakse Bridge, Laos) (VSL International)

REFERENCES

Figure 5.8: Bouygues-Construction
Figures 5.1 to 5.7, 5.9 to 5.11, 5.13: VSL International
Figure 5.12: Freyssinet International

Bridge Deck Erection Equipment
ISBN 978-0-7277-6193-4

https://doi.org/10.1680/bdee.61934.063

Chapter 6
Precast segmental span-by-span erection by launching gantry

6.1. Introduction

Decks that are made self-supporting for self-weight span by span are by far the most common structural systems used for precast segmental viaducts. Unlike other systems, the contractor has little choice than to use overhead or underslung Launching Gantries (LG) when erecting such viaducts, unless it is possible to utilise ground-supported falsework. Decks are built span by span by stressing all segments of a span together with longitudinal prestressing cables, which run the full span length (refer to Section III, Chapter 15 for explanations about prestressing). Since such launching gantries need to be able to support the weight of an entire span, they are among the heaviest erection equipment used for bridge construction.

The method by which segments are joined and stressed together to make self-supporting spans has a direct impact on the LG's performance criteria and configuration. The permanent work designer should understand the detail of how such bridges are built, since many design and detailing aspects are governed by construction considerations and not by in-service requirements. The contractor, when ordering a launching gantry, needs a good understanding of how the bridge they have to build works structurally, to enable them to order the right erection equipment (for details, refer to Section III, Chapter 15).

The content of this chapter is an extended version of a piece written for *Structural Engineering International* (Meyer, 2011).

6.2. Application range

Precast segmental bridge decks, which are built span by span, are used for spans from 20 to 60 m, deck widths from 5 to 24 m and span weights from 2000 to more than 20 000 kN. Typical range of application is between 30 and 50 m, with span weights between 2000 and 10 000 kN and segment weights between 300 and 1500 kN. These bridges can be classified according to how they are made self-supporting. As explained in Section III, Chapter 15, we can differentiate between three systems (Ganz and Meyer, 1997). The three systems are shown in Figure 6.1 with segmentation of the deck, the sequence of construction, the location of the construction joints and the layout of the longitudinal prestressing (in red).

6.3. General arrangement of the BDE

For erection of span-by-span precast segmental viaducts, a contractor has four options (A to D), as shown in Figure 6.2.

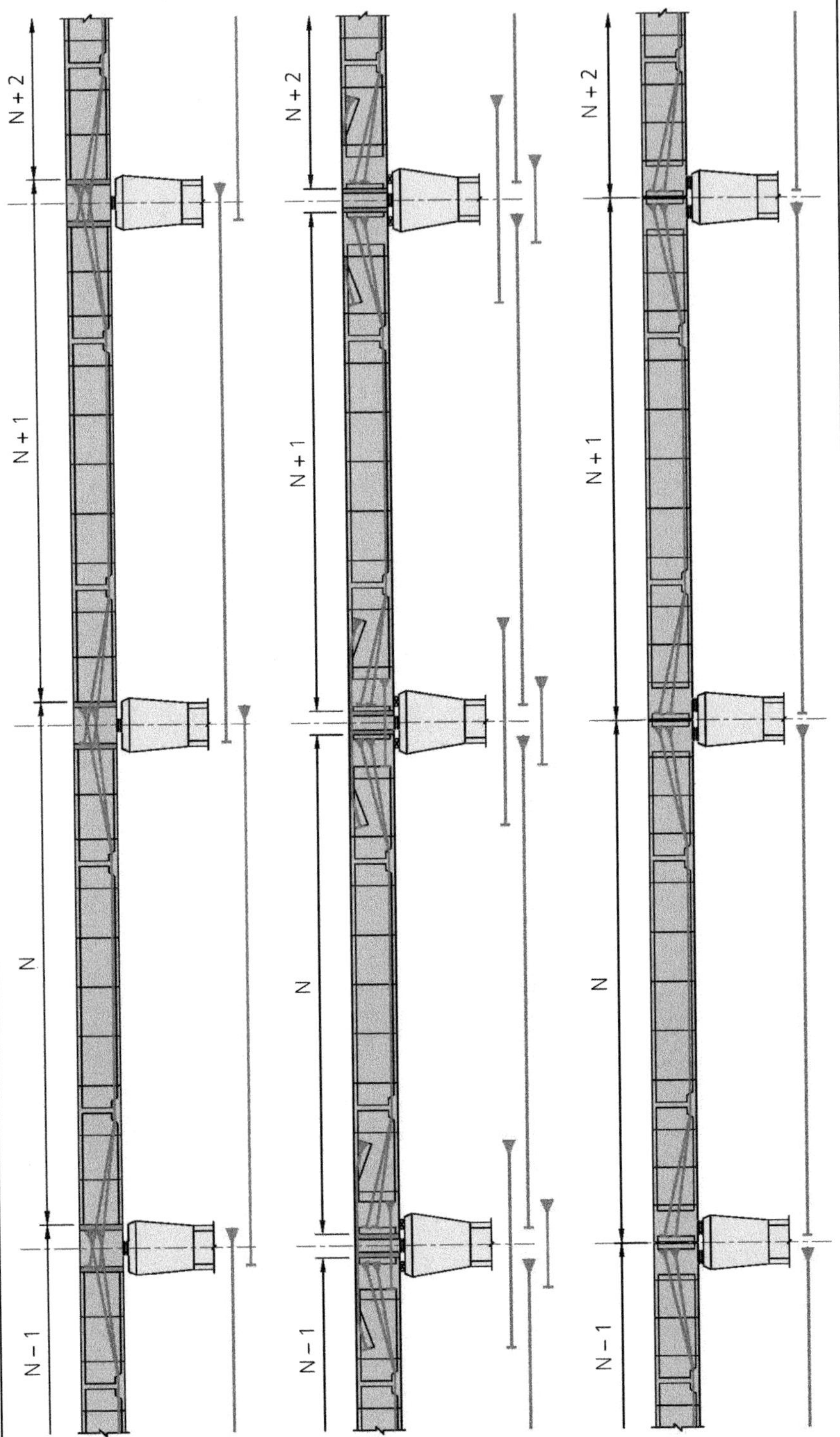

Figure 6.1 The three basic systems for construction of precast segmental bridge decks (VSL International)

Figure 6.2 The four options for construction of precast segmental viaducts (VSL International)

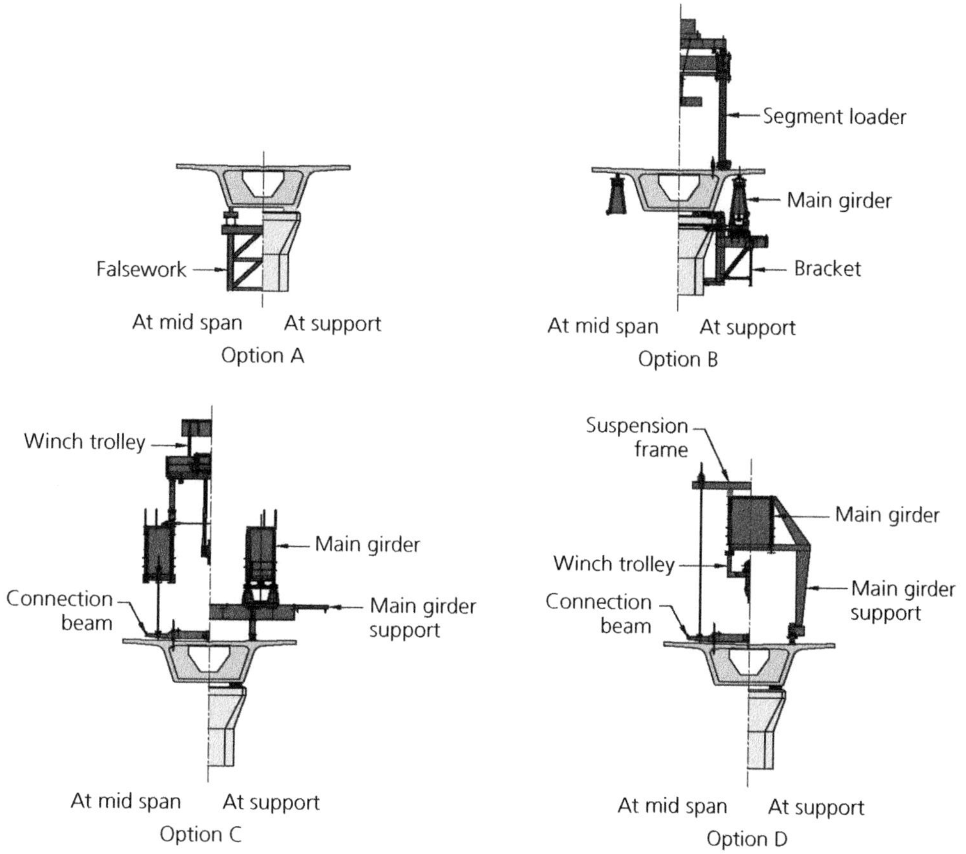

- Option A: ground-based falsework supporting segments underneath bottom slab and crane for handling of segments.
- Option B: underslung gantry with pair of main girders supporting segments underneath their wings and segment loader or crane for handling of segments.
- Option C: overhead gantry with pair of main girders for suspension of segments and for support of winch trolley for handling of segments.
- Option D: overhead gantry with single main girder for suspension of segments and winch trolley for handling of segments.

Since the subject of this book is Bridge Deck Erection Equipment (BDE), the following explanations focus on underslung and overhead launching gantries and only a few comments are given about ground-based falsework. Ground-based falsework may be a valuable option for very small projects and for starting, end and odd spans on projects where the majority of spans are erected by gantries. Often such falsework systems need to be equipped with a pair of sliding beams to allow longitudinal movement of segments along the span to get segments into their final position, or at least for fine adjustment during joining.

Figure 6.3 Isometric view of a typical underslung launching gantry (VSL International)

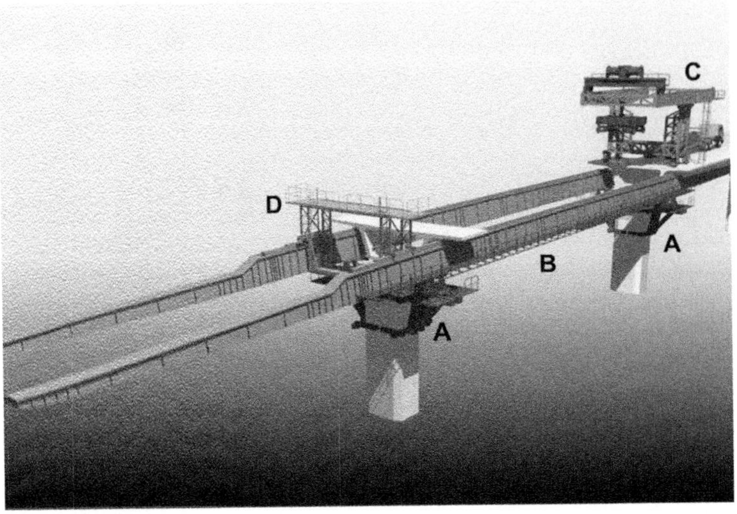

A. Brackets
B. Pair of main girders with segment trolleys and segment support jacks or screw jacks
C. Segment loader with winch trolley and lifting beam
D. Stressing platform

The rest of this subsection describes the main components of underslung and overhead launching gantries. The typical kinematics for these types of LGs are presented in subsection 6.4, 'Kinematics'.

6.3.1 Underslung launching gantries

Underslung LGs were originally developed for the erection of series of spans made up of box-girder segments with distinctive wing slabs, which are supported on single pier shafts with reasonably straight alignment in plan.

A typical underslung gantry is made up of the following main components (for illustration, refer to Figure 6.3).

A. Brackets: Pairs of brackets are attached laterally to the piers supporting the span to be erected. Normally a minimum of three sets are required per gantry. They are generally relocated by a crane. They are clamped against the pier shafts with high-tensile bars for transfer of bending moments and preferably transfer shear to pier by bearing and not by friction. Therefore often a pair of small recesses is needed at the pier top to enable such shear transfer.
Brackets have mechanical components, which allow vertical, transverse, longitudinal translations and rotation about horizontal and vertical axes between the main girders and brackets.

B. Pair of main girders: They support precast segments of a span underneath their wings until they are stressed together to create a span. Main girders have a centre part, on which segments are moved along and then supported by, and a nose at each end, used for launching. Noses are tapered at the front to allow launching onto pre-placed brackets. The rear nose is sometimes omitted by provision of a pair of C-frames at the rear of the main girder. They run on the completed deck N when launching the gantry into span N + 1. In this case, the bridge deck has to be checked for the corresponding moving point loads and there is the need for running tracks on top of the deck. The deck also has to be free of starter bars for, for example, cast-in-situ rail plinths. Since noses are relatively light, the provision of main girders with noses at both ends is the standard case.

The main girders of underslung gantries are generally equipped with segment trolleys, which run along the centre portion of the main girder's top chords. They are used to transport segments lowered onto the main girders either at the front or the rear end to their final position and are equipped with vertical and transverse hydraulic jacks to position the segment correctly during joining of segments. Hydraulic jacks (with safety rings) or screw jacks are used to support the segments on the main girders.

Main girders of underslung gantries not only have to carry the weight of a span by simple beam action back to the brackets, but also need to provide access for the operating crew to receive segments, move them along the girders and join them against each other. The classical underslung gantry gives ideal conditions in this respect, because the support of the segments is on top of the main girders, easily accessible from platforms attached to the outer side of the two main girders.

Since the main girders of underslung gantries need to run beside the piers and support the segments directly, their use is basically restricted to box-girder segments, which have cantilevering wings. Further restrictions are tight curvature in plan, bifurcations (change, for example, from twin-track segments to a pair of single-track segments and vice versa) and need for adequate vertical clearance below the wing slab in the case of a deck supported on cross heads or portals. To extend the use of underslung gantries beyond the normal application range with regard to curvature in plan, the gantry's main girders can be designed with hinged centre portions. With such articulation, spans with a radius as small as 75 m have been erected. The main girders of these gantries need to be positioned beside the segment's deck slab, since for the movement of the segments along a kinked centre portion a three-hinged portal crane is required using the pair of main girders as rail beams. Compared with the typical underslung gantry, these main girders do not, between piers, encroach on vertical clearance below the viaduct they erect. However they occupy a wider corridor, the wing slabs are subjected to bigger transverse positive bending moments and the main girders, which need to cope with significant torsion, have to be restrained on the brackets against overturning. Special care is required for design and detailing of the centre hinges (refer to Figure 6.16).

C. **Segment loader:** Sitting at the front end of the previously completed deck, a segment loader is used to load segments delivered over the completed part of the viaduct onto the main girders. Alternatively, if space permits and the lifting height and the segment weight are not too big, this task can also be achieved with a ground-based crane. A segment loader is generally equipped with a winch trolley, which has a segment- lifting beam attached to its hook block for rotation and fine adjustment of the segments when lowering them onto the segment trolleys.

D. **Stressing platform:** For underslung gantries, the access to the front face of the span, where the stressing anchorages of the longitudinal prestressing cables are located, is unobstructed. Stressing platforms are simply supported between the pair of main girders.

6.3.2 Overhead launching gantries

Overhead launching gantries were developed to enable deck erection for viaducts with relatively small radii in plan, with deck slabs without significant wings, with limited headroom below the deck or which are supported on portals.

A typical overhead gantry is made up of the following main components (for illustration, refer to Figure 6.4).

Figure 6.4 Isometric view of a typical overhead gantry (VSL International)

A. Support of the main girders
B. Main girders
C. Winch trolley with segment-lifting beam
D. Segment hanger bars and Connection Beams (CB)
E. Bracket
F. Temporary legs
G. Stressing platform

A. **Support of the main girders:** Depending on the bridge deck system, the main supports of the gantry's main girders are placed on the pier segments for system 1, on the piers for system 2 or on both pier segment and pier for system 3 (with the rear support on the previously erected deck and the front support on the front half of the leading pier). For more details about geometrical detailing of piers and pier segments, refer to Section 3, Chapter 15. When designing and detailing the front support for system 2- and 3-type bridge decks, it is very important to allow space within the front support to have adequate access for stressing the longitudinal prestressing cables. In addition, the longitudinal clearance between a support on a pier and the decks supported on the same pier needs to be sufficient. Enough space is required to enable the lifting or lowering of the last segment into its final position with allowance for the protruding part of the shear keys, the maximum deck gradient and reasonable construction tolerances for the cast-in-situ piers and the precast segmental deck. The supports of the main girders include mechanical components, which allow vertical, transverse, longitudinal translations and rotation about horizontal and vertical axes between the main girders and the supports.

B. **Main girders:** For overhead launching gantries there are two options: either a single main girder or a pair of main girders. Girders have a centre part, from which segments are suspended during span erection, and generally front and rear noses for launching. These noses are heavier than those needed for underslung gantries. The rear nose may be used to pick up segments delivered by a trailer over the deck. The front nose may be needed to place the pier segments for system 1 bridge decks. Both noses are often utilised for relocation of the gantry's main supports during the launching by the winch trolley or by a dedicated longitudinal winch. *Single main girder:* The main girder needs more vertical clearance against the deck compared to the option with a pair of main girders, since the winch trolley is running below the girder. One suspension frame per segment is required on top of the girder from which the segment's hanger bars are suspended. The single main girder is specifically suitable for gantries without a rear nose, when it is possible to roll or slide the gantry's rear support over the just-completed span during the launching, since a flexural connection of its rear support to the main girder is straightforward.

Pair of main girders: Segment hanger bars are generally directly anchored at the centre of the girders. Alternatively, hanger bars can, for special cases, also be placed either inside or outside of the main girders. Such arrangements give the option to use rectangular plate girders for the centre portion of the main girders and to park the hanger bars vertically, when not used. Parking the hanger bars vertically is, however, also possible for bars anchored at the centre of the main girders, if such girders are rectangular trusses.

As with underslung gantries, the main girders of overhead gantries do not only have to carry the weight of a span by simple beam action back to the supports, but also need to provide access for the operating crew to suspend the segments and need to provide the running tracks for the winch trolley. Often trusses and not box girders are used for the main girders, since trusses offer the possibility to locate working platforms needed for the suspension of the segments inside the trusses.

C. **Winch trolley:** In cases of a pair of main girders, the winch trolley runs on top of the main girders, whereas in cases of a single main girder the winch trolley runs below the main girder. A segment- lifting beam connected to the winch's hook block allows for adjustment of the segments in space, needed when joining the segments to each other.

D. **Segment hanger bars and connection beams:** Since the segment attachment is not only to the winch trolley's lifting beam but also to the hanger bars, each segment is equipped with a connection beam attached to the segment at four points. The lifting beam and the segment hanger bars are then indirectly connected to the segment through these connection beams. For segments with U-sections and top flanges, which cantilever inwards, it is possible to omit connection beams. Hanger bars can pass directly through holes in these flanges and the segment-lifting beam has transverse telescopic arms, which are extended underneath the top flanges to pick up the segment.

E. **Bracket:** A bracket is needed for system 1 bridge decks, since the pier segments have to be placed by the launching gantry during the first stage launching. For all other systems, brackets are generally needed only for relocation of the launching gantry's main supports during the self-launching of the launching gantry. Alternatively, to avoid the need for brackets, gantries for erection of system 2 and 3 bridge decks can be equipped with a third auxiliary main support, allowing relocation of the main supports, when the main girder is in front and rear cantilever mode respectively.

F. **Temporary legs:** Overhead gantries with front and rear noses have, in general, temporary legs at the front and at the rear end, which are utilised during the launching for the relocation of the gantry's main supports and for the placing of the pier segments for system 1 bridge decks. They are equipped with hydraulic jacks to compensate for the main girder's deflection when the main girder is in front or rear cantilever mode for the relocation of the supports. The temporary legs become redundant if the gantry is, instead, equipped with a third auxiliary main support, as mentioned under E above. This is, however, only possible for erection of system 2 and 3 bridge decks, where the launching gantry does not have to sit on the pier segment for the erection of the field segments.

G. **Stressing platform:** The stressing platform can be attached to the brackets or to the front support.

Table 6.1 gives a comparison between the features of underslung and overhead launching gantries and rates how well these types of gantries cope with deck erection aspects.

Table 6.1 Comparison between the features of underslung and overhead LGs

Activity	Underslung LG: Typical MGs	Underslung LG: MGs with centre hinge and portal crane for handling of segments	Overhead LG: Pair of MGs	Overhead LG: Single MG
Assembly and relocation	A	B	B	C
Launching	Simple, but need for three sets of brackets and crane for relocation		Fully self-launching or with bracket relocated by crane	
Lifting of segments	Crane or lifting frame	Portal crane on top of MGs	By LG (winch trolley)	
Segment support/ suspension	Under wing slab next to web: A for big curvature in plan, C for tight curves	Under tip of wing slab: A with regard to alignment in plan, C with regard to transverse bending in wing slab	Need for connection beams (except for U-sections): B	Need for suspension frames and connection beams: B
Load transfer	A, but C if LG is very flexible		A	A
Deck support on portals	C		A	
Variation in deck's cross-sectional geometry (width at base, inclination of webs)	C		A	
Vertical clearance above deck	A	A	B	C
Vertical clearance below deck	B	A between piers, C at piers	A	A
Area to be blocked off during erection	Area below deck: A	Area below deck and MGs: C	Area below deck: A	Area below deck: A

A: ideal; B: average; C: not ideal; MG: main girder

6.4. Kinematics

Figure 6.5 gives kinematics for the launching and erection of a system 3 deck with a typical underslung launching gantry and with segment delivery over the completed deck.

- Stage 1: Span N – 2/N – 1 completed; relocation of bracket from pier N – 3 to pier N by crane/trailer; launching of segment loader from deck N – 3/N – 2 to deck N – 2/N – 1.
- Stage 2: Launching of underslung gantry's main girders from span N – 2/N – 1 to span N – 1/N.
- Stage 3: Placing, alignment and joining of segments of deck N – 1/N on main girders.

Figure 6.5 Kinematics for a typical underslung LG for the erection of a system 3 deck (VSL International)

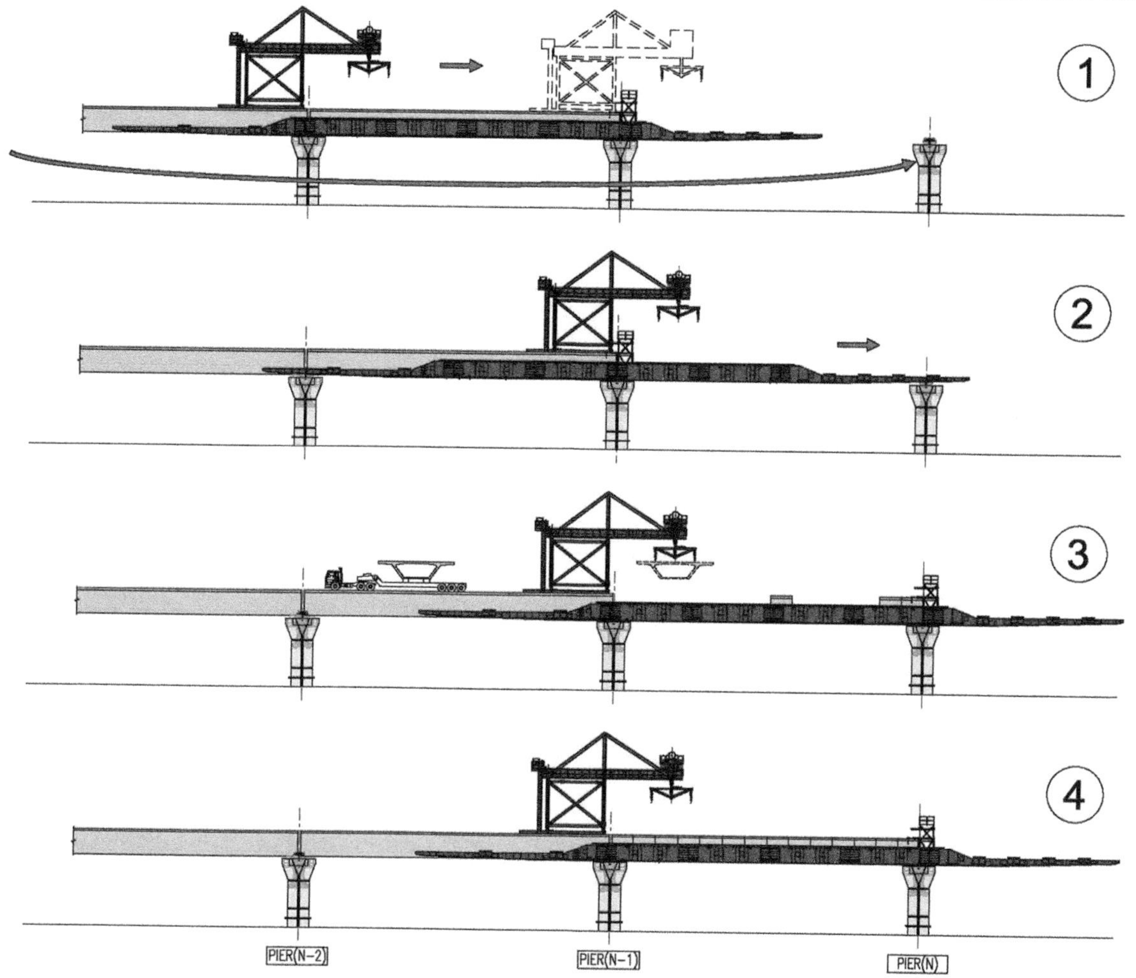

- Stage 4: Installation/stressing of longitudinal prestressing cables and load transfer of deck from gantry onto deck's support on pier N – 1 and N.

Figure 6.6 gives kinematics for the launching and erection of a system 1 deck with a typical overhead launching gantry and with segment delivery over the completed deck.

- Stage 1: Span N – 2/N – 1 completed; suspension of pier segment N in LG.
- Stage 2: First-stage launching of LG until temporary front leg can be activated on bracket attached to pier N; placing and nailing of pier segment N.
- Stage 3: Activation of temporary rear leg; relocation of LG's main support from pier N – 2 to pier N; removal of bracket at pier N – 1; deactivation of temporary rear leg and second-stage launching of LG, until LG is centred over span N – 1/N.
- Stage 4: Suspension of field segments of span N – 1/N in LG in two layers to allow moving in and rotation of last field segment.

Figure 6.6 Kinematics for a typical overhead LG for the erection of a system 1 deck with segment delivery over the deck (VSL International)

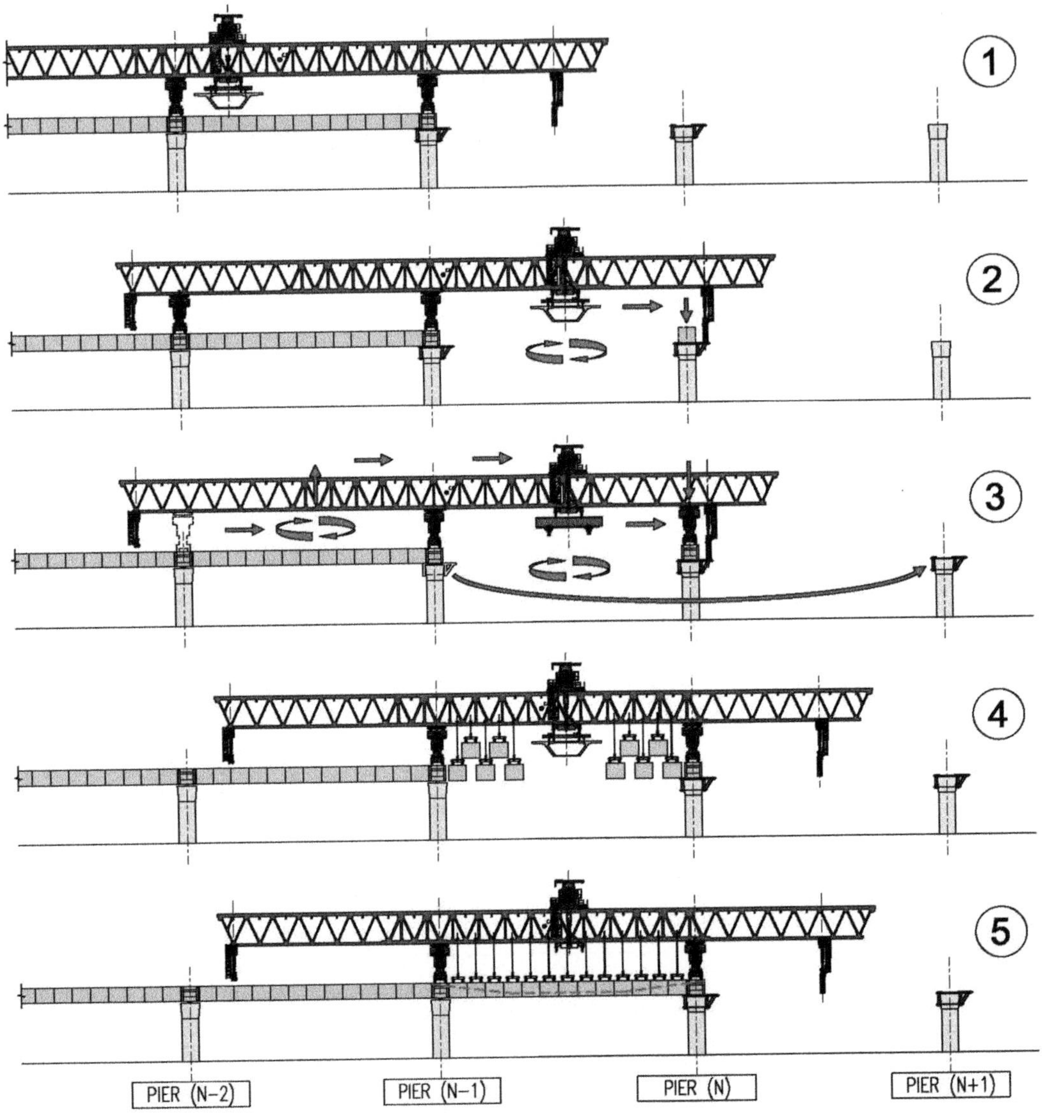

- Stage 5: Alignment and joining of field segments; preparation and casting of wet joints between field and pier segments; installation/stressing of longitudinal prestressing cables and release of field segments from the LG.

6.5. Permanent work interfaces

6.5.1 Segment support

Depending on which type of LG is being used, the segment attachment for handling and load introduction when supported on or suspended from the LG is different (Figure 6.7).

Figure 6.7 Attachment of segments for handling and load introduction when supported on or suspended from the launching gantry (VSL International)

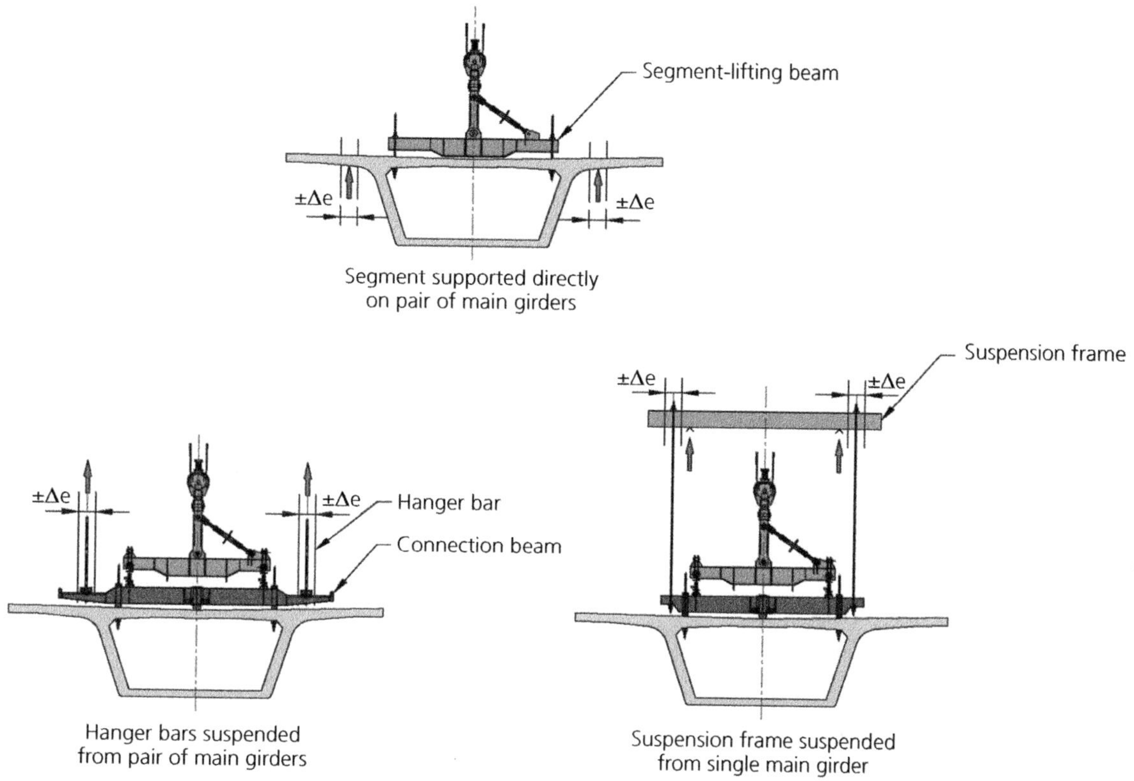

In the case of an underslung LG, inserts/holes are required in the deck slab for the segment handling. In addition, the wing slabs need to be checked as to whether they can take the segment weight when resting on the main girders, especially if the deck slab is transversely prestressed. When using an overhead LG, each segment has a Connection Beam (CB), which is fixed to the segment's deck slab at four points. The winch trolley's lifting beam is connected to this CB, as well as the two or four hanger bars used for the suspension of the segment from the LG's main girders.

6.5.2 Load transfer

The load transfer takes place when a series of segments making up a span, individually suspended from or supported on the main girders of an LG, is converted into a self-supporting concrete girder by stressing the longitudinal prestressing cables. Since the flexural stiffness of the LG's main girders (made of steel) is considerably softer than the flexural stiffness of the span being erected (in prestressed concrete), hogging up of the concrete deck from prestressing will always be significantly smaller than the deflection of the steel girders when supporting all the segments of a span. Figure 6.8 shows a typical overhead gantry unloaded (top diagram) and loaded by the segments (bottom diagram). The midspan deflection of the gantry is indicated by Δv.

Hence if no provisions are made, the launching gantry remains locked to the concrete deck after the load transfer. In the best case, the LG can be freed by de-tensioning individually the segment hanger

Figure 6.8 Load transfer from the launching gantry to temporary or permanent supports (VSL International)

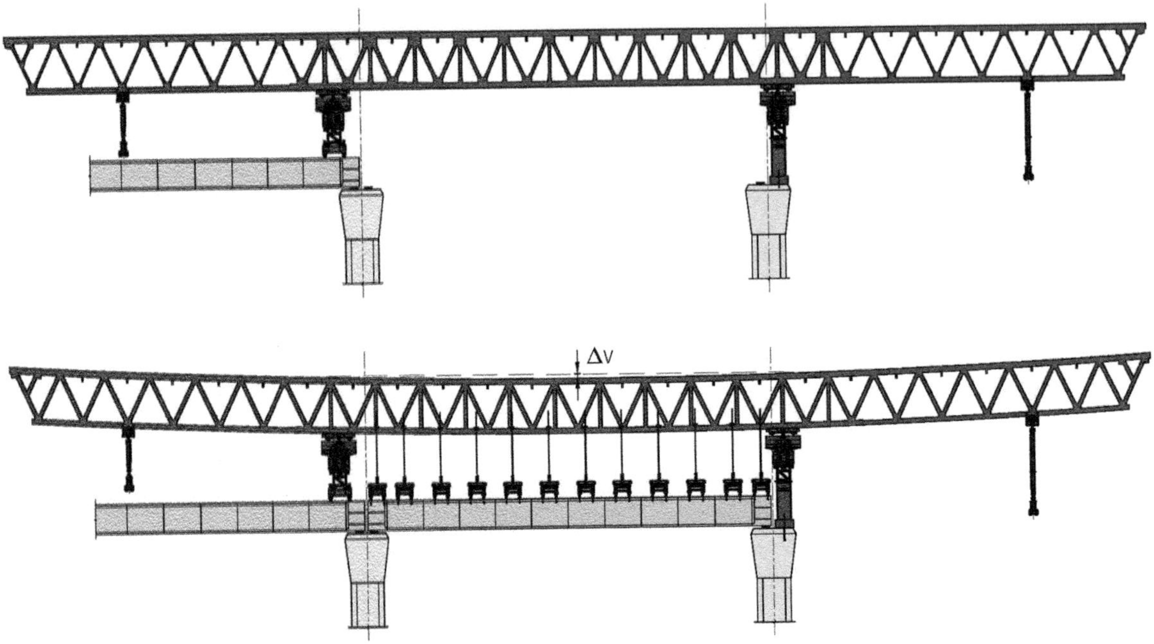

bars or by lowering the supports, if they are equipped with jacks of adequate capacity. In the worst case, glued segment joints may open up. Due to overloading during the load transfer, the hanger bars of an overhead LG nearest to the supports may even fail or the wing slabs of a deck erected with an underslung LG may crack. Such risks can be eliminated by supporting the main girders on hydraulic jacks and not on rigid supports. If the span is erected as a simply supported girder (systems 2 and 3), it is possible to load transfer the span onto temporary span support jacks, which can be lifted instead of the gantry being lowered during the load transfer.

6.5.3 Reactions to permanent work

(Refer to Section III, Chapter 15, for more on this topic.) One of the advantages of the classical underslung LG is the fact that the vertical support reactions are introduced during the span erection and the launching concentrically into the piers. For overhead LGs, the situation is different. Whereas for system 1 and 2 bridge decks the LG's main supports are always sitting on top of the pier segments and at the centre of the piers respectively, the LG's reactions are not centred for system 3 bridge decks. The permanent work needs to be checked for these load cases. In general, however, such load cases are mostly not controlling for the design of the deck, the piers and the foundations. It is also important to verify that the leading pier is able to take the longitudinal forces from the launching, especially when the LG is sliding (on material with relatively high friction coefficients, but which allows higher bearing stresses) and not rolling over its supports, and the piers are high.

6.6. Cycle times

In Tables 6.2 and 6.3, two cycles are shown for the erection of a 45-m-long span made up of 15 single-cell box girder segments of 12 m width, which are stressed together longitudinally with external PT. The soffit of the bridge deck is 6 m above ground. The first cycle, shown in Table 6.2, is for a system 1 span

Table 6.2 Cycle for a system 1 span erected with an overhead launching gantry

Activity	Duration in hours	Day 1				Day 2				Day 3				Day 4				Day 5			
First-stage launching of LG and placing of downstream pier segment of span N – 1/N	5		X																		
Alignment and nailing of pier segment	5			X																	
Second-stage launching of LG	5			X																	
Suspension of field segments of span N in LG (14 No)	10 + 4						X	X			X										
Alignment and gluing of segments	6 + 8											X			X	X					
Preparation and casting of wet joints	3															X	X				
Curing of wet joint concrete	12																X	X	X		
Installation of longitudinal PT	7																		X		
Stressing of longitudinal PT and load transfer	3																			X	

Table 6.3 Cycle for a system 3 span erected with an underslung launching gantry

Activity	Duration in hours	Day 1				Day 2				Day 3			
Launching of LG	6		X										
Loading of segments of span N – 1/N onto LG (15 No)	4 + 3			X			X						
Alignment of segments	7						X	X					
Installation of longitudinal PT (external)	7										X		
Stressing of longitudinal PT and load transfer	3											X	

erected with an overhead launching gantry. The second cycle, shown in Table 6.3, is for a system 3 span erected with an underslung launching gantry. Shown cycles are for illustration only and the overall cycle times should not be used for actual projects unless verified against project-specific site constraints.

6.6.1 Cycle for a system 1 span erected with an overhead launching gantry

Assumptions are as follows:

- starting point: span N – 2/N – 1 completed, launching gantry ready for launching
- one pier segment and 14 field segments
- use of connection beams

- epoxy-glued joints between field segments and unreinforced wet joints between field and pier segments
- use of early-strength concrete for the wet joints (12 hours to reach minimum strength required for load transfer)
- load transfer to permanent bearings
- one shift per day (10 hours per shift)
- segment delivery from below and no delay with segment delivery.

6.6.2 Cycle for a system 3 span erected with an underslung launching gantry
Assumptions are as follows:

- starting point: span N – 2/N – 1 completed, launching gantry ready for launching
- two pier segments and 13 field segments
- dry joints
- load transfer to span support jacks
- one shift per day (10 hours per shift)
- segment delivery over deck and no delay with segment delivery.

6.7. Risks
The list of risks given in Table 6.4 is by no means comprehensive, but addresses risks which are particular to gantries used for span-by-span deck erection with precast segments. For a specific project, a main contractor needs to carry out a full risk assessment of the operations performed with the erection gantries, considering the actual features of the gantry, the erection sequence and the site constraints.

6.8. Performance criteria
The following items need to be addressed when drafting a performance specification for a launching gantry.

- Segment
  - segment characteristics (per segment type): weight, dimensions, centre of gravity, thickening of pier segment's wing slabs (for underslung LG)
  - delivery: from below, from behind, both
- Alignment: gradient, curvature in plan and cross-fall variations
- Permanent work interfaces
  - permissible deck loading (standing on deck during erection, rolling over deck during launching)
  - permissible temporary loading for permanent bearings (overhead LG's rear support during erection of span N – 1/N on downstream pier segment of span N – 2/N – 1)
  - temporary support of decks on piers (location of span support jacks; load introduction into pier segment {beside bearing locations}, etc.)
  - permissible eccentric loading of piers
  - permissible wing loading (segment support for underslung gantry)
  - access for stressing of PT cables (particularly for system 2 and 3 decks erected with overhead gantry; lateral clearances against masking walls for system 3 decks)
  - space constraints for supports of gantry on front pier and rear pier (between pier segments for system 2 decks erected with overhead gantry): permanent bearings, span support jacks, seismic buffers/shear keys; permissible recesses for support of an underslung gantry's pier brackets; longitudinal clearance of an overhead LG's front supports against span being erected (adequate longitudinal clearance for lifting/lowering of the last field segment into its final position)

Table 6.4 List of risks specific to the operation of launching gantries for span-by-span erection with precast segments

Activity	Risk	Mitigating measures	Applicable for	
			Underslung gantry	Overhead gantry
Installation of handrails on top and inside of segment delivered on trailer	Vertical access and work at height: climbing up and down, work performed at height (at least 3 to 4 m above ground)	Access ladder fixed to stable base or access staircase, safety line fixed to segment and winch's segment-lifting beam	X	X
Attachment of Connection Beams (CBs) to segment delivered on trailer			N/A	X
Attachment of winch's SLB to segment delivered on trailer			X	X
Transfer to hanger bars	Longitudinal and vertical access and work at height: worker to stand on segment suspended from winch (some of the segments vertically staggered); worker above in main truss installing washers and nuts, which can fall down	Safety line fixed to connection beams, self-engaging hanger bars; storage boxes for nuts and washers	N/A	X
Storage of hanger bars	Hanger bars can fall down through longitudinal slots between main truss's bottom chords	Storage racks	N/A	X
Free-hanging hanger bars exposed to wind	Bars may turn out of couplers/nuts and subsequently fall down due to wind-induced oscillation about bars' vertical axis	No vertical parking of hanger bars, or provision of lock nuts	N/A	X
Working on top of and inside of segment	Falling down from segment	Provision of temporary handrails around all exposed edges on top and inside of segment, installed prior to lifting of segment from trailer	X	X

Table 6.4 Continued

Activity	Risk	Mitigating measures	Applicable for	
			Underslung gantry	Overhead gantry
Moving from segment to segment when temporarily suspended in or supported on LG prior to joining and gluing	Falling at or through longitudinal gap between segments	Safety line fixed to connection beams (top) and to each segment (inside)	X	X
Segment erection over live traffic	Drop of segment (unlikely), falling of debris, tools (possible)	No segment handling over live traffic	X	X
Segment suspension (stationary) over live traffic	Drop of segment (unlikely)	Permitted with back-up suspension system (e.g. suspension with four hanger bars with two bars diagonally across able to hold the segment)	N/A	X
Epoxy gluing of segment joints	Dropping epoxy	No traffic if over road, plastic sheets or similar below joints to catch epoxy if over water	X	X
Epoxy gluing of segment joints	Closing of gap between segments due to swinging segments	Use of timber shims, which are supported on top of the deck slab	N/A	X
Support of LG	Failure at support's contact area	Adequate edge distances, so that support's contact area is within concrete confined by reinforcement; shimming with similar material over entire footprint of support's contact area	X	X
Longitudinal movement of main girders	Over-launching or loss of control of longitudinal movement → loss of equilibrium → overturning of LG's supports and loss of entire LG	Back-up fixity, nominal tie-down of supports, presence of LG engineer and LG supervisor	X	X

  - support conditions on portals and half joints
  - deck geometry at bifurcations and ramps
  - starter bar location and dimensions (for link slabs, cast-in-situ part of masking walls, rail support plinths, etc.)
- Follow-up work (closing of lifting holes, removal of swift lift studs, etc.).
- Clearances: vertical below and above deck (overhead power lines, overhead bridges etc.); horizontal (houses, power lines, etc.)
- Cycles (double or single shift, restrictions for segment delivery, restriction for noisy type of work {strand pushing}, etc.)
- Mechanisation (labour costs, skill level, etc.)
- Longitudinal fixity (redundancies)
- Access: into LG (from deck, front pier) and inside LG during different stages of operation; details of temporary safety barriers for segments being erected (on top, inside); access from ground to piers (ahead of LG's arrival, during deck erection, for follow-up work) and onto and into deck
- Shipment costs (initial, if reused)
- Reusability
- Method, schedule and crane definition for assembly, relocation, dismantling.

In addition, the buyer needs to give a time frame for fine-tuning of the LG concept, the support reactions/inserts and delivery and commissioning of the LG.

6.9. Photo gallery

Figure 6.9 System 1 road deck built with overhead LG (Shenzhen Western Corridor, Hong Kong): spans activated as single-span frames, pair of main girders, connection beams (in red), segment delivery from ground (and from behind) (VSL International)

Figure 6.10 System 1 road deck built with overhead LG (Bandra Worli Sealink, Mumbai): single main girder with suspension frames, segment delivery from behind and from below (from sea) (VSL International)

Figure 6.11 System 3 rail deck built with overhead LG (Metro Monterrey, Mexico): vertically staggered double main girders (rear main girder with suspension frames sits at its front end on front main girder), U-section with hanger bars attached to top flanges, segment delivery from ground (DEAL)

Figure 6.12 System 3 rail deck built with overhead LG (Dubai Metro): pair of main girders and hanger bars (outside of main girders) anchored at top in suspension frames and at bottom attached to U-section's top flanges, segment delivery from ground (DEAL)

Figure 6.13 System 2 road deck built with overhead LG (Humin Road, Shanghai): pair of main girders, 24-m-wide multi-cell box girder deck, delivered from ground (VSL International)

Figure 6.14 System 1 rail deck built with overhead LG (BTS, Central Station, Bangkok): pair of main girders equipped with 14 upper cross beams supporting winches/SMUs to achieve full deck erection during one night shift, segment delivery from ground (VSL International)

Figure 6.15 System 2 rail deck built with underslung LG (Westrail, Hong Kong): final structural systems are single-span frames, segment delivery from ground lifted onto LG with mobile crane, internal prestressing (VSL International)

Figure 6.16 System 1 rail deck built with underslung LG with centre hinges and portal crane (LRT Sengkang, Singapore): segment delivery from ground, lifted onto LG with portal crane (NRS)

Figure 6.17 System 3 road deck built with underslung LG (Windsor Flood elevation, Australia): segment delivery from ground, lifted onto LG with crawler crane (VSL International)

Figure 6.18 System 3 rail deck built with underslung LG (Airport Rail Link, Bangkok): segment delivery over deck placed onto LG with segment loader (VSL International)

REFERENCES

Meyer ME (2011) Under-Slung and Overhead Gantries for Span by Span Erection of Precast Segmental Bridge Decks, *Structural Engineering International* **21(2)**: 399–405.

Ganz HR and Meyer ME (1997) *Segmentbauweise*, *Schweizer Ingenieur und Architekt* **26**: 4–9.

Figures 6.1 to 6.10, 6.13 to 6.15, 6.17, 6.18: VSL International
Figures 6.11 and 6.12: DEAL
Figure 6.16: NRS
Table 6.1: VSL International

Bridge Deck Erection Equipment
ISBN 978-0-7277-6193-4

https://doi.org/10.1680/bdee.61934.087

Chapter 7
Precast girder erection by launching gantry

7.1. Introduction

Bridge decks which are made from precast beams (girders) supporting a cast-in-situ deck slab are one of the most common types of bridge deck construction systems in the world, for both road and railway bridges.

Precast beams are mostly erected by a pair of mobile or crawler cranes, ground based or mounted onto barges. Launching Gantries (LGs) are usually employed where the use of cranes is not possible.

7.2. Application range

I-beams are the most used and also the most standardised type of precast girders, with typical spans between 20 and 40 m, which can even be increased up to 65 m for exceptional cases. Bulb Tees and Deck Bulb Tees cover the same range of span lengths as the I-beams. The increased width of the top slab reduces the amount of formwork needed to cast in situ the deck slab and contributes significantly to increasing the transverse moment of inertia and therefore the transverse stability. This is important for the handling of very long beams. I-beams and Bulb Tees need to be temporarily stabilised during erection, until the transverse diaphragms and deck slab have been cast. This operation is usually on the critical path.

U-beams, when resting on four bearings, are stable on their own. Such beams can therefore be erected faster. Furthermore, the width of these beams can be increased significantly compared with the other types of beams (up to 4 m), which simplifies the construction of the deck slab greatly.

The weight of I-beams, Bulb Tees and Deck Bulb Tees can vary from a minimum of 200 kN up to around 2500 kN. If precast girders have to be transported on public roads, the girder's weight usually has to be limited to 1000 kN, which gives a maximum span length in the range of 40 m.

For more explanations about detailing aspects of bridge decks made up of precast girders, refer to Section III, Chapter 15.

Longer and heavier girders have to be precast on site in a temporary precast yard, which is normally located on the alignment of the motorway/railway or nearby in order to facilitate transportation to the erection front. A convenient way to handle beams in such cases is with a pair of rubber-tyred straddle carriers. These carriers remove the girders from the moulds, store them temporarily for additional curing of the concrete and sometimes for the completion of the Post-Tensioning (PT) activities, and transport them at the end to the erection front.

Use of LGs is suggested, and sometime unavoidable, when

- the precast girders are heavier than 1200 to 1500 kN

- the piers are higher than 40 to 50 m
- the access conditions do not allow the use of mobile or crawler cranes with sufficient capacity and reach to lift the girders in a tandem lift directly from the ground onto the cross heads
- working in urban areas where the impact on traffic must be minimised.

7.3. General arrangement of the BDE

There are three basic configurations of the LG

- type 1: with multiple supports (beam delivery several spans back or beside span N, where beams are being placed)
- type 2: with two supports and legs (beam delivery one or two spans back or beside span N)
- type 3: compact (beam delivery always beside span N).

Figure 7.1 shows an LG **type 1**, which is designed to pick up the precast girders several spans behind span N. Hence the follow-up work after placing of the precast girders to complete the deck is not affected by the gantry's operation. LG type 1 does not have front and rear legs, but has an auxiliary support and multiple intermediate supports in addition to the main front and rear supports. The supports are relocated by means of the service hoists located at the main girder tips.

Figure 7.1 Arrangement in elevation of an LG type 1: with multiple supports (without legs) (DEAL)

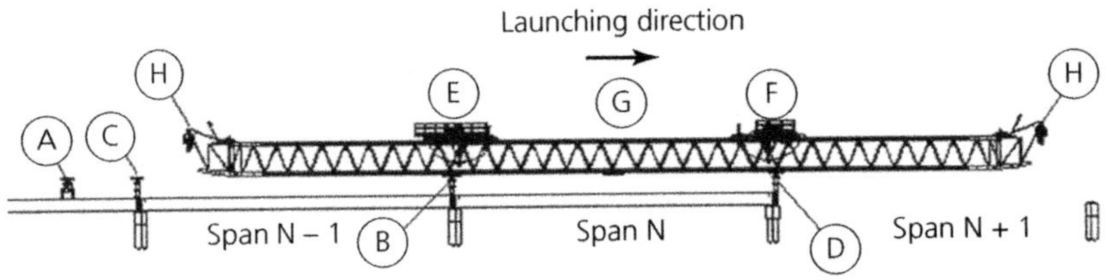

A. Auxiliary support
B. Rear support
C. Intermediate support
D. Front support
E/F. Rear/front winch
G. Main girder
H. Service hoist

Figure 7.2 shows an LG **type 2**, which is designed to pick up girders either beside span N or delivered from the back over the already completed deck by means of transporters. Hence the follow-up work

Figure 7.2 Arrangement in elevation of an LG type 2: with two supports and legs (DEAL)

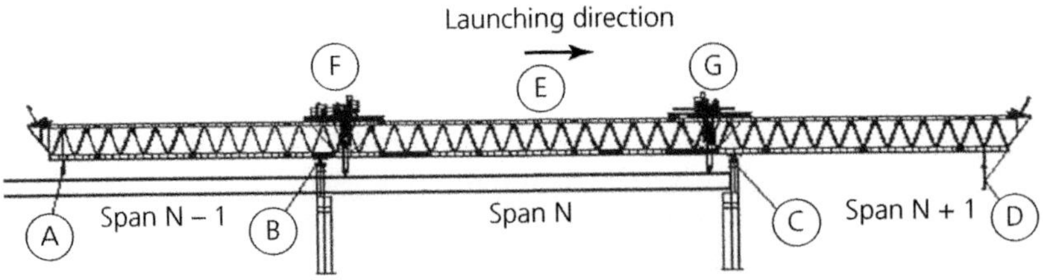

A. Rear leg
B. Rear support
C. Front support
D. Front leg
E. Main girder
F/G. Rear/front trolley

Figure 7.3 Typical section of an LG type 1 and 2, with pair of main girders (DEAL)

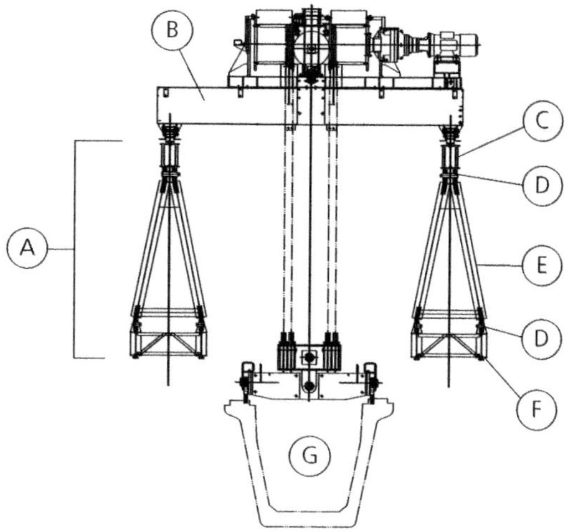

A. Main girder
B. Winch trolley
C. Upper chord
D. Pinned connection
E. Diagonals
F. Lower chord
G. Precast beam

after placing of the precast girders is either not affected at all or can be minimised by transporting girders with rail-bound bogies over the partially completed decks. It is also possible to configure the LG so it can pick up girders in span N – 2 and not N – 1. The LG type 2 has a front and rear support and front and rear legs, which are equipped with hydraulic support jacks.

Since the LG types 1 and 2 are usually designed to receive girders delivered over the deck, they generally have a pair of main girders (Figure 7.3).

The LG **type 3** is a simplified LG type 2. It is an economical option, when:

- girders are always delivered from the side
- the cross heads are basically parallel to each other
- the span length varies by only a few metres.

The main girder of an LG type 3 is a single truss with lifting devices positioned directly inside the main truss or above it. Such single truss gantries can be self-launching (Figure 7.4) or not self-launching (Figure 7.5).

The weight of the main girder and the trolleys is greatly reduced when compared to the main girders of LG types 1 and 2.

It is often possible to relocate an LG type 3 from span to span by a crawler crane. In such cases, the LG does not have to be self-launching and the weight of the equipment is further minimised as the launching noses can be omitted. This helps to further reduce the equipment's costs. Figure 7.5 shows such an arrangement, with winches located in a longitudinally adjustable platform above the single main girder.

Figure 7.4 LG type 3: single main truss; self-launching with fixed centre leg activated for relocation of the supports (VSL International)

Since the crawler crane is required for a limited time only for the relocation of the gantry, it can also carry out other tasks on site, such as construction of the foundations, the piers and the pier cross heads.

One typical use of an LG type 3 is for construction of bridges over water when all activities are performed from a service trestle located parallel to the bridge being built. The precast girders are transported on the service trestle beside the span where they have to be placed. The use of an LG type 3 avoids the need for heavy cranes to erect the beams, which could control the design of the service trestle. For illustration, refer to Figure 7.6.

Figure 7.5 LG type 3: single main truss; not self-launching (DEAL)

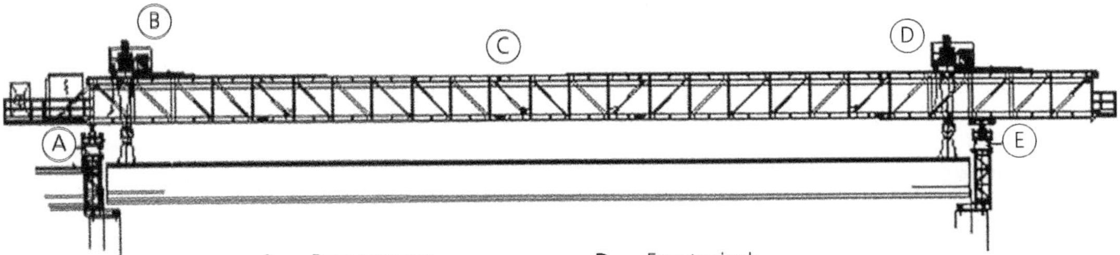

A. Rear support
B. Rear winch
C. Main girder (single truss)
D. Front winch
E. Front support

Figure 7.6 LG type 3 (DEAL)

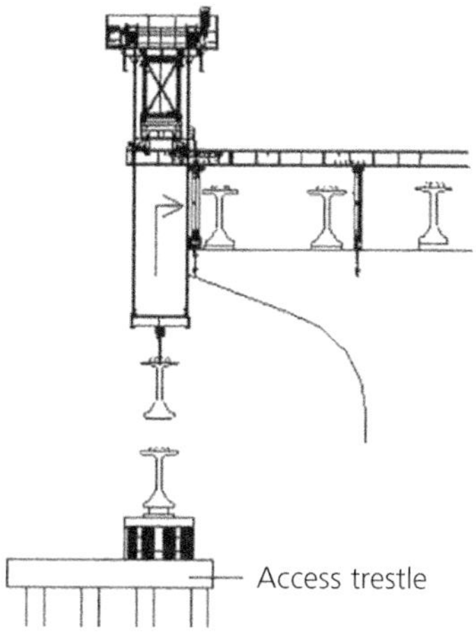

Cross-section at support

Overall view

In its simplest configuration, an LG type 3 does not lift the girders up to the cross head level but is used only to shift the girders along the cross heads into their final positions. This is, of course, only an option if mobilisation of a pair of crawler or mobile cranes for the lifting of the girders is viable and if there is enough space to position such cranes. The lifting devices are then generally just a pair of long stroke jacks. For illustration of this LG type 3 configuration, also called a 'beam shifter', refer to Figure 7.7.

Figure 7.7 A beam shifter, the simplest configuration of an LG type 3 (DEAL)

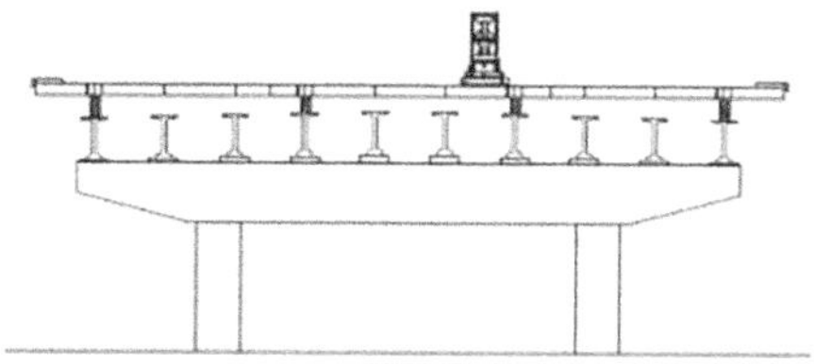

Rear support

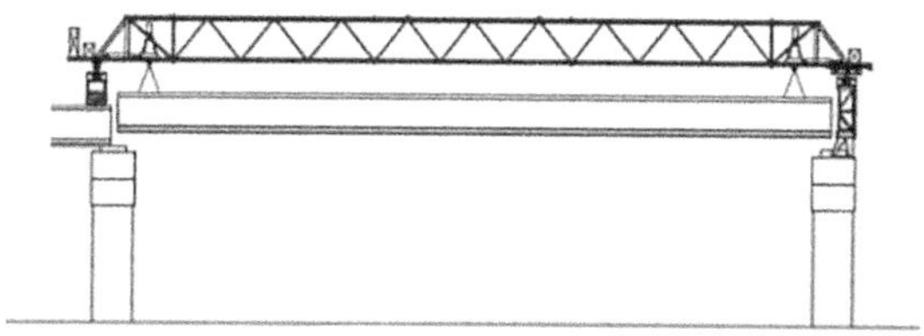

Longitudinal view

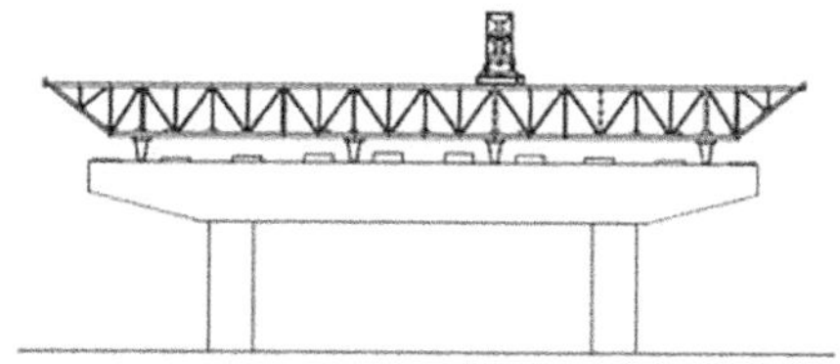

Front support

Overall view

7.4. Launching gantry elements

The main elements of launching gantries for erection of precast girders are

- main girders
- winch trolley units
- supports (front, rear; auxiliary support and multiple intermediate supports if any)
- front and rear legs (if any)
- service hoists (if any).

7.4.1 Main girder

Except for an LG type 3, the main girder is a pair of two parallel trusses (refer to Figure 7.3), most frequently of triangular shape, preassembled to elements of a length usually within the transportation limits (i.e. approximately 12 m).

The elements of the truss (comprising of upper/bottom chords and diagonals) may be welded or bolted, depending on the dimensions: welded when the total height of the truss is within the 2.3 m limit for transport by closed containers (for LGs for placing of girders weighing up to 1000 kN and span length of 35 m approximately) or bolted when exceeding such limits (up to 2500 kN and 60 m). In this case, the connection between diagonals and chords is usually made by means of pins.

The two trusses are connected at the end by connection frames usually arranged at the top chord level to maintain maximum clearance between the trusses.

The total length of the main girders of self-launching gantries exceeds slightly the sum of the two longest adjacent spans. For LG type 1, the bottom chords of the end elements of the main trusses are curved upwards to compensate the cantilever deflection so that the main trusses can easily get onto the support rollers in front and off the support rollers at the rear.

7.4.2 Winch trolley units

Two winch trolley units (in other chapters also referred to as Upper Cross Beams/UCBs) are installed: a master winch trolley and a slave winch trolley. Each trolley is composed mainly of a supporting frame, mounted on wheels, which run on the upper chords of the main trusses. The lifting winches are installed on the supporting frames. The trolley where the electric generator and electric switch box are installed is called the master winch trolley. For ease of access for maintenance and refuelling, it is usually the rear of the two trolleys.

Quite often the identification name of the equipment is composed by the maximum weight of the girders followed by the maximum length of the span: for instance, equipment name LG 120/40 means 120 ton heavy precast girders over 40 m span length.

The capacity of each lifting winch is at least half of the maximum girder weight to be handled. The movement range of the winch's hook depends on whether the girders are delivered from below or from behind over the deck. The hook must therefore be able to travel over the following distances:

- for delivery from the ground, over a range which is equal to the highest level of the deck above ground plus reasonable tolerance
- for delivery from the deck, over a range which is equal to the depth of the segment plus the height of the transporter and a reasonable tolerance.

The target erection cycle derived from the overall construction programme defines how many precast girders have to be erected during a shift. The lifting speed of the winches is affected by the way the girders are delivered: slow speeds, usually in the range of 1 m/min, are common when girders are fed from the back and faster speeds, up to 2–3 m/min depending on the maximum level of the deck, are used when they are fed from the ground.

If the erection cycle permits, lifting winches can be replaced by strand-moving units. For feeding of the beams from the back it is possible to use hydraulic long stroke jacks. For details about various options for lifting devices, refer to Section III, Chapter 17. In this chapter, the term 'winch' is used for the main lifting devices, even though alternative devices can and are frequently used.

7.4.3 Longitudinal movement system for the winch trolleys

If girders are delivered along the deck, the design of the longitudinal movement of the winch trolleys deserves particular consideration. The arrangement is dictated by the fact that most of the work is performed by winches at a fixed distance from each other. A change of distance between the winch trolleys may be necessary to cope with the variation in span length along the project, but this happens infrequently in a well-designed bridge. The simplest approach is to have completely independent trolleys whose movements must be synchronised by the operator when moving the precast girders along the main trusses. For LG type 1, however, which picks up the precast girders several span behind the LG, it is more effective to optimise the longitudinal movement system to work with the trolleys at a 'mechanically fixed distance' with the option of changing the distance when necessary. This can be achieved by means of one capstan winch with a rope path involving both winch trolleys in the same loop plus, as an option, a small capstan winch to change the distance between trolleys when necessary. The reeving pulleys of the capstan winches are installed onto the end connection frames of the main girders (for information on the basic capstan winch principle, refer to Section III, Chapter 17).

This movement system is used for both the coupled movement of the trolleys along the girders and for moving the main girders above their supports during the self-launch. During launching, one trolley is anchored to the fixed support by the anchor frames plus the 45° anchor ropes (see Figure 7.8).

Figure 7.8 Use of the trolley's longitudinal movement system for the launching of the gantry's main girder (DEAL)

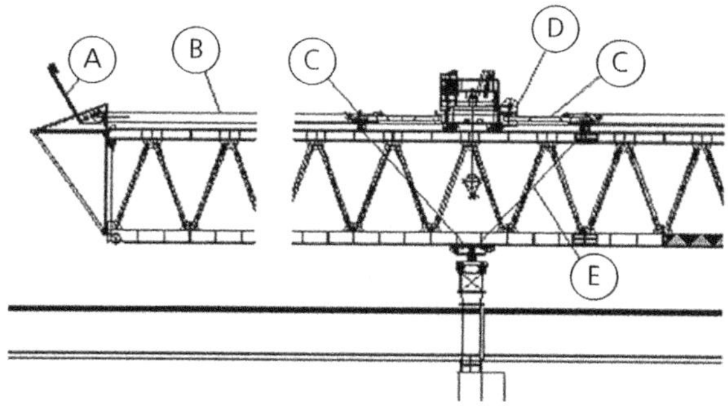

A. Counterweights
B. Capstan ropes
C. Anchor frame
D. Capstan winch on trolley
E. 45° trolley anchor ropes

7.4.4 Anchorage system of the main girder to the supports

An independent longitudinal fixity between the main girder and the supports is also required. Such a system is, for example, based on an anchor rope installed along the bottom chord of the main girder and one or more anchor clamps to be engaged/disengaged to a support when required.

7.4.5 Supports

The supports are composed of a pair of transverse beams with welded rails on top (also called transverse runway or lower cross beams) and several support columns. They are supported on the deck or the pier's cross head at several points. They are installed onto the front pier's cross head (front support) and onto the rear pier's cross head or the already erected precast girders (rear support). The support columns are anchored to the concrete by HT bars. As the transverse beams must always be kept horizontal whatever the cross slope and longitudinal gradient of the bridge are, the height of the support columns has often to be variable. This is achieved by using sets of bolted steel shims and screw jacks for fine adjustment. The screw jacks enable construction tolerances to be taken up.

In the event of feeding the beams from the side, the runways need to have side cantilevers to allow the pick-up of the girders on the side of the cross head. This leads to heavier runway sections and heavier anchoring arrangements (see Figure 7.9).

On top of the transverse beams run the 'roller groups' formed by 'under rollers', 'rotating bases' and 'upper rollers' (in other chapters this is referred to as Universal Lower Roller Support – ULRS). For details, refer to Figure 7.10. The upper rollers allow the longitudinal movements of the main girder over the supports, the under rollers allows the transversal movement, and the rotating base allows the variation of the plan angle between the centre line of the main trusses and the centre line of the supports. Under rollers are motorised to facilitate the transverse movement. In this case, for safety reasons and for accuracy of placing the beams, the advised movement speed should be not greater than 3 m/min.

Figure 7.9 Transverse beams (runways) with side cantilevers for girder feeding from the side (DEAL)

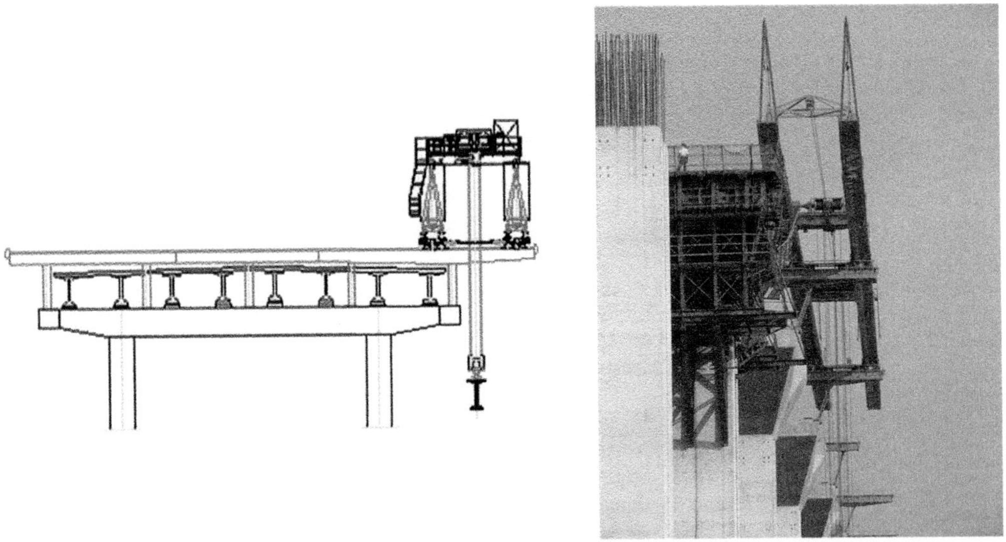

Figure 7.10 Roller group (ULRS) (DEAL)

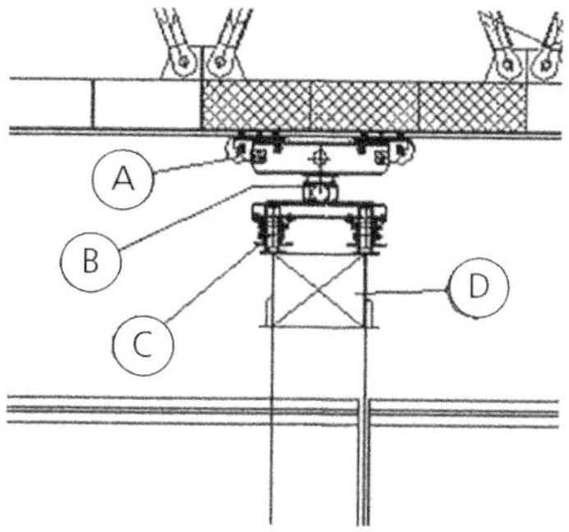

A. Upper rollers
B. Rotating base
C. Lower rollers
D. Transverse beam (runway)

When cross heads are, for example, short or the transverse movement of the beams is not critical for the erection cycle, the under rollers can be replaced by sledges which are moved by hydraulic long stroke jacks along the transverse beams.

For an LG type 1, the number of supports depends on the organisation of the deck construction activities, particularly on the pouring and curing of the cast-in-situ deck slab, but is at least five. For an LG type 2 the number of supports is always two, whereas for an LG type 3 the number of supports is three. If an LG type 3 is made self-launching by legs, the number of supports can be reduced to only two.

When the beams are fed into an LG type 1 from the back, the last back support (called the auxiliary support) has to allow for the delivery of the beams. The solution depends on the mode of transportation: straddle carrier, low-bed transporter, etc.

7.4.6 Service hoists

An LG type 1 needs front and rear service hoists (Figure 7.11) to allow the relocation of the supports by the equipment itself. Such service hoists can be manually or electrically operated rope or chain hoists.

7.4.7 Legs

If LG type 2 and LG type 3 are configured as self-launching, they require rear and front legs for the relocation of the supports. Depending on the gantry's kinematics, these legs may also act as support during the placing of the girders. The legs are equipped by hydraulic support jacks that are extended to put them in force (Figure 7.12). The stroke of the jacks must be enough to manage also the deflection of the main girder in maximum cantilever condition.

7.5. Gantry kinematics

In this subsection, the kinematics for LG types 1 and 2 are shown.

7.5.1 Basic kinematics of LG type 1

The longitudinal movement of the main trusses is obtained by alternative movements of the winches along the main girder and of the main girder over the supports. Figure 7.13 shows the main steps for

Figure 7.11 Service hoists on an LG type 1 (DEAL)

Figure 7.12 Legs engaged to ground on an LG type 2 (DEAL)

Figure 7.13 Steps for longitudinal movement of the main trusses by one span (DEAL)

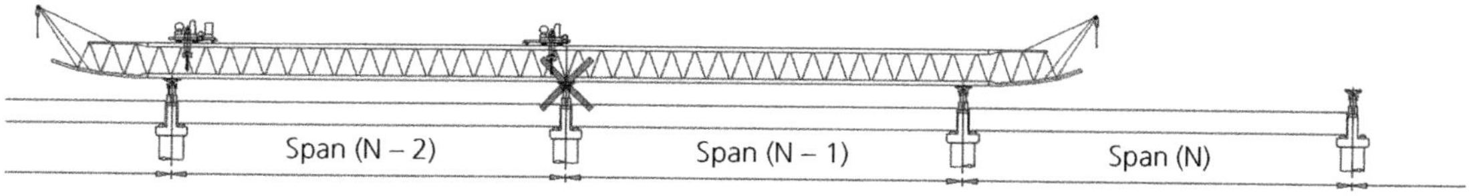

Step 1: The main trusses are anchored to one support as indicated.

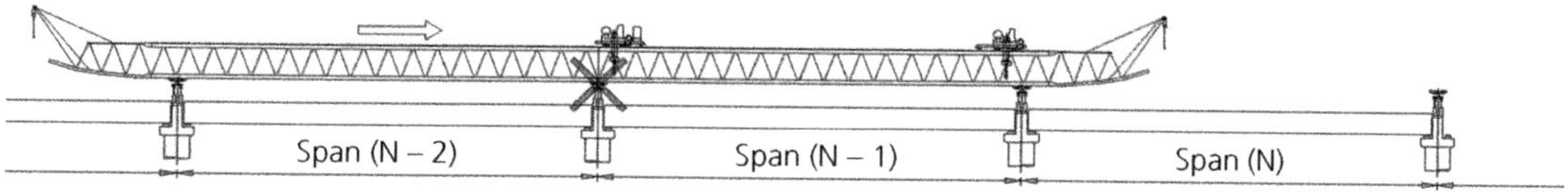

Step 2: Move forward the winch trolleys by one span.

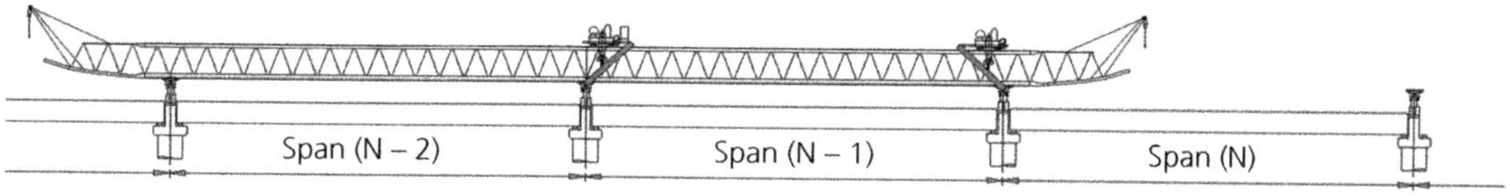

Step 3: Anchor the winch trolleys to one support each (as indicated) and disengage the main truss anchorages from the support.

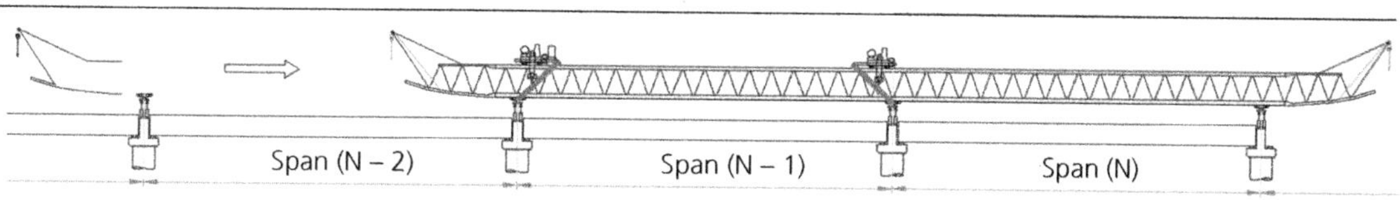

Step 4: Move the main trusses forward by one span.

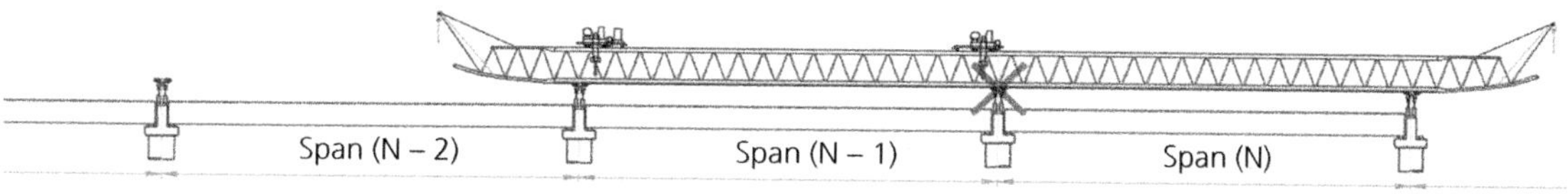

Step 5: Anchor the main trusses to one support as indicated and disengage the winch trolley anchorages. The gantry is in the same configuration as at the start of step 1, but now over span N – 1 and N.

launching the LG by one span forward. The same operation can be performed with or without a beam suspended from the winch trolleys and can be combined with the relocation of the supports by the service hoists.

Figure 7.14 (on pages 98 and 99) shows the placing in span (N) of a precast beam delivered over the deck in span (N – 3). The cycle also includes the relocation of the supports, which is generally required only after all beams of a span have been placed.

Any transverse movement of the main trusses along the transverse beams is performed only when the main trusses are resting on two supports in order to have a statically determined arrangement during transversal move. The same is the case when the precast beam is placed at its final position.

Figure 7.14 Placing of a precast beam delivered over the deck and relocation of the supports (DEAL)

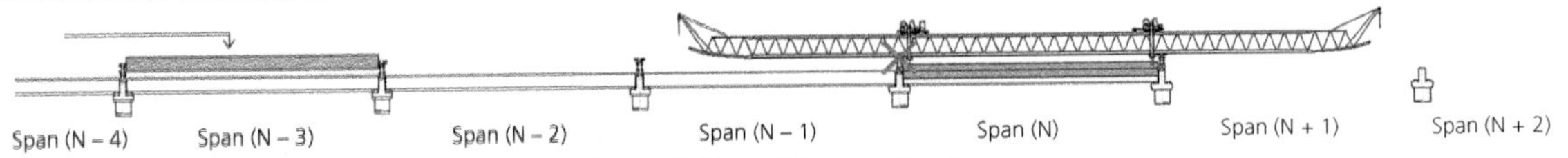

Step 1: Erect the second-last beam of span N, beam delivered in span N – 3.

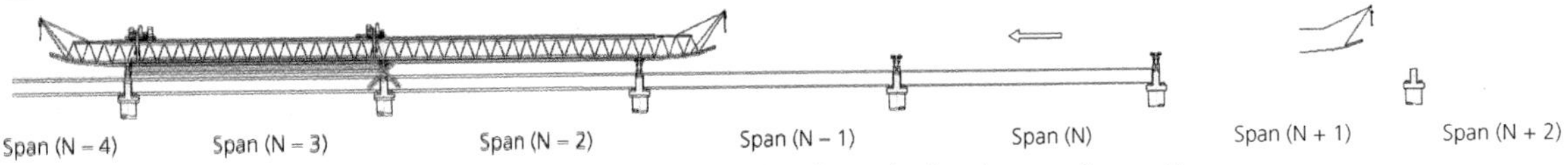

Step 2: Move backwards the main trusses on span N – 3 and pick up the last beam of span N.

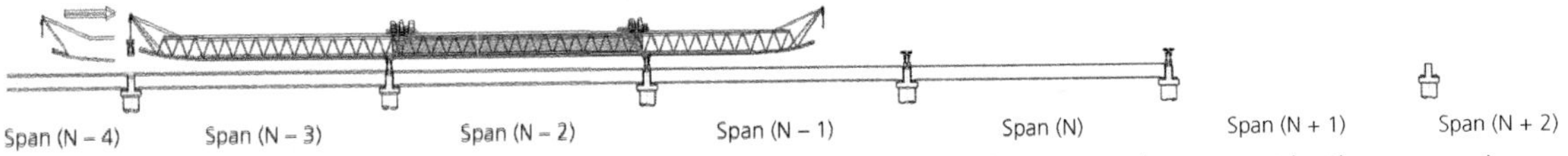

Step 3: Move the main trusses forward together with the beam and pick up the rear service support by the rear service hoist.

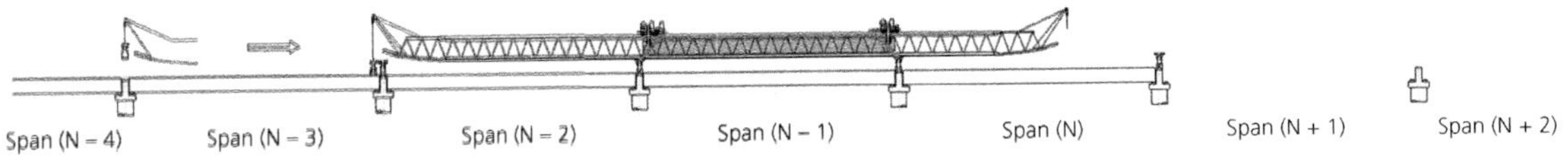

Step 4: Move the main trusses forward and relocate the auxiliary support.

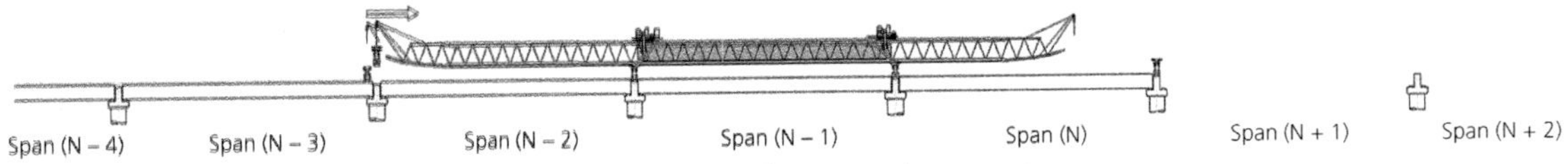

Step 5: Move the main trusses slightly forward and pick up the rear main support.

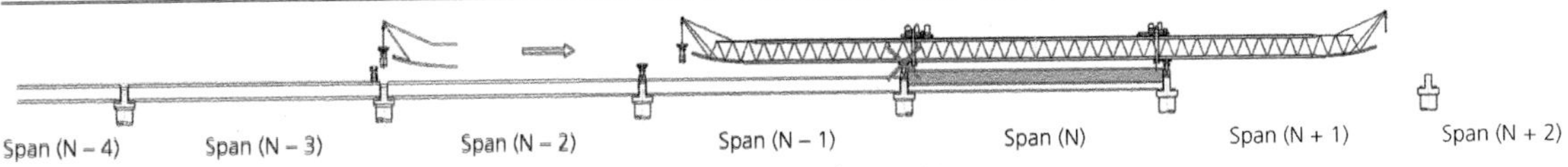

Step 6: Move the main trusses forward and erect the last beam of span N.

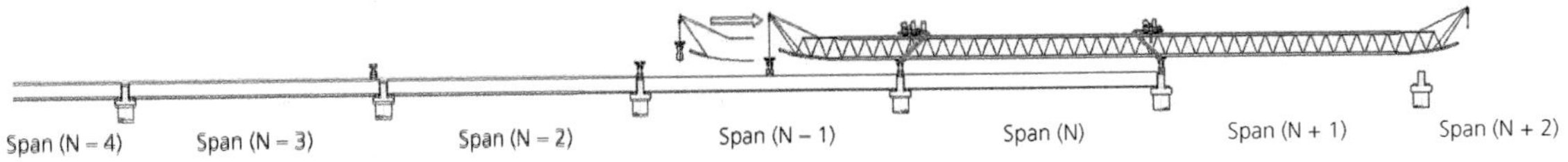

Step 7: Move the main trusses slightly forward and park the support on the deck.

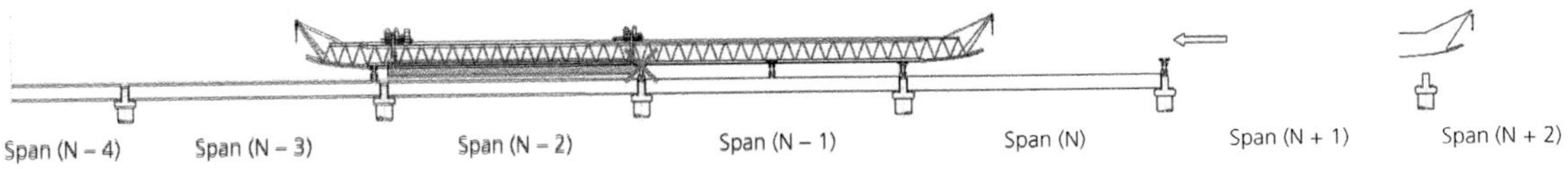

Step 8: Move back the main trusses on span N – 2 and pick up the first beam of span N + 1.

Figure 7.14 Continued

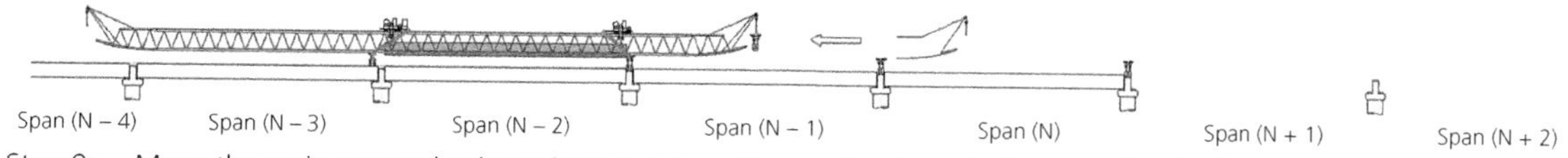

Step 9: Move the main trusses backwards and pick up the previously parked support by the front service hoist.

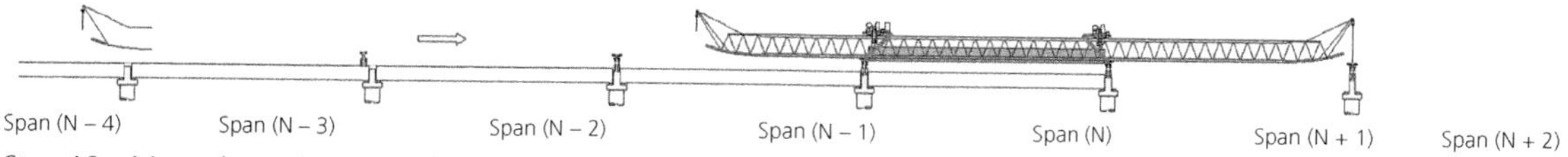

Step 10: Move the main trusses forward together with the beam and place the support on pier N + 1.

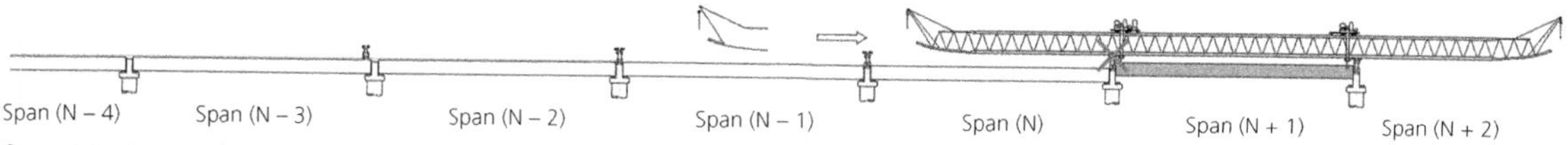

Step 11: Move the main trusses forward and erect the first beam of span N + 1.

The '4 plus 1' supports shown in the above kinematics (four front/rear supports and one auxiliary support) is the minimum number of supports required to allow the relocation of the supports by the LG itself. The number of supports varies according to the maximum distance from the girder delivery point to the 'erection' span. When a bridge has only a few spans, it is convenient to arrange a sufficient number of supports to equip all the piers. With such an arrangement, the follow-up activities related to the deck slab construction can be completely independent of the girder erection activities.

The kinematics shown in Figures 7.13 and 7.14 are based on cross heads with a half-joint cross-section (for different types of support configurations of beams and cross heads, refer to Section III, Chapter 15). The preferred type of cross-section for the cross head when erecting girders with LG type 1 is clearly the half-joint type, since each support is used as rear, intermediate and front support.

7.5.2 Basic kinematics of LG type 2

Figure 7.15 (on pages 100 and 101) shows a full erection cycle, including support relocation, with girder delivery from the side in span N.

LG type 2 can, like LG type 1, also be used to pick up girders delivered over the deck. The girders are in this case transported over the already completed deck generally by a prime mover and a bogie. If the girders are transported on rail bogies they can also be transported over partially completed spans where only the beams have been placed as yet.

In the simplest configuration for beam delivery over the deck, the main trusses are, for placing of the beams in span N, simply supported and centred on supports placed over the front and rear piers of span N. After the front of the beam has been lifted off the bogie by the front trolley's winch, the beam is moved forward by synchronised movement of the prime mover on the deck and the front trolley on the trusses until the rear trolley's winch can pick up the beam at its rear end. After the beam has been centred longitudinally on the main trusses over span N, the main trusses are moved transversely to the beam's final position, where the beam can then be lowered down.

If the deck follow-up work going on in span N – 1 should not be obstructed during placing of the beams for span N, a different gantry configuration can be used as follows: the beams, which are to

Figure 7.15 Full cycle with support relocation, with beam delivery for span N from the side (DEAL)

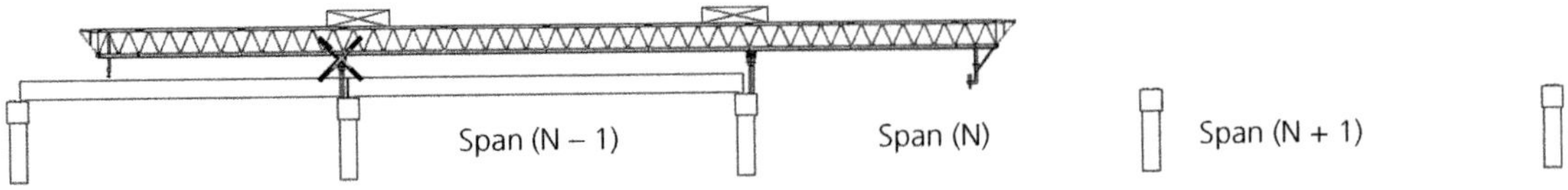

Step 1: Span N – 1 erection is completed.

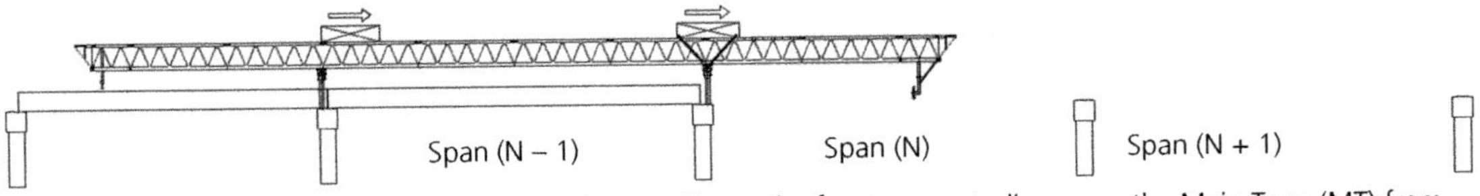

Step 2: Move the winches forward, anchor the front trolley to the front support, disengage the Main Truss (MT) from the rear support.

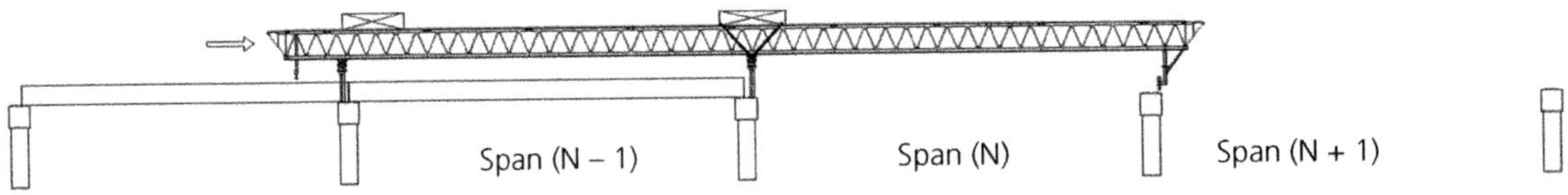

Step 3: Move the equipment forward.

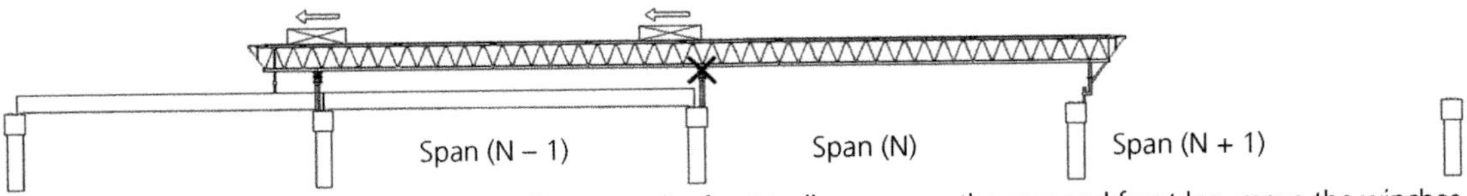

Step 4: Anchor the MT to the front support, disengage the front trolley, engage the rear and front leg, move the winches back, pick up the rear support by the rear winch.

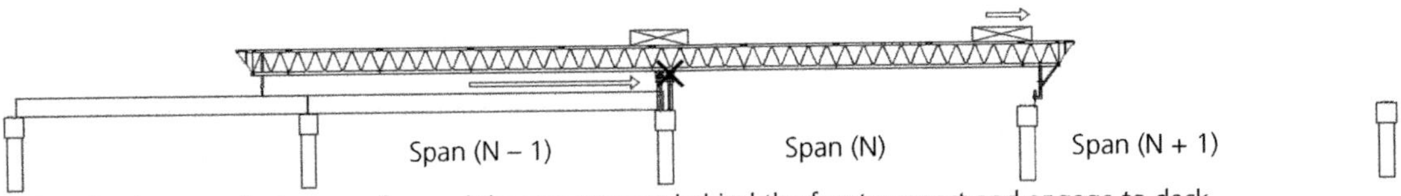

Step 5: By the rear winch, move forward the rear support behind the front support and engage to deck.

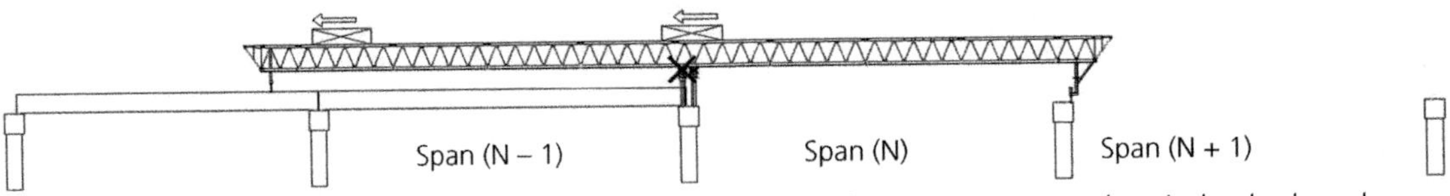

Step 6: Anchor the MT to the rear support, disengage the MT from the front support, move the winches backwards, pick up the front support by the front winch.

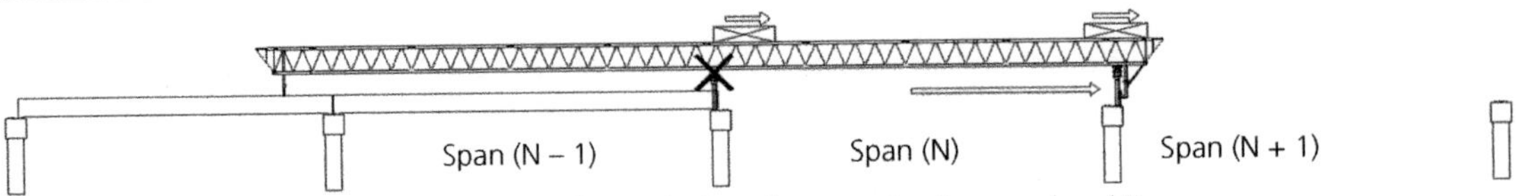

Step 7: Move the front support forward by the front winch and engage to pier cross head N.

Figure 7.15 Continued

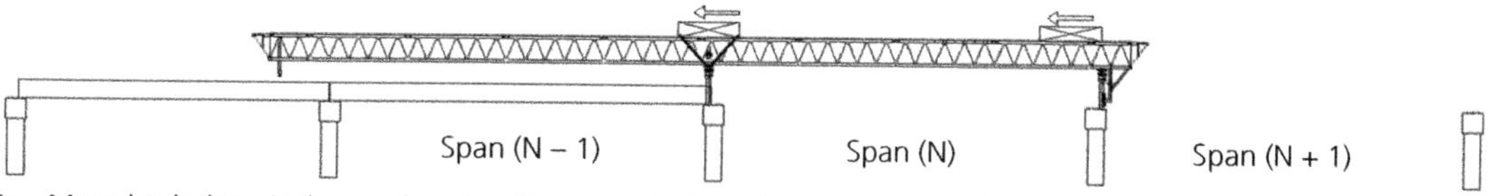

Step 8: Move back the winches and anchor the rear winch to the rear support, disengage the MT from the rear support, disengage front and rear legs.

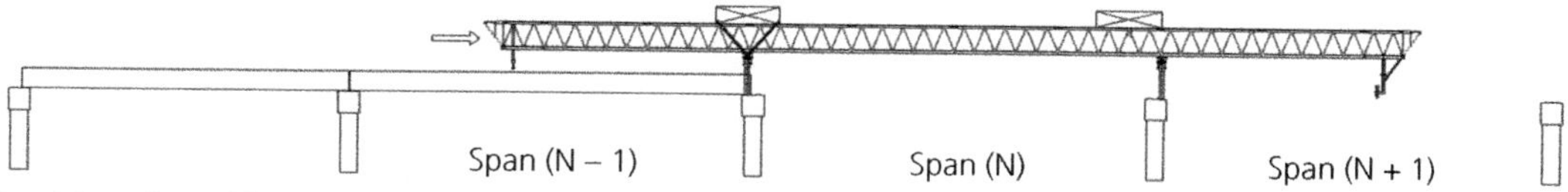

Step 9: Move the MT forward.

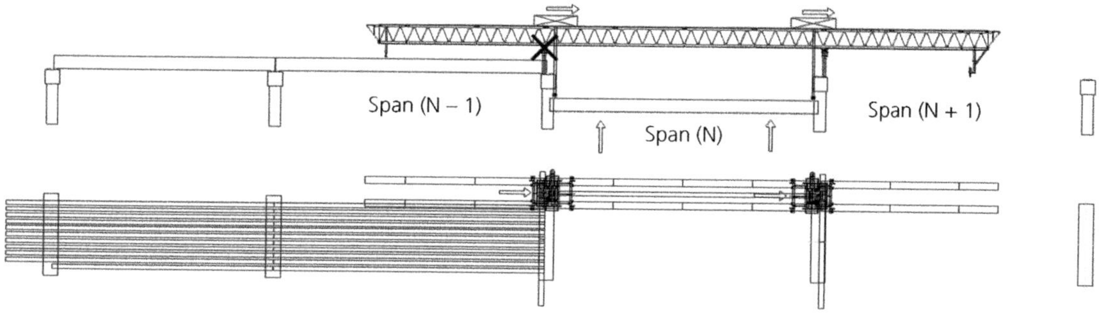

Step 10: Anchor the MT to the rear support, disengage the rear winch from the rear support, move the winches forward, pick up the first girder of the span from the side of span N.

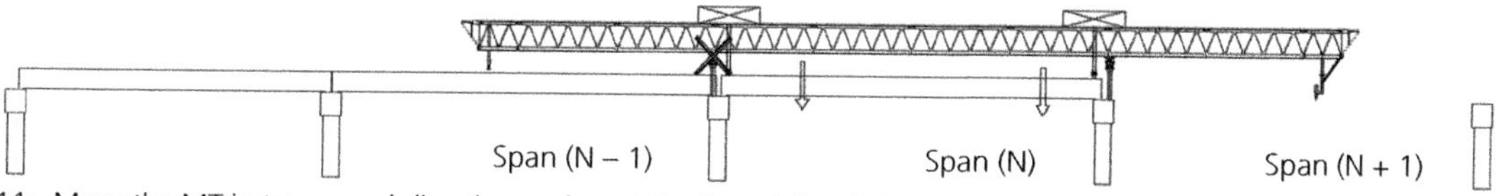

Step 11: Move the MT in transversal direction and erect the first girder of the span. Repeat for all the girders of the span N.

be placed in span N are picked up by the LG in span N – 2. The LG's rear support is therefore placed over the rear pier of span N – 1 and its front support over the rear pier of span N. During pick-up of a precast beam, the main trusses are transversely centred over the deck. Once the precast beam, suspended from the trolleys, has been moved longitudinally over span N – 1, the main trusses are moved transversely. Subsequently the main trusses, still with the beam suspended and kept over span N – 1, are launched forward until the front leg can be activated on the front half of span N's front cross head. Now the beam can be moved forward over the main trusses to its final position, where it is lowered down. The big advantage of this kinematic is that minimum support area is required on span N's front cross head during placing of the beams in span N, particularly of interest if the cross head has a rectangular cross-section. (For illustration of the different support conditions, refer to Section III, Chapter 15.) Furthermore, the LG's winches can also be utilised to assist with the follow-up work in span N – 1, whenever the LG is waiting for the arrival of the next beam (e.g. placing of precast planks between the beams). The main trusses, front leg and ULRS need, of course, to be designed and detailed for the corresponding static forces.

Figure 7.16 LG type 2 with a single box girder as main girder; segment delivery over completed deck to span N − 1 (NRS)

For an LG type 2, it is also possible to use a single box girder or truss as the main girder (for an example, refer to Figure 7.16).

7.6. Permanent works interface

The main interfaces with the permanent structure can be summarised as:

- available space on the pier's cross head to accommodate the front support, Front Leg (FL) and provide support area for the beams to be placed (upstream side)
- fixation of the front support on the pier's cross head
- placing and fixation of the rear and, if applicable, intermediate supports on the erected beams
- temporary stabilisation of the just-erected beams (before diaphragms and deck slab have been poured)
- fixing of the entire system for the 'out of service' wind conditions.

In addition, the permanent work verifications required are:

- the capacity of the cross head and pier to carry the construction load due to unsymmetrical loading by the precast beams and by the LG during the beam erection and the launching (transverse and longitudinal eccentricities)
- the capacity of the (individual) precast beams to carry the rear support, Rear Leg (RL), intermediate and auxiliary supports and, if applicable, the reactions from beams being transported over the not-yet-completed decks
- the capacity of the permanent bearing, if applicable, to take the loading during construction.

7.6.1 LG's front support

Different support conditions exist for the front support of the LG depending on whether the precast beam and the cross head have full or half-joint configurations (for possible configuration, refer to Section III, Chapter 15).

From a deck erection point of view, the longitudinal dimensions of a cross head should not only be determined by the final design conditions but should also consider the combined footprint of:

- the precast beams to be placed (upstream side)
- the front support/transverse beam (LG type 1)
- the front support and the FL (LG type 2, general)
- the FL only (for an LG type 2 when the FL is used during placing of girders).

If the cross head has a half-joint arrangement, hence its top surface is not used to support the girders, the length of the cross head at the top should be about 800 mm in general and up to 1200 to 1500 mm for an LG type 1.

If the cross head has a full section, it is very likely that temporary brackets have to be attached to the downstream side of the cross head to provide more longitudinal space for the LG's front support (and FL if applicable) during relocation of the front support and the placing of the precast beams.

The situation is much more favourable when the beams are placed with an LG type 2 with activation of the FL on the leading cross head during the placing of the beams instead of the front support. An FL will hardly require more space than is needed to support the beams.

7.6.2 LG's rear support

The rear support and RL are generally sitting on already erected beams for self-launching LGs type 2. In the case of an LG type 1, only the auxiliary support is sitting on beams, provided, of course, that the cross heads have half joints.

7.6.3 Gradient and cross fall

As mentioned earlier in the equipment description, the transverse beam of the supports must be kept horizontal. Therefore there is the need to adjust the height of the transverse beam's columns according to the cross fall and gradient. This issue is particularly sensitive in cases of significant cross fall together with particularly wide bridge decks where columns may reach heights of up to one metre. These columns need to be properly designed considering vertical and horizontal forces. Provision of such columns cannot be left to the 'creativity' of the site personnel.

7.6.4 Temporary stabilisation of the precast beams

A proper design also has to be performed for temporary stabilisation of the precast beams, before the deck slab is poured, taking into consideration the loads from the LG. The necessary connections to anchor the temporary stabilisation of the beams have to be provided both in the beams and in the cross head.

7.6.5 Beam delivery

The capacity of the bridge to carry the construction load due to feeding of the beams over the recently completed bridge deck has to be checked.

When beams are fed from the ground beside the bridge deck, the transverse beam has to cantilever beyond the cross head and therefore suitable HT bars must be foreseen to assure stability of the system. The transverse beam's support reactions may be significant and may control the local design and sizing of the cross head in this area. In addition, the global stability of the cross head must also be checked.

7.7. Cycle time

The cycle time depends mainly on the mode of feeding the beams: sideways from the ground or from the back via the already completed deck. When beams are fed from the ground, erection activities are independent from the follow-up activities for the deck after the beams have been placed. In contrast, when beams are fed from the back via the already completed deck, follow-up activities, including the concrete curing of the diaphragms and the deck slab, are on the critical path. For simplicity, only the latter is considered in this subsection and is shown for an LG type 1.

The deck slab is permitted to support the transporter delivering the beams only when its concrete has reached the required minimum strength. Frequently, specifications require the production of sufficient test results to confirm that the deck's concrete has reached the 28-day characteristic strength by the proposed time of loading.

All of the above means that usually the curing time is in the range of one/two weeks at the minimum. Therefore the number of supports has to be selected mainly on the basis of the time required to form and pour the deck slab and the subsequent curing. For example, the kinematics shown in subsection *7.5.1, 'Basic kinematics of LG type 1'* (4 + 1 supports), allow one entire cycle each for erection of span N, for deck pour of span N − 1 and for curing of span N − 2. Span N − 3 is finally ready to carry the weight of the transporter delivering the beams.

With the addition of one extra support, it is therefore possible, for instance, to have two entire cycles for curing the deck slab and so on. Table 7.1 shows the steps of a typical cycle with the assumption of 4 + 1 supports.

7.8. Risks

The list of risks given in Table 7.2 is not intended to be comprehensive, but addresses risks which are specific to LGs used for precast beam erection. For a specific project, a main contractor needs to carry out a full risk assessment of the operations performed with the BDE, considering the actual features of the BDE, the actual erection sequence and the site constraints.

7.9. Performance criteria

The following items need to be addressed when writing a performance specification for an LG.

- Precast beams
  - general data (per beam type): weight, dimensions (width including transverse starter bars, height including vertical starter bars), centre of gravity
  - delivery: from below, from behind, both
  - position of lifting points: fixed distance, variable
  - arrangement of lifting points: bars, strand loops, other
  - temporary stabilisation needs of the beams (if any).
- Alignment: gradient, curvature in plan and cross fall variations.
- Environmental conditions: maximum 'in service' wind speed, maximum 'out of service' wind speed.

Table 7.1 Steps of a typical cycle with 4 + 1 supports

		TYPICAL CYCLE 4+1 SUPPORTS					
		SHIFTS					
Activity	**Min**	1	2	3	4	5	6
LAST BEAM ERECTION ON SPAN (N) AND SUPPORT RELOCATION							
Move back on span (n-1)	20						
Move back on span (n-2)	20						
Move back on span (n-3)	20						
Pick up the last beam	10						
Move forward on span (n-2) and pick up the auxiliary support	60						
Move forward on span (n-1) and place the auxiliary support	60						
Pick up the main support	90						
Move forward on span (n)	20						
Place the beam on span(n)	30						
Park the main support onto the deck	30						
Move back on span (n-1)	20						
Move back on span (n-2)	20						
FIRST BEAM ERECTION ON SPAN (N+1) AND SUPPORT RELOCATION							
Pick up the beam	10						
Move back and pick up the parked support	30						
Move forward on span (n-1)	20						
Move forward on span (n)	20						
Move forward and place the main support	120						
Move forward on span (n+1)	20						
Place the beam on span(n+1)	30						
BEAM ERECTION FOR BEAM 2 TO 6							
Step 1 Move back on span (n)	20						
Step 2 Move back on span (n-1)	20						
Step 3 Move back on span (n-2)	20						
Step 4 Pick up the beam	10						
Step 5 Move foward on span (n-1)	20						
Step 6 Move foward on span (n)	20						
Step 7 Move foward on span (n+1)	20						
Step 8 Place the beam on span(n+1)	30						
Repeat steps 1 nto 8 for beam 3	160						
Repeat steps 1 nto 8 for beam 4	160						
Repeat steps 1 nto 8 for beam 5	160						
Repeat steps 1 nto 8 for beam 6	160						

- Lifting strokes: maximum pier height to fix hook height of the winches if the delivery is from below.
- Speeds of movements: speed of individual movements or time for individual tasks.
- Permanent work interfaces
  - permissible deck loading; magnitude, location and contact area (if beam delivery is over deck and/or if LG's rear support is on beams and not on cross heads)
  - permissible cross head loading: magnitude, location and contact area (especially on side edges if delivery is from side)
  - geometry detail of cross heads, including beam layout envelope.
- Clearances: vertical below and above deck (overhead power lines, bridges, etc.); horizontal (houses, power lines, etc.).
- Access: into LG (from deck, from front pier) and inside LG during different stages of operation; access from ground to piers.

In addition, the buyer needs to give a time frame for fine-tuning of the LG's concept and support reactions/inserts definition, along with a time frame for the delivery and commissioning of the LG.

Table 7.2 List of risks specific to operation of LGs for precast beam erection

Activity	Risk	Mitigating measures
Working on top of precast beams Attachment of hook to precast beams delivered on trailer	Falling down of the worker from the beams	Provision of temporary lifeline along all beams installed on top (between the two lifting points), installed prior to lifting the beam
Beam erection over live traffic	Dropping of beam (unlikely), falling of debris, tools (possible)	No beam handling over live traffic
Beam installed over live traffic	Dropping of beam (unlikely)	Traffic permitted only once beam has been placed and secured against lateral overturning
Lowering of the beam on its supports on the cross head	Crushing of people working in narrow space on cross head	Define by clear instruction where workers can stand while a beam is suspended from the winches. Worker and operator to communicate by radio receiver
Temporary stabilising of beams	Failure of temporary props to cross head or between beams	Properly designed, simple to install, adjustable
Installation of formwork or precast planks between precast beams and installation of reinforcing steel bars	Falling down of materials, tools (possible)	Provision of temporary protection at soffit level of beams if live traffic/pedestrian circulation cannot be stopped
Support of LG	Concrete failure at support's contact area	Adequate edge distances, so that support's contact area is within concrete confined by reinforcement; shimming with similar material over entire footprint of support's contact area
Longitudinal movement of LG's main girders	Over-launching or loss of control of longitudinal movement → loss of equilibrium → overturning of LG's supports and loss of entire LG	Back-up fixity, provision of nominally stressed anchor bars at supports, presence of LG engineer and LG supervisor during launching stages
Relocation of the supports	Crushing/falling down of people working on deck/cross head	Suspend construction work on deck during support relocation over the span. Define by clear instruction where workers can stand during support relocation and main girder's self-launch

7.10. Photo gallery

Figure 7.17 LG type 1, girder delivery over deck (Sang Pyung Bridge Project, South Korea) (DEAL)

Figure 7.18 LG type 3, relocated by crane operating from temporary trestle (Indian Street Bridge, Florida, USA) (DEAL)

Figure 7.19 LG type 3, with SMUs as lifting devices, relocated by ground-based crane (East Coast Parkway, Singapore) (VSL)

Figure 7.20 LG type 3, self-launching (Southern Outer Bangkok Ring Road Project, Thailand) (VSL)

Figure 7.21 LG type 2, girder delivery over deck, with SMUs as lifting devices, in two-span configuration for placing of girders (with activation of FL) (Bukit Brown Viaduct, Singapore) (VSL International)

Figure 7.22 LG type 2, girder delivery over deck, LG also used for construction of substructure (Washington Bypass Project, North Carolina, USA) (DEAL)

REFERENCES

Figures 7.1–7.3, 7.5–7.15, 7.17, 7.18, 7.20, 7.22: DEAL
Figures 7.4, 7.19, 7.21: VSL International
Figure 7.16: NRS
Table 7.1: DEAL

Bridge Deck Erection Equipment
ISBN 978-0-7277-6193-4

https://doi.org/10.1680/bdee.61934.111

Chapter 8
Precast full span erection by launching gantry

8.1. Introduction

The Full Span Method (FSM) system has seen a huge rise in popularity during the past 15 years for the construction of high-speed railway line viaducts in many parts of the world.

With this method, entire bridge spans of a viaduct are produced in a casting yard located at the beginning of the viaduct. The Precast Spans (PS) are then transported, usually along the already erected viaduct, and placed at their final position by Bridge Deck Erection Equipment (BDE).

The structural requirements for high-speed railway lines are stringent: big horizontal radii, small gradient and small settlement of embankment. This often results in great portions of the track being put on viaducts. These viaducts can be kilometres long and construction time becomes a critical issue.

For traditional erection methods, cycle times vary from a minimum of two days (i.e. span-by-span segmental erection) up to 14 days (cast in situ with a movable scaffolding system), resulting in the need for a significant number of BDE on the same viaduct. With FSM, a deck erection cycle time of less than one day per span with one BDE is common. Also, two spans per day is an achievable result. Hence by using FSM it is possible to reduce the number of construction fronts for the deck construction. Construction of the substructure and piers remains, however, a time-consuming activity, which can be sped up only by increasing the number of construction fronts. Furthermore, FSM can only be utilised to its full potential if the substructure and the piers can be made ahead of the arrival of the BDE. Simply omitting a span where there is a problem with foundations is not possible.

Very often delivery distance and related transportation time control the entire construction cycle.

Initial investment costs for FSM are high. In addition, a lot of space is required for the prefabrication facilities. FSM becomes, therefore, economically viable only if the viaducts are longer than 3 to 4 km.

In recent years, FSM launching equipment has also been designed and fabricated for highway bridges with spans of up to 60 m and a weight of up to 18 000 kN. The initial investment costs are very high, due to the large dimensions and the huge weight of the BDE and the need for sophisticated Self-Propelled Transporters (SPT). Use of FSM for such applications is economically viable only for viaducts of at least 10 km in length.

If the PS is delivered over the deck, reactions on the deck are significant. In order to avoid construction loads governing the design of the permanent structure, the use of the full span method is usually limited to:

- high-speed railways, where the superimposed permanent loads (i.e. the ballast) and the live loads are very high
- light rail with twin tracks, where it is possible to spread the loads of the PS between the two decks during transport
- motorways with twin decks, where it is possible to spread the loads of the PS between the two decks during transport.

For light rail, if weight and dimensions of the PS allow transportation on public roads, the erection may be done by conventional cranes and/or by a BDE. For offshore viaducts it may be possible to deliver the PS by barge. Erection may be done by a floating crane and/or by a BDE.

As the combined length of viaducts on a project where FSM is viable is usually several kilometres, it is possible that the viaducts are interrupted by embankments, bridges erected in advance by other methods and tunnels. In the case of tunnels, the BDE has to be designed so that it can be launched through the tunnels. Often the PS's wing slabs cannot be fully precast to allow transportation through the tunnels.

8.2. Application range

The full span construction method is applied for both motorway and railway bridges. However, the type of FSM and BDE and therefore the FSM's application range varies significantly depending on whether the viaduct supports a high-speed railway track, a light railway track or a highway. For details, refer to Tables 8.1 to 8.3.

Table 8.1 High-speed railway viaducts

Average span length	25 to 35 m	m
Type of viaduct	Usually double-track viaducts	
Typical cross section	U or box section	
Span weight	6000 to 10 000 kN	ton
Horizontal curve:		
for erection	≥3000 m	m
for transportation of the PS	≥200 m	m
Gradient	≤3%	%

Table 8.2 Light rail viaducts

Span length	20 to 40 m
Type of viaduct	Usually parallel twin decks or twin viaducts
Typical cross section	U or box section
Span weight	2000 to 4500 kN
Horizontal curve:	
for erection	≥400 m
for transportation of the PS	≥200 m
Gradient	≤3%

Table 8.3 Highway viaducts

Span length	50 to 60 m
Type of viaduct	Always parallel twin viaducts
Typical cross section	Box section
Span weight	13 000 to 18 000 kN
Horizontal curve:	
for erection	≥3000 m
for transportation of the PS	≥200 m
Gradient	≤3%

For more details about cross sections for the FSM, refer to Section III, Chapter 15.

8.3. General arrangement of the BDE

Different types of BDE for FSM have been developed. If the PS is delivered over the deck, the BDE can be classified into three main types.

- Type 1: Launching Gantry (LG) plus Self-Propelled Transporter (SPT). The LG is able to unload the Precast Span (PS) from the SPT and place it at its final position. The BDE type 1 is designed to work on single-deck viaducts only.
- Type 2: Launching Gantry (LG) plus Dual Self-Propelled Transporter (DSPT). The LG is able to unload the PS from the DSPT and place it at its final position. Unlike BDE type 1, the BDE type 2 is configured to work across twin decks or viaducts. It erects both decks in span N before it is launched into the next span N + 1.
- Type 3: Straddle Launching Carrier (SLC) plus Self-Launching Support Beam (SLSB). The SLC picks up the PS directly at the precast yard, transports it to the erection front over the erected deck and places it in its final position by moving over the SLSB.

8.3.1 Type 1 – Launching Gantry (LG) + Self-Propelled Transporter (SPT) on single viaduct

The BDE (refer to Figure 8.1) consists of a Launching Gantry (LG) and a Self-Propelled Transporter (SPT). The PS is loaded onto the SPT at the cast yard by means of a pair of dedicated lifting equipment (e.g. portal cranes). Then the PS is driven to the erection front over the erected part of the deck and/or the embankment. At the erection front the LG is able to unload the PS from the SPT, move it longitudinally over and lower it down to its final position on the piers. After the erection stage, while the SPT returns back to the cast yard to load the following PS, the LG self-launches to the following span.

Figure 8.2 shows the main components of the BDE shown in Figure 8.1. For individual projects, the configuration of the BDE may be altered.

Advantages of a BDE type 1 are:

- Low impact on superstructure from the LG during self-launch: the total weight of the LG is in fact divided almost equally between the rear portal moved over the deck and the launching leg moved over the support beam.
- The transporter is not used during the placing of the PS, so it can be moved back to the casting yard as soon as the PS has been unloaded by the LG. This reduces the transport cycle time.

Figure 8.1 Example of BDE type 1: LG + SPT on single-deck viaduct (NRS)

Disadvantages of a BDE type 1 are:

- Requires additional heavy lifting portal cranes in the casting yard to load the PS onto the SPT.
- Cannot be used for twin viaducts, since the wide rear portal of the BDE working on one viaduct does not permit the transport of a PS over the adjacent parallel viaduct.
- Cannot pass through tunnels without dismantling.
- Significant conflicts with the finishing works (i.e. parapets, sidewalk, etc.): all the auxiliary equipment needs to be removed from the whole already erected deck from the precast yard to the erection front in order to allow the passage of the SPT.
- Cannot erect viaduct spans immediately before or immediately after tunnels (needs to erect at least one to two spans before and after tunnels by other means).
- Needs dismantling and/or additional equipment to allow for transportation of the whole equipment from one viaduct to the next over an embankment.

8.3.2 Type 2 – Launching Gantry (LG) + Dual Self-Propelled Transporter (DSPT)

The BDE (refer to Figure 8.3) consists of a Launching Gantry (LG) and a Dual Self-Propelled Transporter (DSPT) that always needs to be customised. The system is configured to work on twin bridges or twin decks and the construction load during transport (the weight of the PS plus the self-weight of the DSPT) is substantially distributed over four spans: two longitudinally adjacent spans on the two parallel viaducts.

The PS is loaded onto the DSPT in the casting yard by means of a pair of dedicated portal cranes. Then the PS is driven to the erection front over the already erected part of the deck. At the erection front the LG is able to unload the PS from the DSPT, move it longitudinally over and lower it into its final position on the piers. After the erection stage, while the DSPT returns to the casting yard to load the following FS, the LG self-launches to the following span.

Figure 8.2 Main components of a BDE type 1 (NRS)

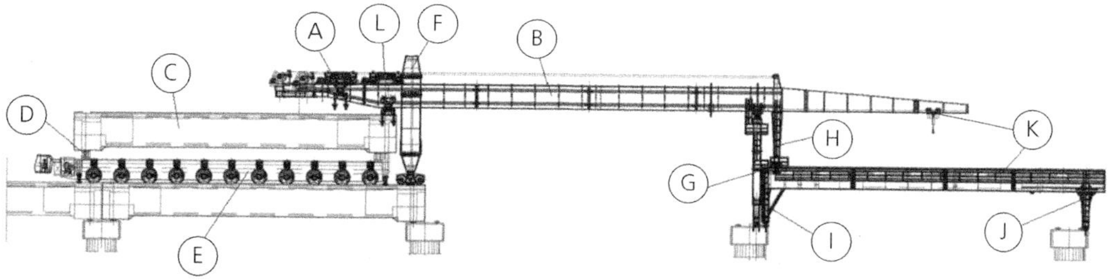

A. Rear winch
B. Main Girder (MG)
C. Precast Span (PS)
D. Trolley on SPT
E. SPT
F. Rear portal
G. Front leg
H. Launching leg
I. Fixed leg of SB
J. Movable leg of SB
K. Support Beam (SB)
L. Front winch

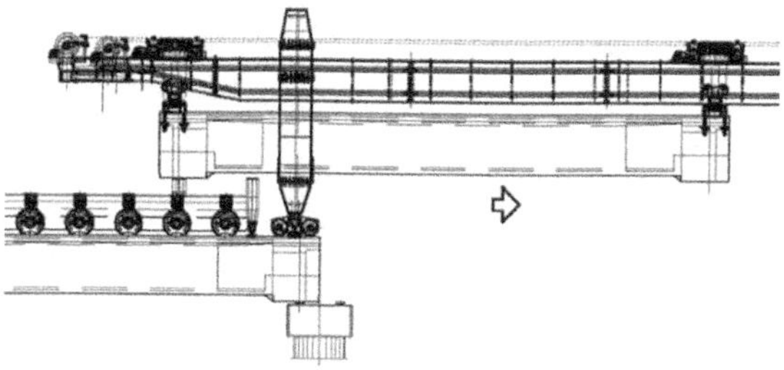

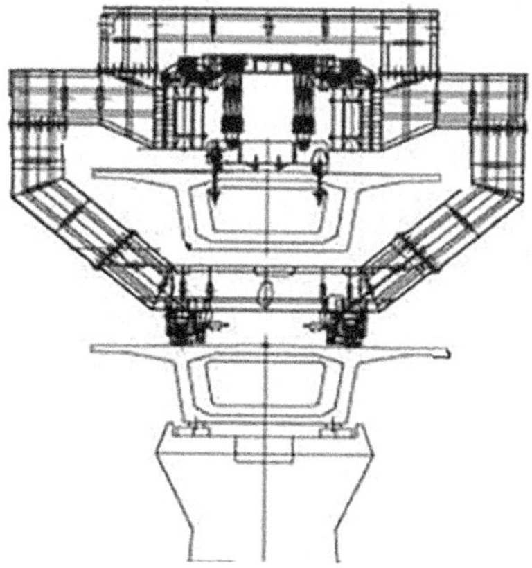

Details of the rear portal
The rear portal is a specific feature of the shown equipment. As the ratio between the width of the deck and the distance between the PS's webs increases, the design of the rear portal becomes more challenging and heavier.

Despite the size difference between highway and light railway viaducts, the erection system is very similar, varying only in size and capacity.

Figure 8.5 shows the main components of the BDE shown in Figure 8.3. For individual projects, the configuration of the BDE may be altered.

Figure 8.4 shows the main components of the delivery system (DSPT).

Figure 8.3 Example of a BDE type 2: LG on twin-deck viaduct (DEAL)

Advantages of BDE type 2 are:

- Very stable launching system thanks to a wide support on both rear and front piers.
- Reduced space requirements on piers for anchoring.
- The transporter is not used during placing of the PS, so it can be moved back to the casting yard as soon as the PS has been unloaded by the LG, reducing transport cycle time.
- Limited load on the deck during launching of the LG.

Disadvantages of a BDE type 2 are:

- Requires additional heavy-duty portal cranes in the casting yard to load the PS onto the DSPT.
- Can only work for deck erection of parallel twin viaducts or twin decks, where it reduces the loading during transportation of the PS to a level permissible for the deck.
- Needs dismantling and/or additional equipment to allow for transportation of the whole BDE from viaduct to viaduct over an embankment.
- Cannot pass through tunnels.

8.3.3 Type 3 – Straddle Launching Carrier (SLC) plus Self-Launching Support Beam (SLSB)

The BDE (refer to Figure 8.6) consists of a Straddle Launching Carrier (SLC) and a Self-Launching Support Beam (SLSB). The SLC transports the PS from the casting yard to the erection front and places it directly in its final position over the piers. Once the SLC reaches the end of the viaduct, the SLC's front wheel group drives over the rear end of the SLSB, which cantilevers at its rear end slightly over the last-placed span. The SLC's front reaction is subsequently transferred to a trolley on top of the SLSB. By synchronised longitudinal movement of this trolley over the SLSB and the rear wheel group over the viaduct, the SLC transports the PS over its final position. The SLSB is then launched forward underneath the stationary SLC and PS to free the span for lowering of the PS to its final level. Figure 8.6 shows the stage when the SLC and PS are moved over the span.

Figure 8.4 Type 2 – DSPT (DEAL)

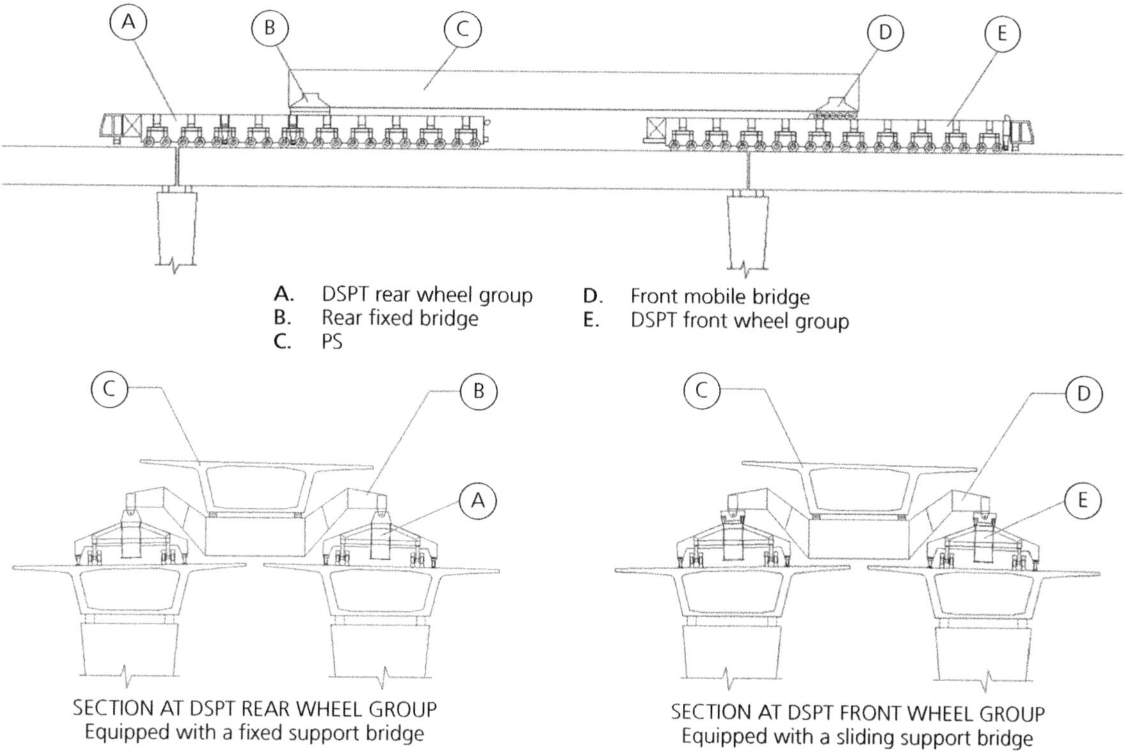

Figure 8.7 shows the main components of the BDE shown in Figure 8.6. For individual projects the configuration of the BDE may be altered.

Advantages of BDE type 3 are:

- Low impact on superstructure from special carrier during transportation and launching: the total weight of the SLC and PS is being distributed equally on two adjacent spans directly over the deck's webs through two or four massive rubber wheels per support point (no local bending in deck slab).
- The special launching carrier can lift and transport the PS directly from the casting yard, theoretically without need of additional heavy-duty portal cranes to load the PS on a transporter (however, for an ambitious construction programme, the casting and erection cycles need to be made independent from each other, for which provision of dedicated heavy-duty portal cranes is essential).
- Reduced conflicts with the finishing works (i.e. parapets, sidewalk, etc.): all the auxiliary equipment may be parked in the middle of the bridge (i.e. between the webs) to allow the SLC to pass, even loaded, since the SLC transports the PS with significant vertical clearance over the deck and its wheels load the deck only over its webs.
- Can work on twin bridges and twin decks with a dedicated SLSB per deck.
- Can pass over special spans erected ahead of arrival of the BDE.
- Can pass through tunnels without dismantling.

Figure 8.5 Main components of a BDE type 2 (DEAL)

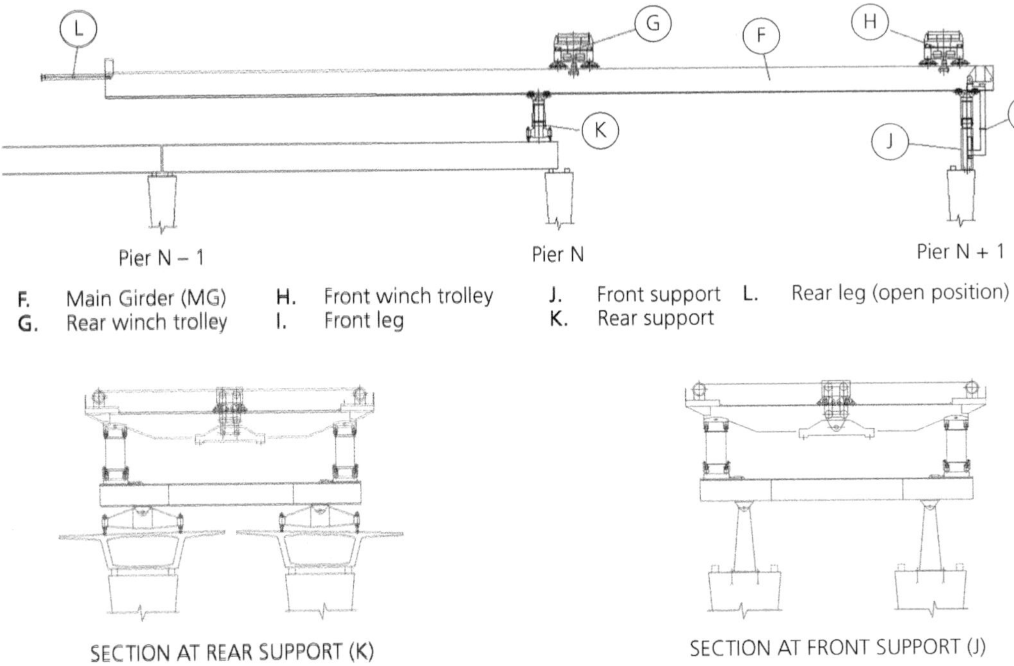

Figure 8.6 Example of a BDE type 3: SLC + SLSB (DEAL)

Figure 8.7 Main components of a BDE type 3: SLC + SLSB (DEAL)

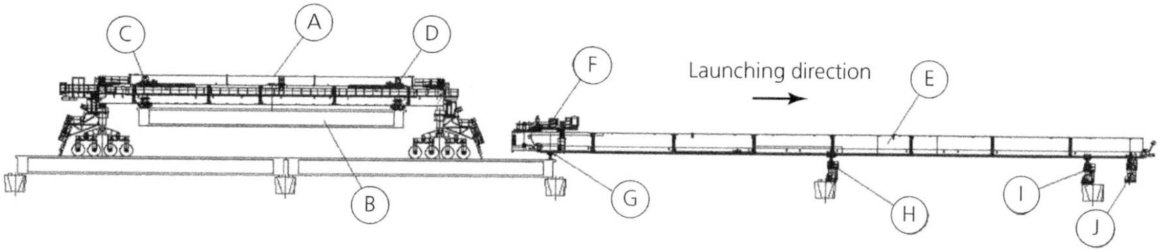

A. SLC
B. PS
C. Rear winch
D. Front winch
E. SLSB
F. Trolley
G. Rear support
H. Front support
I. Front auxiliary support 1
J. Front auxiliary support 2

- Can erect viaduct spans immediately before or immediately after a tunnel (even inside).
- Easy to relocate from one viaduct to the next.

Disadvantages of BDE type 3 are:

- Requires significant anchoring on piers (a need for large pier heads).
- Transporter cannot be moved back to casting yard until almost the full erection cycle is completed; only relocation of the SLSB supports may be performed out of the critical path.
- Limitation in span length: above about 40 m, the size of the SLSB would become too deep, making the overall height of the SLC on top of the SLSB too big for stability.

8.4. Kinematics

8.4.1 Kinematics of type 1 (LG + SPT)

The SPT moves the PS from the casting yard to just behind the LG. There, the LG lifts the PS from the SPT, moves it forward and places it at its final location. Once complete, the LG self-launches forward to the next span while the transporter returns to the yard to collect another PS. Figure 8.8 shows the phases in detail.

8.4.2 Kinematics of type 2 (LG + DSPT on two parallel viaducts)

The DSPT moves the PS from the casting yard, running on both parallel bridges or decks, to just behind the LG. There the LG lifts the PS from the DSPT, moves it longitudinally over its final position, shifts it transversely to the side and places it at its final position over the piers. Subsequently the second PS for this span is placed. Once complete, the LG self-launches forward to the next span while the DSPT returns to the yard to collect another PS. Figure 8.9 shows the phases in detail.

8.4.3 Kinematics of type 3 (SLC + SLSB)

The SLC lifts the PS from the storage area, transports it along the embankment and over the already completed deck and places it at its final position by driving onto the SLSB. Figure 8.10 shows the phases in detail for placing precast span N between pier N and pier N + 1 (refer to Figure 8.10).

8.4.3.1 Passing through tunnels

When working in mountainous areas, the BDE often has to be moved from one viaduct to another passing through tunnels. For this situation, customised solutions have been developed to obtain very

Figure 8.8 Kinematics of BDE type 1 (LG + SPT) (NRS)

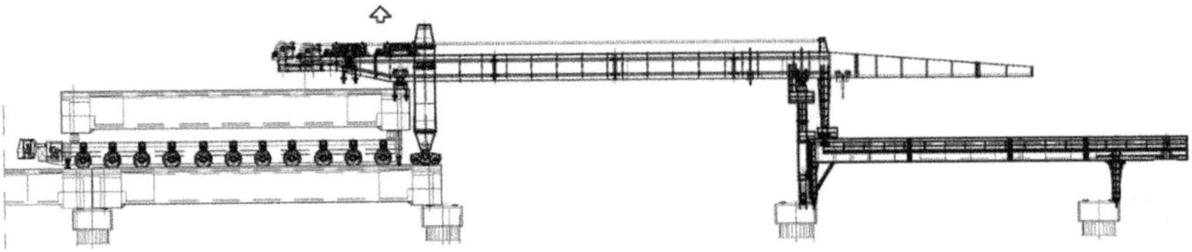

PHASE 1: PS arrives on SPT; PS picked up by front winch and remains on the trolley of the SPT.

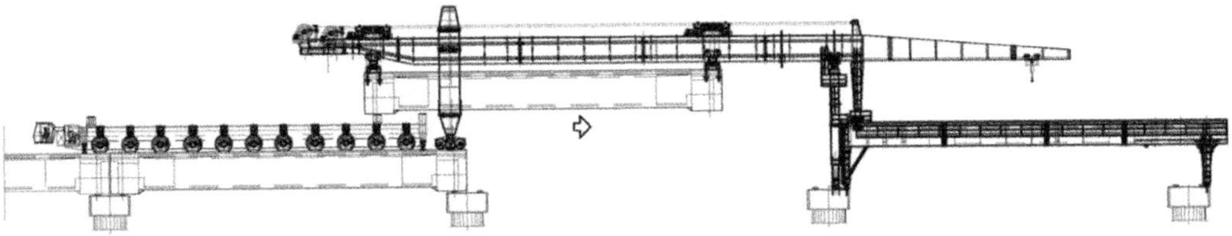

PHASE 2: PS is moved forward through rear portal; rear PS side still supported on the trolley of SPT until the rear side of the PS can be picked up by the rear winch. Then the SPT is free to move back to the PS storage area.

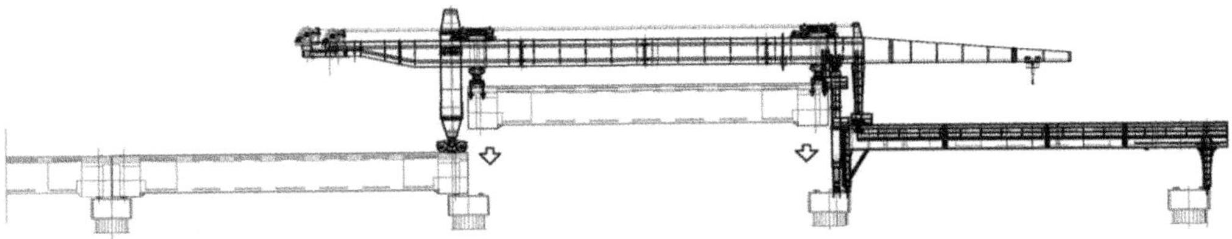

PHASE 3: PS positioning and setting on piers by the two winches.

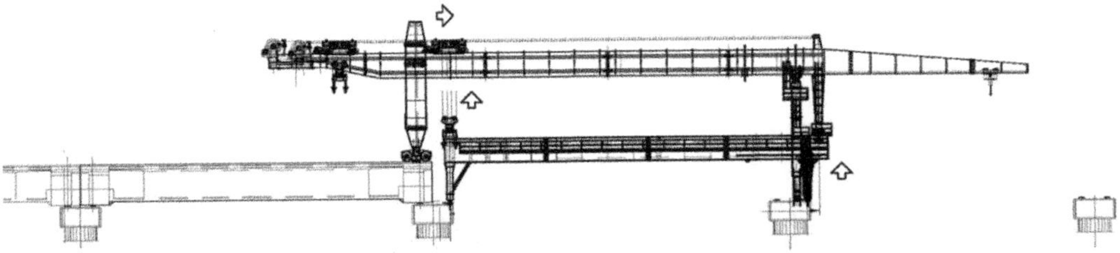

PHASE 4: MG advancing, rear portal running on erected span and the launching leg running on the support beam.

low profiles, leaving room for the PS to be placed right after or even inside the tunnel (refer to Figure 8.11).

8.4.3.2 Working on parallel lines with a BDE type 3

A BDE type 3 allows working on parallel decks with one of the following options:

1 complete one deck, launch backward and erect the parallel deck
2 use two machines working in parallel.

As already mentioned, a type 1 BDE does not allow working on closely spaced twin viaducts in parallel, since its large rear portal blocks the transportation of precast spans for the parallel deck (refer to Figure 8.12).

Figure 8.8 Continued

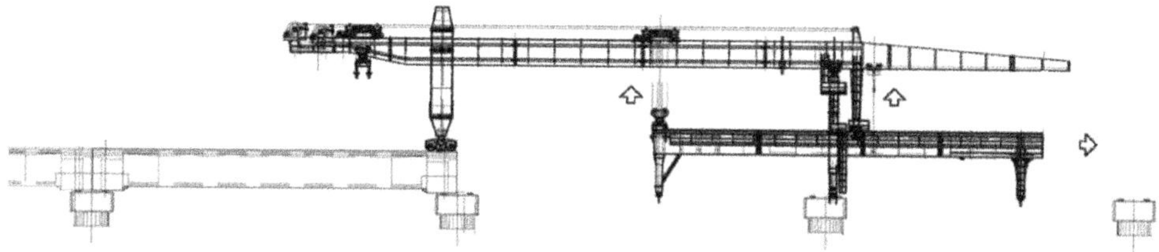

PHASE 5: Support beam partial launching, support beam hanging by front auxiliary winch.

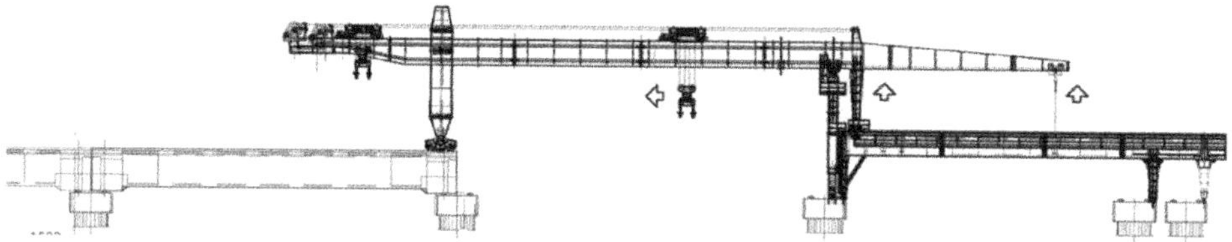

PHASE 6: Completion of the SB launching, adjustment of SB movable leg according to actual pier position and anchoring of both legs of the SB to piers.

8.5. Permanent works interfaces

The main interfaces with the permanent structure can be summarised as:

- adequate space on the pier head to accommodate the front support
- fixity of the front support to the pier by means of vertical HT bars
- activation of the rear and intermediate supports on the erected span (space and vertical fixity provisions by HT bars)
- fixity of the entire BDE system for the 'out of service' condition (wind loading).

The capacity of the bridge to carry the construction loading has to be checked during the basic design and detailing stage. Particular areas for checking are:

deck

- lifting holes reservation to arrange statically determinate lifting system (three points)
- axial load from transporters – SPT and SLC: axle distance, tyre pressure
- axial load from LG (type 1)
- loads on wheel groups (statically determinate distribution on three points achieved by hydraulic systems)

piers/abutment

- longitudinal moment due to eccentricity of support and longitudinal forces
- transverse moment: skewing forces and transversal wind loads, eccentricity due to curve
- anchoring reservations
- bearing points for supports: need for a certain area to distribute the loads within the permissible concrete pressure

embankment

- compacted ground to allow for axial load of transporter

Figure 8.9 Kinematics of BDE type 2: (LG + DSPT on parallel twin viaducts) (DEAL)

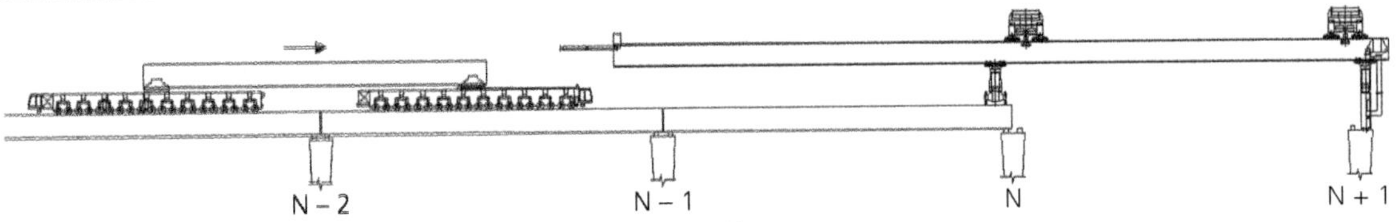

PHASE 1: DSPT arrives with PS, rear leg of the LG is in open position.

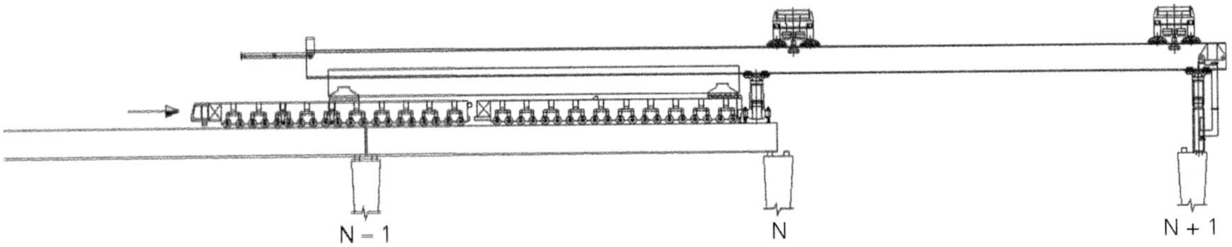

PHASE 2: DSPT enters underneath the LG; once the DSPT's front wheel group reaches the LG's rear support, the front mobile bridge on the DSPT is activated; the PS's front support slides over the front wheel group, whereas at the rear the DSPT's rear wheel group supports and pushes the PS forward. At the end the entire PS is underneath the LG.

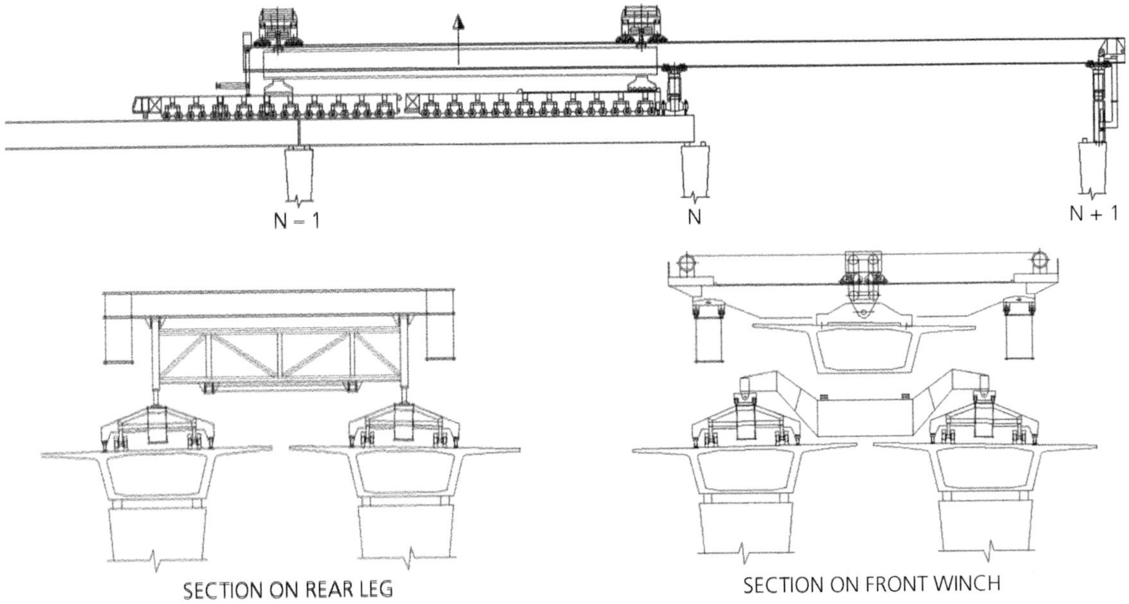

PHASE 3: The rear leg is closed and activated on top of the DSPT's rear wheel group. The winch trolleys are moved over the PS and pick it up

- soil preparation: to avoid mud, holes, etc.
- specific footings for launcher legs/supports for placing of first and final span (may need 'false' pier)
- redistribution of axial load due to transverse moments: skewing forces and wind, eccentricity due to curve.

Figure 8.9 Continued

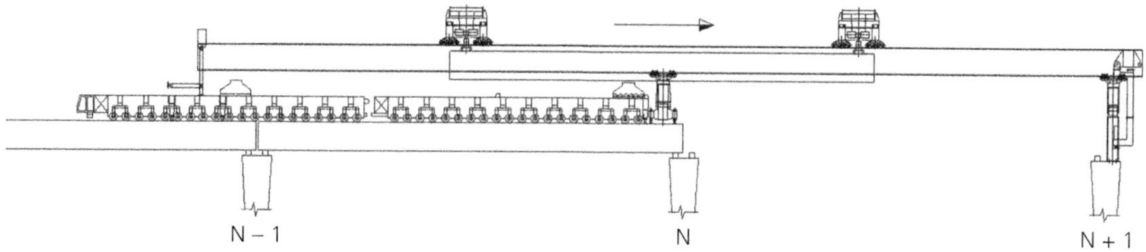

PHASE 4: The PS is moved ahead by the two winch trolleys.

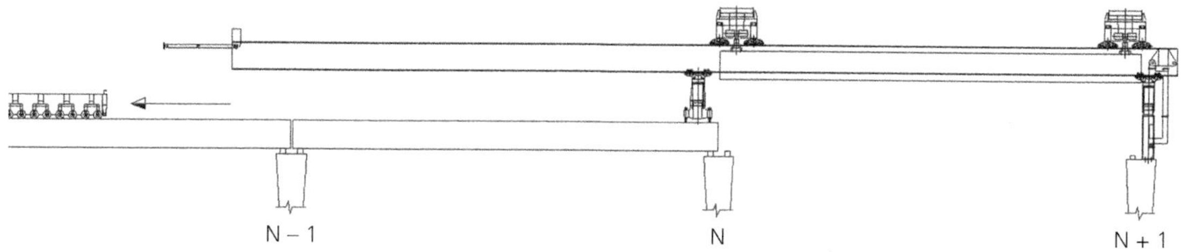

PHASE 5: When the PS has been moved longitudinally over its final position, the LG's rear leg can be disengaged from the DSPT and opened to allow the DSPT to run back to the PS storage area.

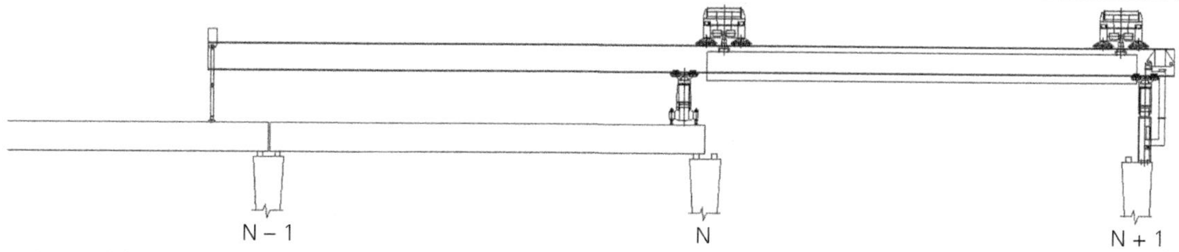

PHASE 6: With the PS still transversely centred in the DSPT, the rear leg is closed again and activated on the deck.

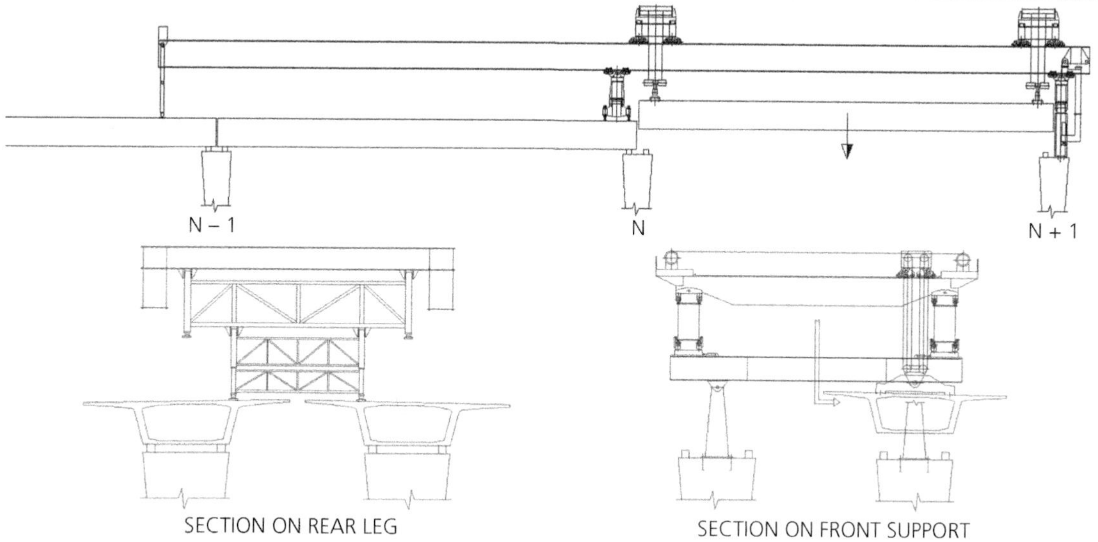

PHASE 7: When the rear leg is activated, the PS can be lowered, shifted transversely by moving the winches on the winch trolleys over its final position and lowered to its final position.

Figure 8.9 Continued

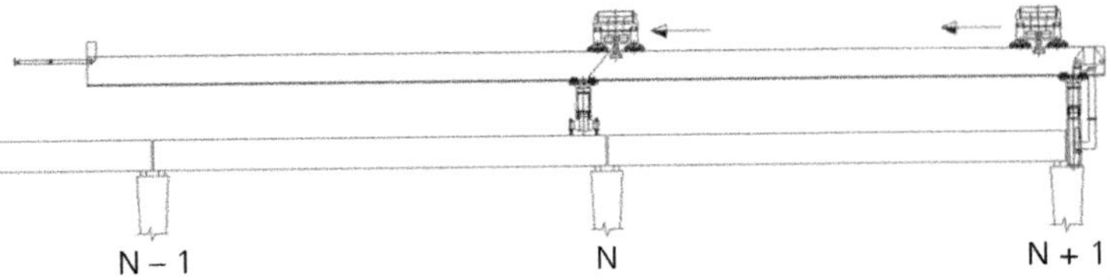

PHASE 8: After PS erection, the rear leg is disengaged, the winches are transversely centred and the trolleys are moved back, so that the rear winch trolley is in position to be anchored to the rear LG's support by a 45° anchor prop.

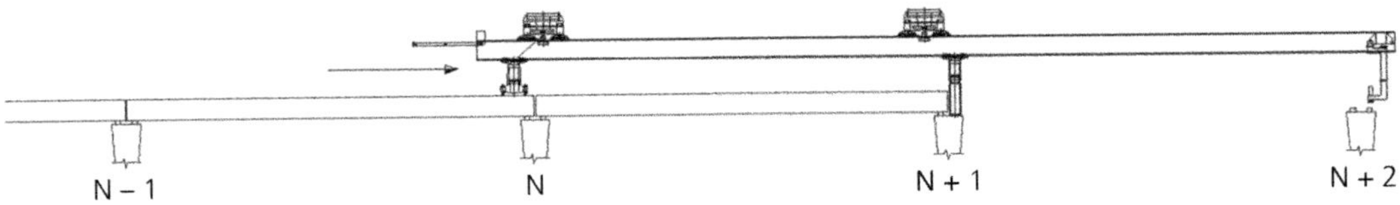

PHASE 9: Once the rear winch trolley is anchored to the LG's rear support, the MG's fixity to the rear support can be released and the MG can be launched forward until the front leg reaches the next pier.

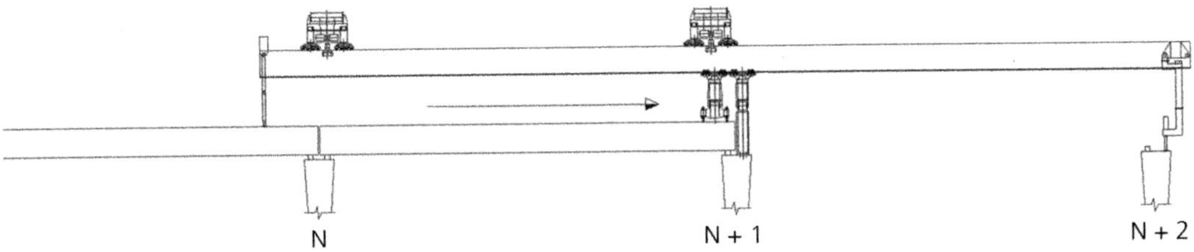

PHASE 10: The rear and front legs are activated on the deck and on pier N + 2 respectively to allow the relocation of the rear support.

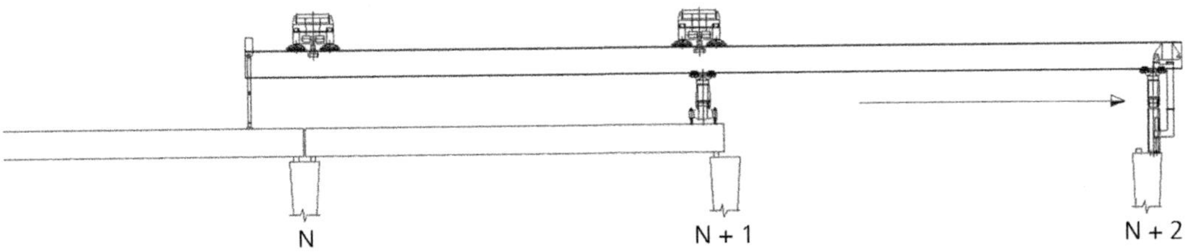

PHASE 11: The rear support is activated on the deck to allow the relocation of the LG's front support to pier N + 2. Once in place, the front support is anchored to pier N + 2 and the front leg is disengaged.

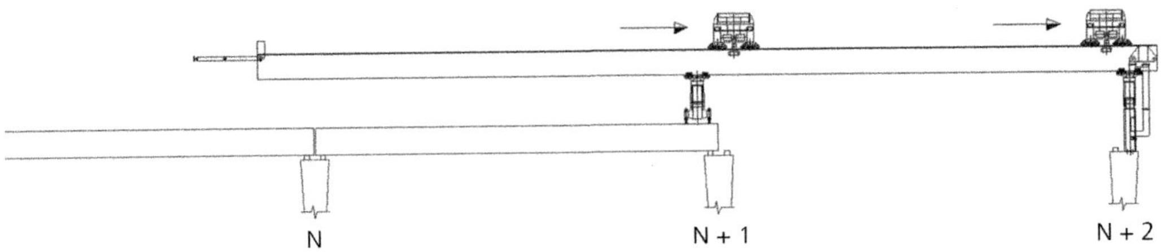

PHASE 12: The winch trolleys are moved forward and the rear leg is opened to receive the next PS.

Figure 8.10 Kinematics of a BDE type 3 (SLC + SLSB on a single viaduct) showing placing of precast span N between piers N and N + 1 (DEAL)

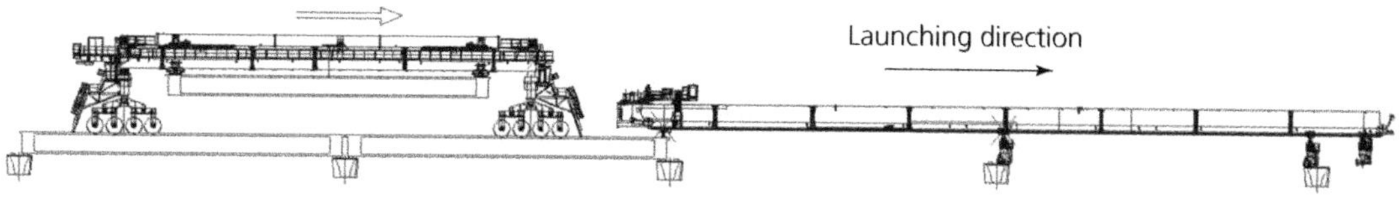

PHASE 1: SLC lifts the PS and transports it to the erection front.

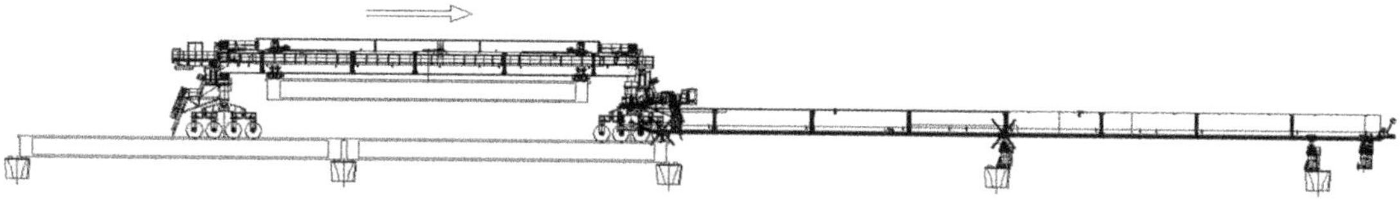

PHASE 2: SLC moves over the rear end of the SLSB, so that the front wheel group is over the SLSB's trolley.

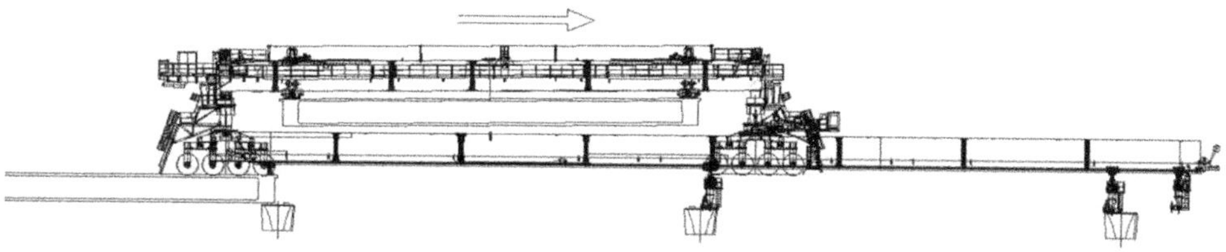

PHASE 3: By the hydraulic jacks of the trolley, the front wheel group of the SLC is picked up on the SLSB; SLC + PS are moved forward while driving on the deck at its rear end and on the SLSB at the front. The movements are synchronised by the control system.

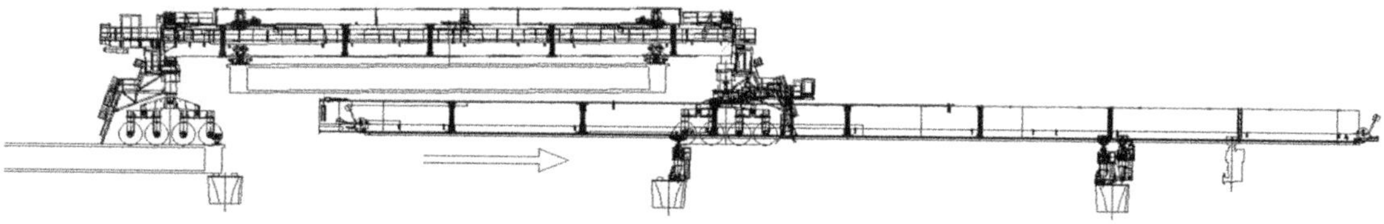

PHASE 4: When the PS is moved in plan over its final position, the movement is stopped. The SLSB is disengaged from the front support and launched forward, rolling over the front support on pier N + 1 and the front auxiliary support I on pier N + 2.

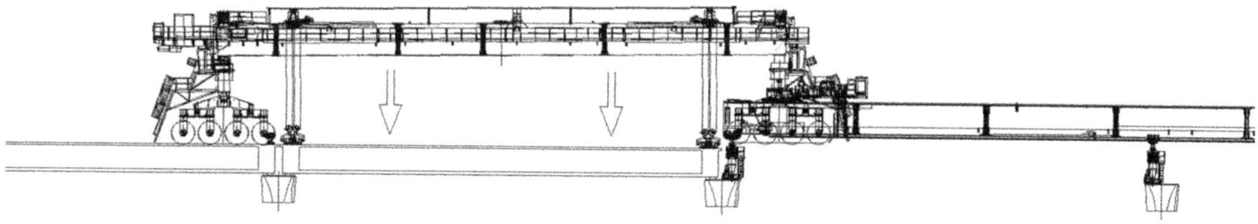

PHASE 5: When the SLSB clears the span, the movement is stopped and the PS is lowered into its final position.

8.6. Cycle time

For all the systems, the transport time and the casting cycle of the PS usually drive the global cycle. If the distance between the PS storage and the erection front is small (less than 3 km), the bottleneck can be the LG's self-launch time. But as soon as the distance increases, the transport time becomes critical.

Figure 8.10 Continued

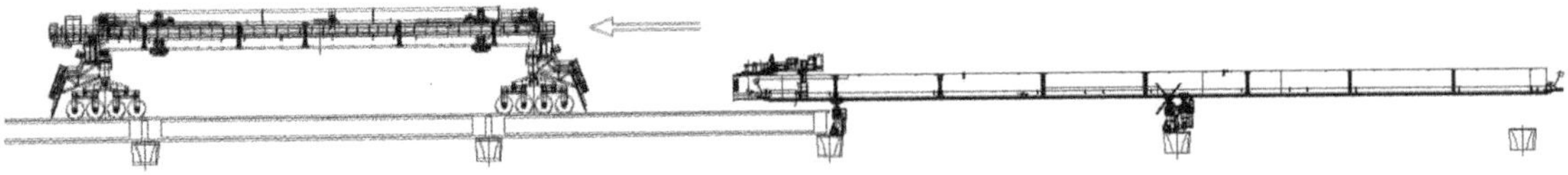

PHASE 6: The SLSB is launched back slightly to cantilever over the just-erected span and to activate its rear support on span N.

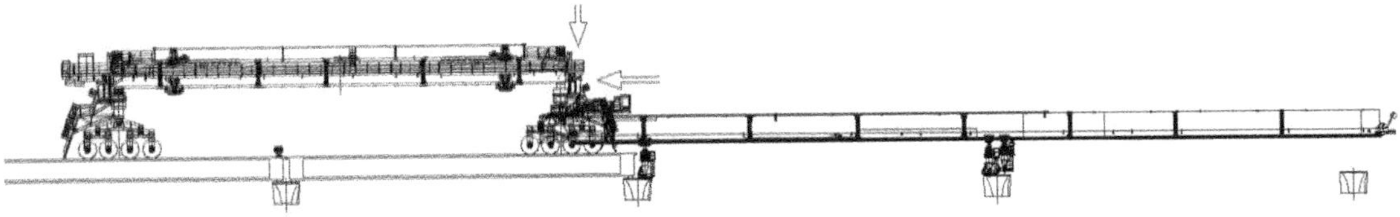

PHASE 7: The SLC is moved back until the front wheel group is again over the deck, so it can be activated.

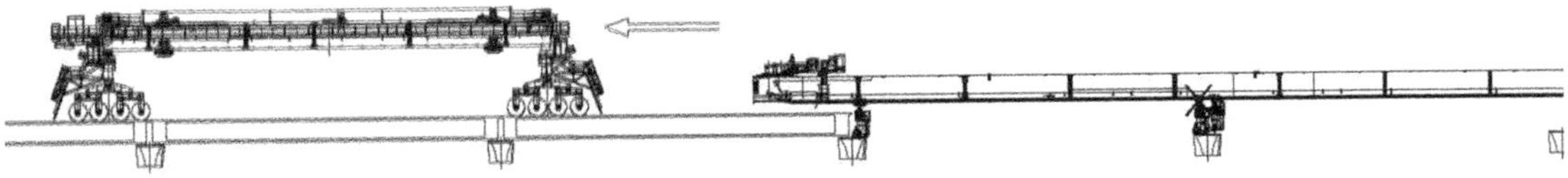

PHASE 8: The SLC can drive back to the PS storage area.

PHASE 9: The front support and the front auxiliary support I are moved from piers N + 1 and N + 2 to piers N + 2 and N + 3 respectively, so that at the end of their relocation the supports are in the same configuration as for phase 1, but now for placing of span N + 1.

Figure 8.11 Special arrangement of a BDE type 3 to work through tunnels (WowJoint Holding)

Figure 8.12 Twin viaducts at close spacing (DEAL)

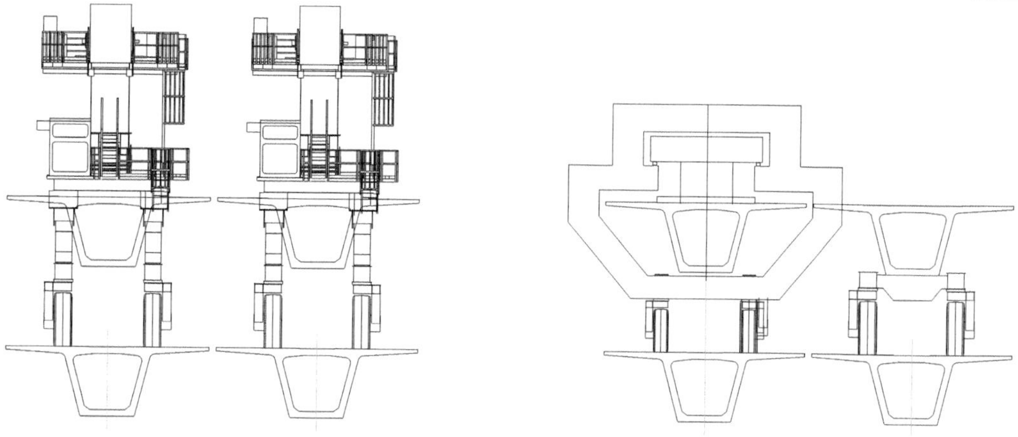

BDE type 3 allows work on parallel decks

BDE type 1 does not allow work on parallel decks

In general it can be said that an erection cycle of two spans a day is feasible with all the FSM systems, as long as the transport of the PS does not become controlling. However, as already explained in the introduction to this chapter, the efficiency of FSM can only be exploited if a sufficient number of construction fronts are opened up for construction of foundations, piers, special viaducts and tunnels. The follow-up work after a span has been placed should start immediately because it is time-consuming. This is particularly important on long viaducts. In such cases, a type of BDE needs to be selected whose transporter does not obstruct this follow-up work.

SPT and DSPT speed range is usually as follows: 2–3 km/h loaded and 4–5 km/h unloaded.

An example cycle time for a BDE type 1 with the PS storage area 2.5 km from the erection front is given in Table 8.4.

8.7. Risks

The list of risks given in Table 8.5 is not intended to be comprehensive, but addresses risks which are specific to launching gantries used for full span erection. For a specific project, a main contractor needs to carry out a full risk assessment of the operations performed with the equipment, considering the actual features of the equipment, the erection sequence and the site constraints.

8.8. Performance criteria

The following items need to be addressed when drafting a performance specification for a BDE for FSM.

- Viaduct type: high-speed railway, light railway, highway.
- Casting yard: in line, besides, same height, etc.
- PS (Precast Span)
  - characteristics: weight, dimensions, centre of gravity, loading capacity
  - delivery: from below, from behind, both.
- Alignment: gradient, curvature in plan and cross fall variations.

Table 8.4 Example cycle time for a BDE type 1; PS storage location 2.5 km from the erection front

ACTIVITY / STEPS	min	DURATION TIME (hours) 1	2	3	4	5
TRANSPORT FS TO ERECTION FRONT	*75*					
- load PS on the SPT *- transport to erection front* *- locate and connect SPT to LG (control/power connection)*						
PICK-UP OF PS BY LG	*45*					
- connect front lifting hook and lift the PS *- move PS forward using front winch and rear movable trolley on SPT* *- connect rear lifting hook and lift the PS*						
PLACE PS ON PIERS	*45*					
- move PS forward by the trolleys *- lower PS in position* *- release lifting hooks*						
MOVE SPT BACK TO CAST YARD	*30*					
- release SPT from LG (power and control) *- drive back SPT to PS loading location*						
LAUNCH LG FORWARD TO NEXT SPAN	*120*					
- release anchor on rear leg *- launch LG forward running on rear leg on deck and on front launching leg on support beam* *- anchor rear leg in position* *- release anchor of the support beam* *- relocate the support beam*						
TRANSPORT FS TO ERECTION FRONT	*75*					

- Permanent work interfaces
  - permissible deck loading (transport of PS over erected deck, standing on deck/pier during erection, rolling over deck during launching)
  - permissible temporary loading for permanent bearings
  - temporary support of PS on piers (location of span support jacks; load introduction into PS's end diaphragm {beside bearing locations}, etc.)
  - permissible eccentric loading of piers
  - space constraints for supports of BDE on front pier and subsequent piers: permanent bearings, span support jacks, seismic buffers/shear keys; longitudinal clearance of a BDE's front support against PS being erected (adequate longitudinal clearance for lifting/lowering into final position considering gradient, etc.)
  - starter bar location and dimensions (for link slabs, cast-in-situ part of masking walls, rail support plinths, etc.).
- Follow-up work: closing of lifting holes, cutting of HT bars etc.; numerous cast-in-situ elements in the case of rail viaducts, which need to be carried out when PS are still being transported over the deck.

Table 8.5 List of risks specific to operation of LGs for full span erection

Activity	Risk	Mitigating measures
Working on top of a PS Attachment of hook to PS delivered on SPT	Falling down of the worker from the PS	Provision of temporary lifeline along all PS on top (between the two lifting points), installed prior to lifting the PS
Transport of PS over viaduct	SPT tyres travel outside the allowed path on top of the deck (above webs)	Mark line on deck, along which the SPT has to drive. Real-time monitoring in the cabin by camera directed on the track line to follow
PS load transfer from SPT to LG or to SLSB	Local overload of support points on deck/SLSB	Strictly monitor pressures on hydraulic jacks of the supports and of the legs. Centralised data-monitoring in operator cabin
For type 3 BDE: longitudinal movement of PS along SLSB while rear wheel group is driving on deck	Pushing/pulling SLSB if movements on SLSB and by wheel group are not synchronised	Strictly monitor pressures on hydraulic drive circuit by real-time comparison with expected values. Centralised data monitoring in operator cabin
PS erection over live traffic	Drop of PS (unlikely), falling of debris, tools (possible)	No PS handling over live traffic
Placing of the PS on the bearings during girder erection	Crushing of people working in narrow space on pier head	Define by clear instruction where workers can stand while PS is suspended from the winches. Worker and operator communicating by radio receiver
Support of launching gantry	Bearing failure of the pier's concrete under the steel support	Adequate edge distances, so that support's contact area is within concrete confined by reinforcement; shimming with similar material over entire footprint of support's contact area
Longitudinal movement of main girders	Over-launching or loss of control of longitudinal movement → loss of equilibrium → overturning of gantry's supports and loss of entire gantry	Back-up fixity, apply nominal stressing of anchor bars at supports, presence of gantry engineer and gantry supervisor during anchoring stages
Relocation of the supports	Crushing/falling down of people working on deck/pier head	Suspend construction works on deck during support relocation through the span. Define by clear instruction where workers can stand during support relocation and main girder self-launch

- Clearances: vertical below, above and beside deck (overhead power lines, overhead bridges, etc.); horizontal (houses, power lines, etc.); very important to define clearances which are required on deck during PS transport over the completed deck to allow follow-up work to start immediately after a span has been placed by the BDE.
- Cycles (double or single shift, restrictions for segment delivery, restriction for noisy type of work, etc.).
- Longitudinal fixity (redundancies).
- Access: to LG/transporter (from deck, front pier) during different stages of operation; access from ground to piers (ahead of BDE's arrival, during deck erection, for follow-up work) and onto and into deck.
- Shipment costs (initial, if reused).
- Reusability.
- Method, schedule and crane definition for assembly, relocation, dismantling.

In addition, the buyer needs to allocate adequate time for fine-tuning the BDE concept and the subsequent confirmation of the support reactions/inserts, and for the delivery and commissioning of the BDE.

8.9. Photo gallery

Figure 8.13 LG type 2 (DEAL)

Figure 8.14 DSPT type 2 (DEAL)

Figure 8.15 SLC type 3 (WowJoint Holding)

Figure 8.16 LG type 1 (NRS)

Figure 8.17 SLC + SLSB type 3 (DEAL)

Figure 8.18 SLC type 3 (DEAL)

Figure 8.19 SLC + SLSB type 3 (DEAL)

Figure 8.20 SLC type 3 (DEAL)

REFERENCES

Figures 8.1, 8.2, 8.8, 8.16: NRS
Figures 8.11, 8.15: WowJoint Holding
Figures 8.3–8.7, 8.9, 8.10, 8.12–8.14, 8.17–8.20: DEAL

Bridge Deck Erection Equipment
ISBN 978-0-7277-6193-4

https://doi.org/10.1680/bdee.61934.135

Chapter 9
Cantilever erection by lifting frames

9.1. Introduction

Construction of bridge decks with fully or partially preassembled segments by the cantilever method is technically possible for any kind of span range and segment weight. The fact that very often cranes and/or lifting frames can be used for the deck erection instead of more costly erection gantries may make precast segmental decks erected by the cantilever method economically viable even for small- to medium-sized projects. For stay-suspended decks of large span bridges, lifting frames or derricks are the standard equipment.

9.2. Application range

Lifting frames (refer to Figure 9.1) can, as an alternative to cranes, be used for:

- cantilever erection of precast segmental decks
- cantilever erection of stay-suspended decks with precast or fully or partially preassembled steel or composite segments
- loading of precast segments onto span-by-span erection gantries (for such applications, the term 'segment loader' is more commonly used than 'lifting frame').

There are no limitations to the use of such frames with regard to span length and segment weight. This characteristic makes lifting frames specifically suitable for medium- to large-span bridges.

Whenever possible, precast segmental decks built by the balanced cantilever method with or without stays are erected by ground- or barge-based cranes. Lifting frames or overhead gantries (for details of this type of BDE refer to Section II, Chapter 5) are selected whenever land- or barge-based cranes cannot be used. This may be the case when:

- match-cast segments have to be joined over choppy water
- there is no space to position cranes on the ground (the viaduct corridor is too narrow, e.g. for a viaduct over an existing at-grade road between houses in urban area; there is running traffic at-grade level next to and parallel to the viaduct under construction)
- ground conditions are not suitable to support cranes (shallow water; shore area; when crossing canals/small rivers, viaducts)
- the crane required would be too large (locally unavailable; too expensive; too big a footprint; too slow to move in and out of site, when erection can only be done during limited hours per day)
- there are headroom restrictions (existing power lines and/or bridges above viaduct; vicinity to airports).

Figure 9.1 Isometric view of a typical lifting frame (type 1) (VSL International)

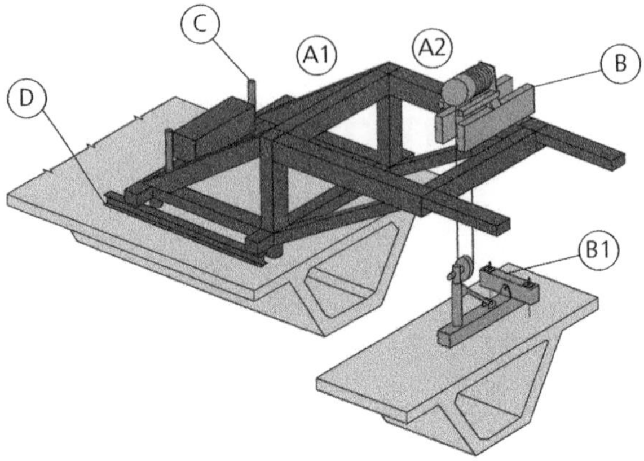

A1. Support frame
A2. Cantilevering main beams
B. Upper Cross Beam (UCB)
B1. Segment Lifting Beam (SLB)
C. Tie-down system
D. Frame movement system

Lifting frames are often used in combination with cranes, which are anyway required for the erection of the pier tables and the relocation of the frames from pier to pier, as follows:

- a lifting frame for erection of the upstream cantilever (towards the previously erected deck, where the crane would have to be positioned beside the deck for the erection of the last few segments, but no space may be available or a too large capacity crane may be needed)
- a crane for erection of the downstream cantilever (where the crane can stand in front of the deck being built).

For composite stay-suspended decks with precast planks, where planks are only placed after the steel grillage has been joined to the deck and after installation and stressing of the respective stays, a lifting frame may be used for the erection of the steelwork and a crane for the placing of the planks.

Lifting frames for composite decks may sometimes have to be designed so they are not only able to lift and erect the fully preassembled steel grillage of a segment but can also assemble the steel grillage in situ for segments, where access from below is blocked for delivery of the fully assembled steel grillage.

An alternative to lifting frames is to place cranes or derricks on top of the deck being erected. In general such arrangements increase the loading the deck needs to handle during erection. For composite decks, cranes and derricks can not only handle the segment's preassembled steel grillage but also the precast planks or even assemble the steel grillage in situ.

A precondition for the use of lifting frames for cantilever erection of precast segmental decks and stay-suspended decks is that the pier table is built by other means (for explanations about detailing of pier tables for concrete bridge decks, refer to Section III, Chapter 15). For precast segmental decks the pier table is either cast in situ or erected by a crane if also precast. If the pier table is cast in situ there is need for wet joints between the pier table and the first pair of field segments. A special suspension system is required to position and to hold the first pair of field segments in place during casting and curing of the wet joints. Such a system can easily be incorporated into lifting frames, but is an extra temporary work structure, if cranes are used for erection. The pier tables need to be long enough to allow for the installation of at least one or even a pair of lifting frames.

Last but not least, lifting frames cannot jump from pier to pier, but need to be relocated from pier to pier by cranes and trailers/barges.

9.3. General arrangement of the BDE

There are three main types of lifting frames (refer to Figure 9.2).

Figure 9.2 Main types of lifting frames (VSL International)

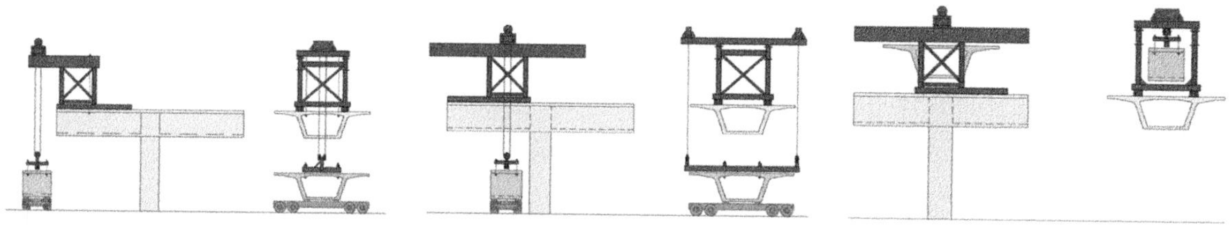

A type 1 lifting frame can only erect segments that are delivered from below their final position. With a type 2 lifting frame, segments can be delivered to anywhere below the cantilever under construction and the lifting frame can move along the deck with a segment suspended below the deck. A type 3 lifting frame can pick up segments delivered to below their final positions on either cantilever side and/or delivered over the deck. It can move along the deck with a segment suspended inside the frame. A segment loader, which is used to place segments delivered over a deck erected span by span with an underslung gantry, is basically a stationary type 3 lifting frame.

The main elements of a typical lifting frame are as follows (refer to Figure 9.1).

A. Main Frame (MF): The MF is made up of a pair of cantilevering main beams, which are mounted onto a support frame. This support frame stands at the tip of the already erected part of the deck and is tied down at its rear end to the deck. Single main beams are possible for type 1 lifting frames with single lifting points and relatively straight deck alignment in plan or for a configuration where the upper cross beam with the lifting gear spans between a pair of main frames, which sit on twin decks.

The height of the MF is for type 1 and 2 lifting frames controlled by:

- the Segment Lifting Beam (SLB), which is different for Lifting Frames (LFs) with single and double lifting points
- the type of segment attachment used
- whether the Upper Cross Beam (UCB) with the lifting gear and together with the SLB has to be moved over the support frame during launching.

Where to position the main frame's two vertical planes on the deck depends on a lot of factors. For concrete decks, the lifting frame's supports have to be over the deck's webs during segment installation. During launching the supports are generally either inside or outside the support corridor used during segment installation (refer to item D below). This means that for a concrete deck these supports are already on the wing slab or the inner deck slab. For steel and composite decks it is often necessary to centre the supports during launching directly over the deck's cross beams, because the orthotropic top slab of a steel deck is too weak to take the big reactions or the planks of a composite deck have not yet been installed or fixed.

Another complication arises for stay-suspended decks. The lifting frame needs to clear the 'vertical' corridor where the stays are installed and located. As explained in Section III, Chapter 15, stays can be anchored to the deck's cross section at its centre or at its sides. If a steel or composite deck has an open section, stays are anchored to the deck either directly to or slightly outside the main girders. The support of a lifting frame directly on such main girders is often not possible, since the space along the deck is only available at intervals and the lifting frame may get locked in after the installation of the stays. For such cases, the lifting frame's supports also have to stand on the deck's cross beam during the segment installation.

Main frames can be set out for segment erection in different ways, as follows:

- option 1: horizontal in longitudinal and transverse direction
- option 2: parallel to gradient, horizontal in transverse direction
- option 3: parallel to gradient and cross fall.

Each of the above options has advantages and disadvantages for the adjustment features of the main frame's supports, the tie-down fixity between the deck and the frame, and the articulation of the lifting gear's sledge. Option 3, for example, is an optimum solution for the tie-downs if the segment lifting holes are used to connect the tie-down bars to a precast segmental deck. Option 1 requires no adjustable articulation for the lifting gear's sledges, etc.

B. Upper Cross Beam (UCB) with the lifting gear and the Segment Lifting Beam (SLB): The UCB is supporting the lifting gear, which is winches, Strand Moving Units (SMUs) or long stroke jacks either in single configuration or in pairs (refer to Figure 9.3). Long stroke jacks can only be used for lifting heights of up to 5 m and operate at a speed of about 1 m/min. Strand moving units have operating speeds between 0.25 to 0.33 m/min and can be used for any load and lifting height. Winches are faster. They operate at speeds of up to 3 m/min. For more details of mechanical and hydraulic components refer to Section III, Chapter 17.

The UCB can travel along the main frame's main beams. Such movement is small for type 1 lifting frames, since movement is only required to lift the segment with a longitudinal clearance against the tip of the deck's cantilever and for fine adjustment when the segment is joined to the deck's cantilever. For type 2 and 3 lifting frames, running tracks for the UCB have to be provided over the full length of the MF's main beams, since when moving the LF with a segment suspended the UCB must be centred over the support frame. For type 3 lifting frames, the main beams may even be symmetrical about the support frame with similar cantilevers at both ends. Such an arrangement is needed if segments can only be delivered from ground under one of the two cantilever arms (as shown in Figure 9.2). By moving the UCB during launching of the lifting frame over the support frame, the centre of gravity of the lifting frame is shifted backwards to within the support frame. This is, however, generally not possible for type 1 lifting frames, where during launching either the rail beams have to be tied down or a counterweight has to be placed at the rear of the support frame to achieve stability (as for the lifting frame shown in Figure 9.1).

Figure 9.3 Different lifting gears (VSL International)

Segment loader with winch

Lifting frame with strand moving unit

Segment loader with long stroke jack

Lifting frames are able to lift a segment without UCBs when two sets of lifting gear are used, the two vertical planes of the main frame are aligned in the same corridor as the segment attachments and the deck is straight in plan. UCBs can also be omitted when a pair of lifting frames with single main beams is used for lifting of a single segment, which again have to be aligned in the same corridor as the segment attachments. Without a UCB the lifting gear can be placed in a cradle and the main frame can be configured in elevation as a rhomboid-shaped truss girder, from whose front tip the cradle is suspended.

Segment Lifting Beams (SLB) are used to connect the lifting gear to the segment. They are generally articulated to allow controlled rotation of the segment about its centre of gravity during lift-off from a barge or a trailer and when joining to the tip of the deck. When segments are lifted-off from barges operating in waters with wave action, the lifting frame, segment attachment, lifting beam and segment support system on the barge need to be designed and detailed accordingly. The provision of buffer elements (sling/rope with free length longer than wave height) between the lifting beam and the segment's attachments allows undisturbed connection of the segment to the lifting beam. If wave action is significant and the speed of the lifting device is too slow to clear the water surface within the wave's return period, the segment supports on the barge can be detailed as shims, which are pulled away once the segment's weight has been taken up by the lifting device. The traffic around a barge can be limited during the lift-off to keep the water calm and the lift-off operation can be restricted to a defined maximum wave height. The lifting frame itself needs to be designed for horizontal loads in line with the permissible horizontal movement of the delivery barge during the lift-off and the lifting height.

C. Tie-down system: Before the main beam's cantilever can be loaded with a segment, the main frame has to be tied down to the deck. In general, high tensile bars are used as tie-downs. Whenever possible they are anchored to precast segmental decks through holes or to segment attachments, which are provided to handle the segments during lifting. The tie-down systems need to be detailed to allow for geometrical variation of the tie-down position and orientation due to:

- variation in nominal segment length
- curvature in plan (segments tapered in plan)
- changes in super elevation and gradient.

They also need to allow for construction tolerances. Tie-downs can be stressed to a force equal to the lifting frame's uplift reaction during segment installation (at Service Limit State/SLS) if the tie-down's working lines are not behind the main frame.

D. Frame movement system: A lifting frame is moved longitudinally from segment to segment by either sliding and being pushed/pulled with long stroke jacks or by self-driven wheels rolling directly over the precast segmental deck. Movement with self-driven wheels requires a higher investment when compared with movement by sliding, but needs less people to operate and is faster. However sliding rails need a smaller footprint on the deck than wheels, which is an advantage particularly for rail viaducts with starter bars for fixity of rail plinths. A lifting frame is usually equipped with vertical jacks, to enable it to switch from the support for launching to the support for segment installation. The much heavier loaded supports during segment installation are generally screw jacks or the vertical jacks themselves (with safety rings), which sit directly on the deck. When a lifting frame is slid, it is slid on top of a pair of rail beams, whose length is a function of the launching jack's stroke. At the end of the jack's stroke, the rail beams can be repositioned for the next launching step by temporarily transferring the lifting frame's self-weight with the vertical jacks from the rail beam to the deck.

E. Stressing platform (for precast segmental decks only; not shown in Figure 9.1): If the stress bars used to temporarily attach a segment are anchored in temporary or permanent blisters, no access to the segment's downstream face is required for installation and stressing of these bars. If such bars are anchored at the segment's downstream face and, of course, for the installation of the permanent cantilever prestressing there is need for a stressing platform. This platform is best suspended from the main frame's cantilever beams whenever not in use. However, when being used it is advisable to attach the platform directly to the segment after it has been temporarily attached to the deck by stressing of the temporary high-tensile bars. If this is not possible, such as when the temporary high-tensile bars are anchored at the segment's downstream face, a back-up suspension system should be provided.

F. Splice platform (for steel and composite decks; not shown in Figure 9.1 as it is not relevant for concrete decks): Steel decks or the main beams of composite decks have to be joined to the already erected part of the deck by either splicing or welding, followed by coating of the splice zone for corrosion protection of the steelwork. Adequate access has to be provided to carry out all this time-consuming work, which is fully on the critical path for the deck erection. In general, splice platforms, which provide this access, are designed and detailed independently from the lifting frame, but need of course to be compatible with the lifting frame and have to respect the space taken up by the stays.

Stressing and splice platforms need particular attention during design, detailing and site installation, considering that many accidents with lifting frames are related to inadequately secured platforms.

9.4. Kinematics

Figure 9.4 shows kinematics for a type 1 lifting frame for erection of a precast segmental deck.

- Stage 1: lifting of segment N
- Stage 2: joining of segment N to deck and stressing of temporary HT bars; release of segment N from segment-lifting beam
- Stage 3: attachment of stressing platform to segment N; installation and stressing of the permanent cantilever prestressing tendons
- Stage 4: moving back of the upper cross beam over lifting frame's support frame; deactivation of lifting frame's tie-down bars; attachment of stressing platform to lifting frame
- Stage 5: launching of lifting frame from segment N – 1 to segment N; tie-down of lifting frame to deck
- Stage 6: move upper cross beam onto main beam's cantilever for pick-up of segment N + 1.

Figure 9.4 Kinematics for a type 1 lifting frame for the erection of a precast segmental deck (VSL International)

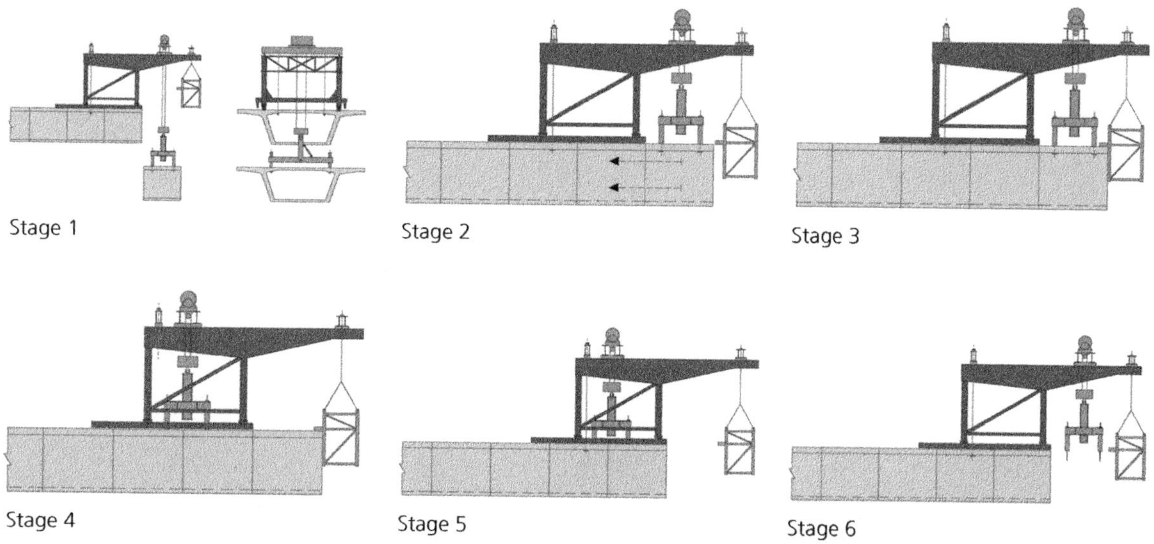

9.5. Permanent work interfaces

9.5.1 Deck reactions during segment installation and during launching of lifting frames

The ratio between the length of the support frame (Lf) versus the length of the main beam's cantilever (Lc) defines the magnitude of the lifting frame's deck reactions during lifting of a segment (with weight G):

- front reaction: $RF = G \times (1 + Lc/Lf)$
- rear reaction (uplift): $RR = G \times Lc/Lf$

LFs with shorter support frames are lighter than those with longer support frames, but induce higher reactions into the deck. A rough rule of thumb is that the weight of a lifting frame is between 30 and 40% of the weight being lifted.

As explained in subsection 9.3, 'General arrangement of the equipment – main frame', the lifting frame's supports during launching are generally either inside or outside the support corridor used during segment installation. It is necessary to verify that the deck slab of a precast segmental deck is structurally able to handle the reactions imposed on the deck outside of its webs. For steel and composite decks the supports during segment installation and launching are often centred over the deck's cross beams. Cross beams and their connections to the main beams have to be checked for the appropriate loading.

9.5.1.1 Rear reactions during installation of precast segments

As mentioned earlier, whenever possible, holes or segment attachments provided for handling the segments should be used to anchor the tie-downs to the deck. The use of two segment attachments would be enough to cope with the reaction from the segment's self-weight, assuming:

- there are four segment attachments
- each attachment is sized and the deck's local zone around the attachment is able to take about 25% of the segment's weight and
- the lifting frame has a ratio between cantilever and back span of 0.5.

However, since the lifting frame's self-weight induces additional out-of-balance moment during lifting, at least four segment attachments have, in general, to be utilised. This means that the lifting frame's tie-down should be centred about the next upstream segment joint. If lifting frames are tied down directly to the webs, making use of dedicated cast-in and lost HT bar anchorages and ducts, much higher tie-down forces can be introduced into the deck through single bars.

9.5.2 Out-of-balance moments ΔM

Out-of-balance moments ΔM during the deck erection are generally not more severe for the erection with lifting frames when compared to the erection with overhead gantries (refer to Section II, Chapter 5), as long as a pair of frames is used and a segment is only lifted after both frames have been launched into similar positions with reference to the pier's centre line. If only one frame is being used in combination with a ground-based crane and this frame is lighter than the segment, ΔM can be controlled by the installation sequence of the segments. After the lifting frame has been launched to the tip of the deck ready to erect the upstream segment N, the crane first installs the downstream segment N, before the lifting frame installs the upstream segment N.

When calculating ΔM and checking whether the deck's fixity to the pier and the pier and foundation are able to handle this moment, Service Limit State (SLS) and Accident Limit States (ALSs) have to be looked at. At SLS, opening-up of gaps in pre-compressed contact area and cracking in piers shall generally be avoided, whereas for ALS the objective is to ensure that the structure does not collapse. Since it is prohibitive to check ΔM for a general load case 'drop of segment' onto the already completed part of the deck, ΔM is generally limited for the accidental load case to an impact factor equivalent to the segment's weight handled by the lifting frame (with a load factor 1). Such an assumption models a scenario whereby a segment is not dropped onto the deck but is suddenly lost at the end of the load transfer from the lifting gear to the tip of the deck.

9.5.3 Stay-suspended decks with open section

For deck erection of cable-stayed bridges with open section, the weight of the frame may be critical to control the magnitude of the negative moment in the deck and the forces in the previously installed stays during lifting of segments and prior to the installation of the next pair of stays.

9.5.4 Actual weight of a lifting frame

It is important to consider the actual weight of a lifting frame for the stage-by-stage analysis of the cantilever construction.

9.5.5 Precast segmental decks – PT arrangement

For optimum safety during construction it is desirable to provide a pair of cantilever tendons for each pair of segments. This avoids the need to rely on temporary prestressing during segment erection. The exception is when the temporary high-tensile bars are anchored in permanent concrete blisters.

9.6. Cycle times

Unlike gantries, lifting frames have to be launched after erection of each segment. Hence there is a critical operation once or twice per day on a pair of frames. The engineer and supervisor in charge of a pair of lifting frames need therefore to be on the lifting frame almost full-time.

Table 9.1 shows an erection cycle for a precast segmental deck (without stays) and Table 9.2 for a composite deck with stays. It is important to note that the cycles shown are for illustration only and that actual cycles need to be worked out for individual projects considering project-specific features and constraints.

9.6.1 Cycle for a precast segmental deck (without stays)

Assumptions made for the cycle shown in Table 9.1 are:

- type I lifting frame
- pair of frames (A and B)
- concrete deck with single-cell box girder segment: 3 m long; cross section 8 m^2
- lifting frame movement: by sliding on rails
- lifting: by winch, height 12 m
- two pairs of cantilever prestressing cables with 19 strands each, 40 m long, anchored at tips of segment pair N, two strand pushers, two stressing jacks (one end stressing)
- start of operation: LFs standing at tip of segment pair N – 2
- both segments N waiting on trailers for unloading at time of segment attachment (need for pair of trailers)
- temporary prestressing anchored in concrete blisters inside of segment
- segment attachment to SLB: with swift lift connectors (no stressing, no need for access inside segment).

Table 9.1 Optimum cycle (after learning curve) for a precast segmental deck (without stays)

	Duration	7–8 am		8–9 am		9–10 am		10–11 am		11 am–12 pm	
■ Parking UCB over MF ■ Pick-up of stressing platform ■ Deactivation of tie-down ■ Launching of rails	30′	A	B								
■ Launching of frame onto segment N – 1 ■ Activation of tie-down	30′		A	B							
■ Lowering of lifting device ■ Attachment of segment N	30′			A	B						
■ Lifting of segment N	5′				A	B					
■ Joining of segment N: epoxy application and temporary PT	30′				A	A B	B				
■ Attachment of SPF to segment N	15′					A	B				
■ Installation of permanent PT anchored at segment N	60′							A B	A B		

Table 9.2 Cycle for a composite deck with stays

	Duration	Day 1		Day 2		Day 3		Day 4		Day 5		Day 6		Day 7		Day 8	
■ Parking UCB over SF ■ Deactivation of tie-down	2 x 30′	X															
■ Launching of LF onto segment N – 1 ■ Activation of tie-down	2 x 60′	X															
■ Third-stage stressing of stays N – 1	120′	X															
■ Lowering of lifting device ■ Attachment of segment N	45′	X															
■ Lifting of segment N	2 x 180′			X													
■ Joining of segment N and splicing by bolting	2400′					X	X	X	X								
■ Installation/first-stage stressing of stays N	900′									X	X						
■ Placing planks, which bridge over segment joint between segment N – 1 and N	240′											X					
■ Cast and cure stitch concrete between planks													X	X	X		
■ Second-stage stressing of stays N	120′															X	

Comments:

- The cycle shown in Table 9.1 is an optimum cycle (after learning curve), whereby two pairs of segments are erected per ten-hour shift. This is possible with the above made assumptions for short- to medium-long cantilevers and for a medium-sized pair of cantilever tendons per pair of segments.
- In addition, this cycle is only possible with decently mechanised lifting frames with:
  - tie-down jacks as part of the lifting frame
  - the stressing platform, when not in use, suspended from the lifting frame's main beams by electrical chain hoists or vertical jacks, which are fixed to a trolley with mechanised movement
  - fully articulated SLB (gradient, super-elevation, rotation).

9.6.2 Cycle for a composite deck with stays

Assumptions made for the cycle shown in Table 9.2 are:

- type I lifting frame
- pair of frames
- composite deck: with open cross section comprising of main edge beams (plate girders, hence with small torsional stiffness) and cross beams in steel supporting precast planks; deck suspended from two planes of stays anchored in deck's main beams; 12 m long, 24 m wide
- lifting frame movement: by sliding on rails
- lifting: by strand moving units (20 m/h lifting with load, 10 m/min lowering unloaded, equipped with motorised coiler), height 60 m; water without significant wave action during segment lift-off (river, no boat traffic around barge permitted during lift-off)
- two pairs of parallel strand stays (with about 61 strands per stay, about 100 m long) per pair of segments, installed and stressed strand by strand
- start of operation: lifting frames are standing at tip of segment pair N – 2
- both segments N waiting on barges for unloading at time of segment attachment (need for pair of barges); planks loaded onto steel grillage
- splicing of main beams by bolting
- segment attachment: SLB attached to segment before being loaded onto barge; connection of SMU's lower anchorage to the SLB by pin.

Comments:

The cycle shown in Table 9.2 is an optimum cycle (after learning curve) based on all work, except for bolting of splices and installation of stays, being carried out during day shifts.

9.7. Risks

The list of risks given in Table 9.3 is by no means comprehensive, but addresses risks which are particular to lifting frames and segment loaders. For a specific project, a main contractor needs to carry out a full risk assessment of the operations performed with the lifting frames, considering the actual features of the lifting frame, the erection sequence and the site constraints.

9.8. Performance criteria

Table 9.4 shows the items that need to be addressed when drafting a performance specification for a lifting frame or segment loader.

In addition, the buyer needs to give a time frame for fine-tuning of the lifting frame's concept, support reactions and inserts and for the delivery and commissioning of the lifting frame.

Table 9.3 List of risks specific to the operation of lifting frames and segment loaders

Activity	Risk	Mitigating measures	Precast segmental deck	Steel or composite deck	Segment loader
Installation of handrails on top and inside of segment delivered on trailer/barge	Vertical access and work at height: climbing up and down, work performed at height (at least 3 to 4 m above ground)	Access ladder fixed to stable base or access staircase, safety line fixed to segment and lifting device's segment lifting beam	X	X	X
Attachment of connection beams to segment delivered on trailer/barge			X	X	X
Attachment of lifting device's SLB to segment delivered on trailer/barge			X	X	X
Working on top of and inside of segment	Falling down from segment	Provision of temporary handrails around all exposed edges on top and inside of segment installed prior to lifting of segment from trailer/barge	X	X	X
Working outside of segment for splicing of segment to deck	Falling down from segment	Provision of platforms with access from top of deck	N/A	X	N/A
Erection: front support of lifting frame on concrete deck	Failure at support's contact area	Support over webs → marking of support area on deck; adequate edge distances, so that support's contact area is within concrete confined by reinforcement; shimming with similar material over entire footprint of support's contact area	X	N/A	X
Erection: front support of lifting frame on steel or composite deck	Failure at support's contact area	Support centred over steel main beams or cross beams → marking of support area on deck	N/A	X	N/A

Erection: rear support of lifting frame	Failure of tie-down bar or support zone	Preloaded tie-down bars; if not preloaded: tie bars with articulated connections to deck and frame and provision of redundancy (two bars per tie-down location with each bar able to take full uplift reaction at ASL); load test of segment connection (prior to start with erection work) to verify its design and detailing	X	X	X
Erection: position of UCB	Lift-off/suspension of segment too far out on main beam's cantilever	Distance marks on main beam; provision of stoppers	X	X	X
Erection: lift-off of segment	Lift-off location in plan outside of permissible range	Marking of trailer position on ground; check of lifting device's verticality after connection to segment prior to lift-off	X	X	N/A
Erection: lift-off of segment from barge	Wave action → barge movement during lifting beam connection and during lift-off	Lifting beam with buffer elements; removable support shims on barge; limitation of traffic around barge; definition of permissible maximum wave height during lift-off	X	X	N/A
Segment erection over live traffic	Drop of segment (unlikely), falling of debris, tools (possible)	No segment handling over live traffic	X	X	N/A
Segment suspension (stationary) over live traffic	Drop of segment (unlikely)	Permitted with back-up suspension system (e.g. suspension with additional hanger bars)	X	X	N/A
Epoxy gluing of segment joints	Dropping epoxy → injury to people, damage to clothes and cars, death for fish	No traffic if over road, plastic sheets or similar below joints to catch epoxy if over water	X	N/A	N/A
Epoxy gluing of segment joints	Closing of gap between segments due to swinging segments	Use of timber shims, which are supported on top of the deck slab	X	N/A	N/A

Table 9.3 Continued

Activity	Risk	Mitigating measures	Precast segmental deck	Steel or composite deck	Segment loader
Segment attachment and release from lifting device	Release from lifting device before segment has been fully fixed to deck (temporarily) → drop of segment	Checklist with hold point	X	X	N/A
Stressing platform	Failure of connection to segment	Back-up system; checklist and hold point	X	N/A	N/A
Placing of precast planks	Crane wrongly positioned locally overloading deck	Marking of crane position on deck	N/A	X	N/A
Placing of precast planks	Interference between steel studs and plank's starter bars	Preassembly of precast plank's reinforcement cage with a template	N/A	X	N/A
Launching: stability of lifting frame	Overturning of lifting frame because UCB not moved back over support frame or rail beam not tied down to deck	Checklist with hold points	X	X	X
Launching: longitudinal movement of lifting frame	Loss of control of longitudinal movement → loss of frame (drop from deck, etc.)	When moving by sliding, active fixity of frame during resetting of jacks; when moving on wheels, provision of back-up brake system	X	X	X
Launching: support points	Local failure of deck at support point due to wrong positioning	Marking of support area/corridor on deck	X	X	X

Table 9.4 List of performance criteria

	Precast segmental deck		Steel or composite deck	Segment loader
	Without stays	With stays	With stays	
■ Segments				
– weight and dimensions	X	X	X	X
– attachment details	X	X	X	X
– segment delivery: from below, from behind, both	X	X	X	X
■ Planks				
– weight and dimensions			X	
– installation method: lifted together or not with steel grillage; handled by LF or by other means; timing/curing requirements for stitch concrete, etc.			X	
■ Stays				
– position of stay's deck anchorages		X	X	
– type and installation method of stays		X	X	
– stressing sequence (one or more stressing stages, first stage prior to or after LF launching, etc.)		X	X	
■ Alignment: changes in gradient and super-elevation (over cantilever being built or span to launch over respectively), curvature in plan	X	X	X	
■ Permanent work interfaces				
– limitations about weight and centre of gravity of LF	X	X	X	X
– permissible deck loading: magnitude, position, contact pressure (during segment installation {front reaction, uplift} and during launching)	X	X	X	X
■ Access				
– installation and stressing of temporary stress bar and PT tendons	X	X		
– installation and stressing of stays		X	X	
– splicing of main girders including corrosion protection of steelwork			X	
– installation of planks			X	
■ Clearances				
– vertical (to comply with airport height restrictions or to be below power lines; to be above deck inserts, etc.)	X	X	X	X
– transverse (to clear starter bars, preinstalled parapets, etc.)	X	X	X	X
– stays and stay deck anchorages (to allow installation, to clear during launching)		X	X	

Table 9.4 Continued

	Precast segmental deck		Steel or composite deck	Segment loader
	Without stays	With stays	With stays	
■ Tie-downs: stressed or non-stressed (with redundancy)	X	X	X	X
■ Cycles				
– segments without stays	X	X	X	X
– segments with stays		X	X	
■ Mechanisation (labour requirement, speed → cycles)	X	X	X	X
■ Shipment costs (initial, if reused)	X	X	X	X
■ Reusability	X	X	X	X
■ Methods, time, crane requirements for assembly, relocation, dismantling	X	X	X	X
■ Follow-up work: closing of holes, etc.	X	X	X	X

9.9. Photo gallery

Figure 9.5 Lifting frame type 1: precast segmental road viaduct (Ipoh, Malaysia) (VSL International)

Figure 9.6 Lifting frame type 1: stay-suspended steel deck (main span of Stonecutters Bridge, Hong Kong) (VSL International)

Figure 9.7 Lifting frame type 1: steel bridge (Haliç Bridge, Istanbul) (FI)

Figure 9.8 Lifting frame type 3: precast segmental road viaduct (High Five, Dallas) (DEAL)

Figure 9.9 Lifting frame type 3: precast segmental road viaduct (Seremban, Malaysia) (VSL International)

Figure 9.10 Lifting frame type 2: precast segmental road viaduct (Lantau, Hong Kong) (FI)

REFERENCES

Figures 9.1–9.6, 9.9: VSL International
Figure 9.8: DEAL
Figure 9.7, 9.10: FI

Bridge Deck Erection Equipment
ISBN 978-0-7277-6193-4

https://doi.org/10.1680/bdee.61934.155

Chapter 10
Progressive cantilever method using stay cables

10.1. Introduction

The Progressive Cantilever Method (PCM) is a cantilever construction method where the precast box girder segments are installed successively, starting at one end of the bridge and progressing towards the other. A temporary staying system is mandatory to relieve the cantilever from the very high bending moments during construction. The staying system consists of stay cables suspended from a temporary movable tower or mast. All segments are delivered over the already constructed deck and lifted and set in position with a movable swivel crane, located at the extremity of the cantilever. Each new segment S is stressed against the previously installed segment S − 1 with HT bars and suspended from the mast by a pair of stay cables.

This chapter presents the progressive cantilever method for bridge decks using temporary stay cables. The bridge deck segments are precast. The chapter does not deal with other uses of temporary stay cables in construction, for example a temporary staying system for arches.

Developed in the early 1970s, the temporary stayed progressive method is well proven and effective. It has been refined on many different projects over the last 40 years (for an example, refer to Figure 10.1).

10.2. Application range

As for permanently cable-stayed structures, there are geometrical and structural limitations to the use of the PCM. The progressive cantilever method using stay cables is suitable for bridges with the features listed below. These features are best practice features. In some cases they can be enhanced, but of course with a proper and complete assessment.

- **Continuous bridge:** The continuity is essential, since the method generates horizontal loads in the deck from the stay cables.
- **Alignment:** Straight or slightly curved bridges can be built using this method. The staying forces generate horizontal transversal thrust due to the curve, which may govern the design of the equipment. The minimum horizontal radius is generally greater than 1300 m for road bridges and 3500 m for rail bridges.
- **Cross section:** The progressive cantilever method is mostly suited to erecting concrete box girders of constant depth. Decks with variable width or cross fall, when properly managed during precasting, can be installed with this method.
- **Span length:** The typical span length is between 40 and 50 m. For road decks, span length can reach 65 m, whereas span length for rail bridges is generally limited to 45 to 50 m. In addition, the span length should remain as constant as reasonably possible, to avoid time-consuming

Figure 10.1 A high-speed railway bridge in southwest France being constructed with the progressive cantilever method using stay cables (VINCI)

changes in the stay cable length. This is not an absolute requirement, but has an impact on the general schedule.

- **Segment size and weight:** Segment size and weight are limited by several factors.
  - Handling and transport considerations are some of the most important factors. The weight controls the capacity of every lifting and handling device in the casting yard, during transport and during erection. Since for PCM the swivel crane is located on the cantilevering part of the deck during construction, the segment's weight has a direct impact on both the deck and the temporary staying mast. When the precast yard is not close to the construction site, transport issues are governed by geometrical dimensions of the elements (length, height, weight).
  - The segments have to pass through the legs of the temporary mast, so their length cannot be greater than the space between those legs. The mast must stand right above the webs of the box girder. The segment length has therefore to be smaller than the distance between the webs by at least 1 m, provision being made for the mast post's width. In a similar manner, the height of the segment (including starter bars, if any) has a direct effect on the mast's structure, since the first horizontal bracing element must be located high enough to let the segment through on its carrier.
  - The typical length of the segments is 2.7 to 2.8 m. The height can vary from 2.5 to 3 m for road decks and up to 4 m for rail decks. Higher segments would have a significant impact on the mast's stability by shifting up the first bracing between the legs.
  - Expectable weight of a segment is 500 kN for road decks, and 700 kN for high-speed rail decks.
- **Pier height:** The progressive cantilever method can easily cope with variation in pier heights and configurations, since no access from the ground is required. In particular, there is no limitation on the height of the piers. Yet reasonably low height from the ground is recommended for the first span, which must be constructed on falsework, as explained in subsection 10.3.5, 'Heavy-duty falsework'.

- **Overall length of the bridge:** No general rule can be issued on the overall length. Obviously, as a general consideration, the longer the bridge, the more cost-effective construction will be.

10.2.1 Advantages

Advantages of the method are as follows.

- Safe: all operations are conducted from the already constructed deck.
- No ground access is required during deck construction, hence it is suitable for the crossing of rivers, roads, railways and steep valleys.
- The method can accommodate a variable horizontal radius.
- Apart from the learning curve period, weekly cycles can be achieved regardless of how long the span is.
- No permanent Post-Tensioning (PT) installation is required during the cantilever erection.
- The special equipment is light and reusable.
- Cable staying provides beneficial compression in the concrete deck, reducing the amount of temporary HT bars.
- The reactions on piers are vertical (no bending moments).

10.2.2 Disadvantages

Disadvantages of the method are as follows.

- The first span has to be erected on falsework.
- Forces in the superstructure during erection are different from those in the completed structure.
- Piers and bearings are subjected to higher loads than only the self-weight of the deck.
- The maximum span is 60 m for road bridges and 45/50 m for rail bridges.
- Requires a minimum number of spans to reduce the impact of the cost of the special equipment on the deck's construction costs. This is a common issue for all methods involving BDE with precast segments.

10.3. General arrangement

10.3.1 Casting yard and storage area

The progressive cantilever construction method offers the full benefit of building decks with precast segments. A precast yard will be located at immediate proximity or further away, provided that road connections allow the transport of the segments. All precasting aspects are dealt with in Section III, Chapter 19.

If the casting yard is far away from the bridge erection site, an intermediate storage area may be necessary. In that case, special additional handling equipment, such as straddle carriers, is required.

10.3.2 Carrier, SPMT

A traditional carrier, or a more advanced Self-Propelled Modular Transporter (SPMT), is required to transport the segments from either the casting yard or from the intermediate storage area to the extremity of the cantilever. The key points governing the choice of the carrier are generally the following.

- Payload: the typical segment weight ranges from 500 to 600 kN (up to 700 kN).
- Geometrical dimensions: the carrier has to pass between the temporary mast's legs. Therefore, its width and height must remain compatible with the space available.

- Nominal speed: the carrier transports segments on the deck from the storage area to the cantilevering end and returns empty back to the yard. For long bridges, the distance travelled can become a governing factor for the cycle time. The use of several carriers may become necessary.

10.3.3 Temporary movable staying system

The temporary staying system is the core equipment of the PCM. It supports the weight of the cantilevering deck. For illustration, refer to Figure 10.2.

The system consists of the following.

- **A steel tower or mast** is made of two columns braced together. Sufficient space between the columns permits the passage of the segment on a carrier. The steel tower is equipped with access platforms movable along a vertical sprocket bar (preferred to a rope-suspended access platform for safety considerations) and winches, necessary for cable installation and tensioning.
- **A stay cable** is made up of several independent strands. The strands are anchored to anchor devices by means of mono-strand barrels, at the bottom directly to the concrete deck and at the top to a tensioning jack connected to the mast. A pair of cables is attached whenever a segment is added to the deck in cantilever mode and removed when the span is complete with all required post-tensioning and before the tower is shifted to the next position. When the span length varies or segment arrangement changes, the length of the stays must be adapted.
- Each stay cable is equipped with its own dedicated **tensioning jack**, which pulls on the stay's bundle of strands simultaneously. Forces in the front cables are balanced by forces in the back cables, which are anchored at the rear to the previously installed deck. The number of jacks corresponds to four times the number of cantilevering segments. After tensioning, the jacks are mechanically secured with two HT bars.
- Two types of **anchorage devices** are necessary for each stay cable. One device, at the bottom end of the stay (see Figure 10.3) is fastened to the concrete deck. The other device, at the top end of the stay (see Figure 10.4), allows for easy connection with the tensioning jack at the mast. These devices are handled many times by the workers. They must be designed to transmit the force, while being as light as possible. They are usually made of high-strength steel. The lower device is anchored to the concrete by means of an anchoring cone and an HT bar. The cone must be designed to avoid concrete spalling.
- **A transfer system** is needed to move the tower from one pier to the other. This system can either be a special trailer or can be incorporated into the tower structure. The transfer system must distribute the tower weight on the deck and must cope with variation of longitudinal gradient and cross fall of the deck.

Typical weight for this BDE, including tensioning jacks and transport tool, ranges from 800 to 1000 kN.

10.3.4 Swivel crane

The swivel crane (refer to Figure 10.5) lifts the segment from the carrier and installs it at the tip of the cantilever. The swivel crane is equipped with a winch and a spreader beam. The orientation of the swivel crane must be controlled to prevent any unsafe situation.

After the segment is set in position, the HT bars are installed and stressed and the swivel crane is moved to the next position. The swivel crane is thus equipped with necessary jacks and a moving system. The stay cables for segment S are installed after the swivel crane has been launched onto segment S.

Figure 10.2 Temporary movable staying system (VINCI Construction)

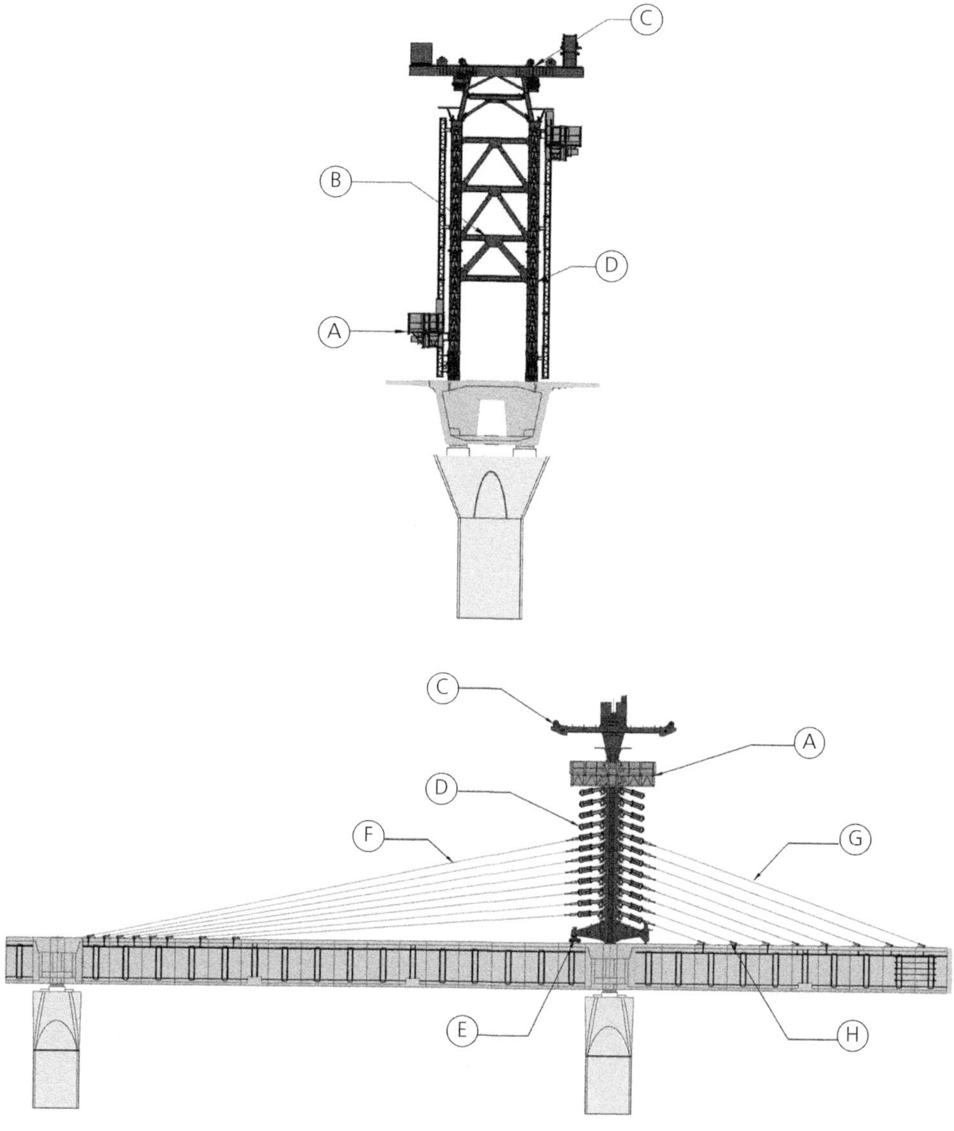

A. Lifting access platforms
B. Steel tower
C. Winches
D. Tensioning jacks
E. Transfer system (bogie)
F. Rear stay cables
G. Front stay cables
H. Anchorage device

The main elements of the swivel crane are as follows.

- A horizontal swivel arm: the swivel arm can be rotated about the crane's vertical post. It supports the upper sheave block of the winch. The sheave block is movable along the swivel arm. The winch drum is fixed on the back cantilever of the swivel arm to balance part of the front cantilever.

Figure 10.3 Bottom anchoring device

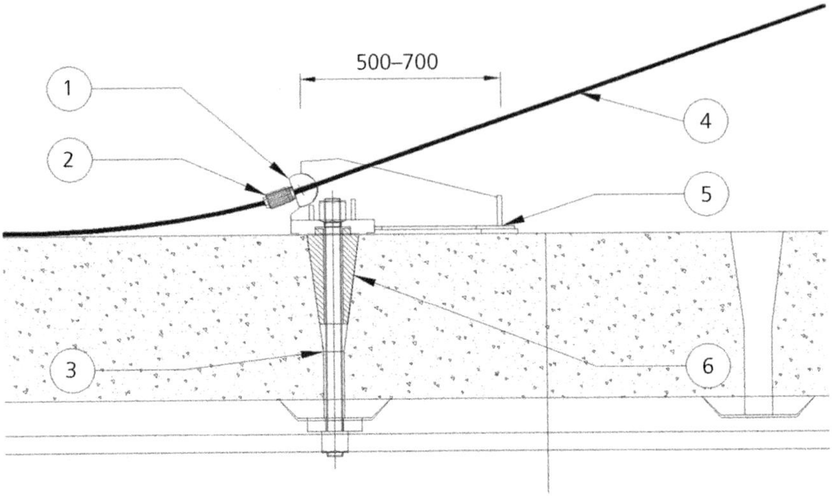

1. Bearing cylinder
2. Mono-strand barrel
3. Tightened HT bar
4. Strands
5. Steel bracket
6. Conical shear key

Figure 10.4 Top anchoring device

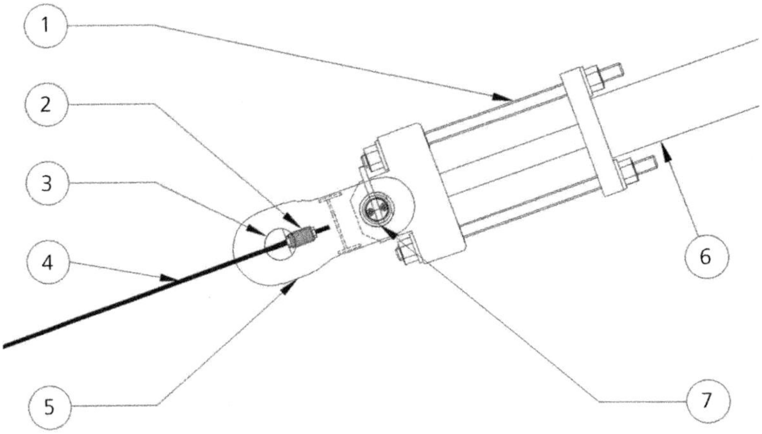

1. Safety HT bars
2. Mono-strand barrel
3. Bearing cylinder
4. Strands
5. Connection plate
6. Tensioning jack
7. Connection pin

Figure 10.5 Swivel crane

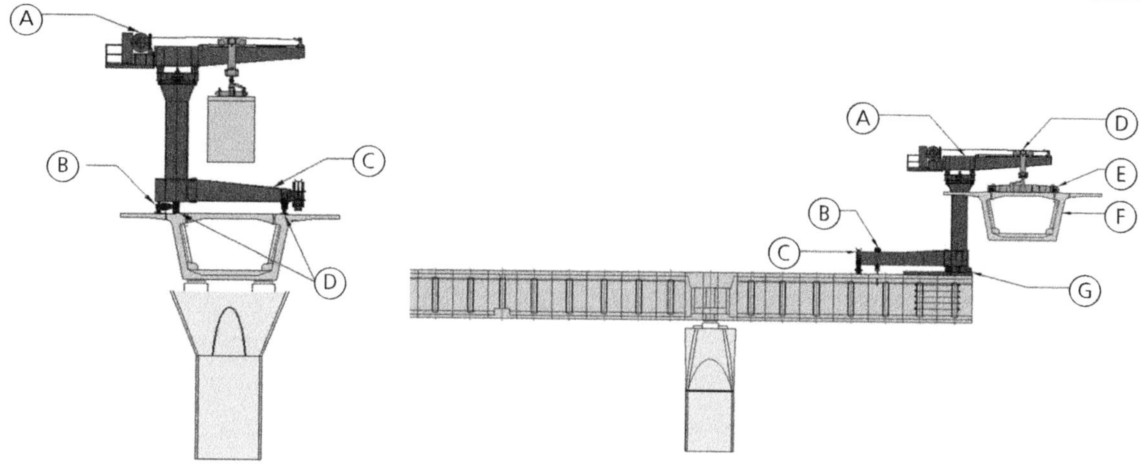

A. Winch
B. Moving system
C. Front leg
D. Swivel crane bearings

A. Horizontal swivel arm
B. Tie-down device
C. Rear leg
D. Trolley

E. Spreader
F. Segment
G. Moving system

- The crane's post must be high enough to let the segment be moved from the delivery position at the rear to the placing position in front.
- The crane is supported on the deck during placing of segments on three points over the webs of the deck
  - directly under the post on a central bearing, properly designed to spread the load on the concrete deck
  - under the tip of the two horizontal legs, which are provided to stabilise the crane against overturning: under the front leg (transverse, perpendicular to the bridge axis) and the rear leg (longitudinal, parallel to the bridge axis)
  - the rear leg has to be tied down to the deck with HT bars, stressed with dedicated jacks.
- A moving system: the moving system consists of wheels with jacks under the tips of both legs and either a wheel or a sledge system with sliding beam near the central bearing.
- Required accesses and platforms: in particular, a platform is needed to apply the epoxy glue over the end face of the segment.

The weight of a typical swivel crane is approximately 400 to 500 kN depending on the segment characteristics.

10.3.5 Heavy-duty falsework

The first span cannot be constructed by the PCM. There is no back span which could balance the weight of the cantilevering span and resist the axial loads in the deck generated by the stay forces. Most commonly, the first span is built on a heavy-duty falsework. The falsework consists generally of two parallel beams supported on heavy-duty towers. The segments are installed on the falsework by crane, either from the ground or from the abutment.

Alternative solutions would require either massive foundation works on the abutment, or the assembly of a temporary counterweight made up of segments fabricated for one of the typical spans. Unless the falsework is too complicated and onerous to be built (in cases of great height, for example), these alternative solutions are hardly cost effective.

10.4. Kinematics

For the kinematics of the PCM using stay cables, refer to Figures 10.6 to 10.14.

Figure 10.6 First span on falsework

The staying system needs back counterweight to balance the loads from the cantilevering part. Hence the first span (N – 1) is commonly built on a heavy-duty falsework, on which all the segments of the first span are placed, joined and prestressed.

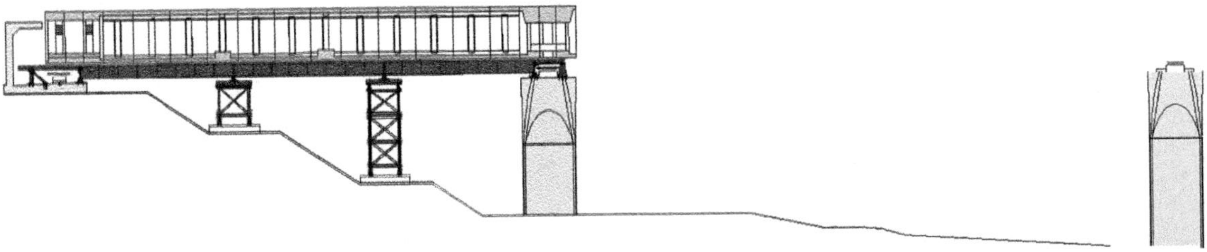

Figure 10.7 Installation of the first few segments

The first three to four segments of span N are attached with the swivel crane in free cantilever mode without stays. These segments are positioned and secured with permanent cantilever PT, one after the other. Special access must be provided for the stressing of the PT tendons.

These initial segments provide the necessary space to have the swivel crane at enough distance from the pier segment for the installation of the tower.

A stitch is cast in situ between the pier segment and the first segment.

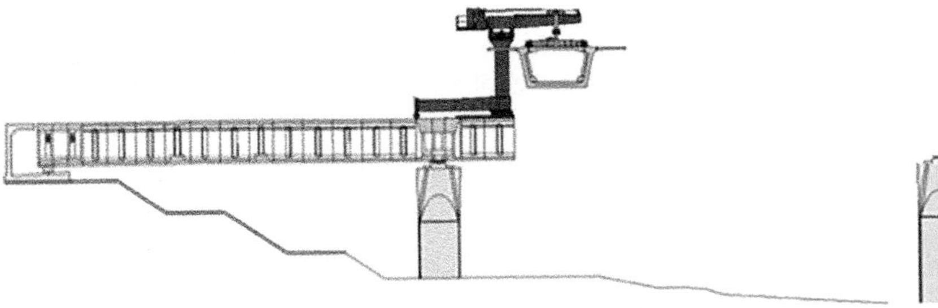

10.4.1 Cantilever loads

The PCM generates static forces due to permanent and moving loads during construction, which are notably different from those generated on the final static system. The cantilever moment due to self-weight is almost balanced by the forces introduced into the deck by the stay cables. The moving loads, mainly swivel crane weight and the segment during its transport on deck, are not equilibrated by the stay cable forces. These loads are taken by both additional tension in the stays and by bending in the

Figure 10.8 Tower installation on pier segment

The tower is moved over and placed on the pier segment.

The first stay cable layer is installed and tensioned. Tensioning of the pair of stay cables in front and at the back of the mast must be as synchronised as possible.

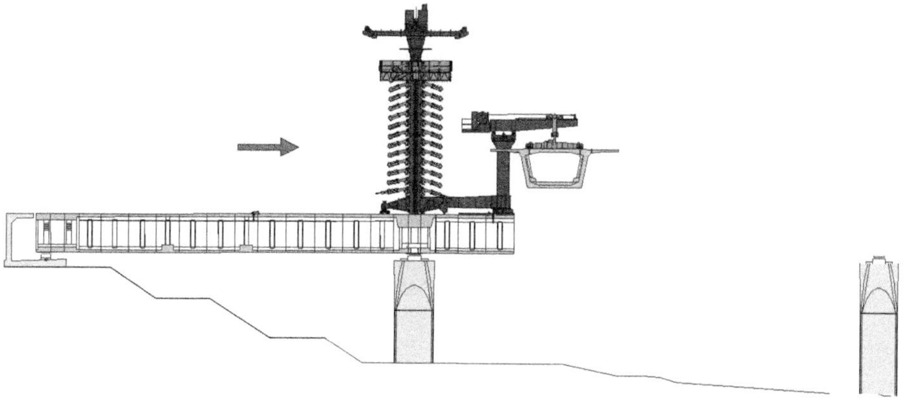

deck, distributed according to the relative stiffness of each of the two components. (See Figure 10.15 for illustration.)

10.4.2 Temporary loads on the final bearings and the piers

Temporary high loads are acting on the piers, as well as on the bearings. These loads are equivalent to twice the weight of the cantilever span being erected, and are significantly higher than the reactions from dead loads in the final structure. However, the increased temporary load is generally not governing, since additional loads (superimposed dead load and live loads) are applied to the structure in its final configuration.

Nevertheless, the impact of temporary high loads should be investigated with care, both for the piers and for the bearings.

10.4.3 Decompression of segments joints

As with any construction method using match-cast precast segment, one constant concern is to avoid tension in the joints between the segments. Generally, a compression equal to 0.2 MPa is recommended in the joints at any time of the construction.

- With the progressive cantilever method, the horizontal component of the staying forces generates additional prestressing in the deck. This horizontal force is beneficial to reduce the decompression in the joints.
- When landing on the next pier, the static scheme changes completely and the inversion of the sign of bending moment can introduce decompression in the joints. This construction phase has to be looked at carefully.
- The thermal gradient loads are changing during a day. During cantilevering, the consequence will be a change in the deflection of the deck, with modifications of the forces in the cable stays. When reaching the next pier but before the final prestressing has been applied, the thermal gradient changes can generate decompression of the joints.

Figure 10.9 Segment lifting and positioning

The typical segment is transported over the deck with the carrier behind the swivel crane. It is lifted by the swivel crane from the carrier and positioned in front of the already cantilevering part of span N.

While lowered, epoxy is applied to the segment's matching face.

The segment is tied to the cantilever with stressed temporary HT bars, from inside the deck. The number of HT bars depends on the size of the segment. The resulting force must generate and maintain enough pressure between two adjacent segments to squeeze the glue and avoid loss of compression in the joint in the later phases.

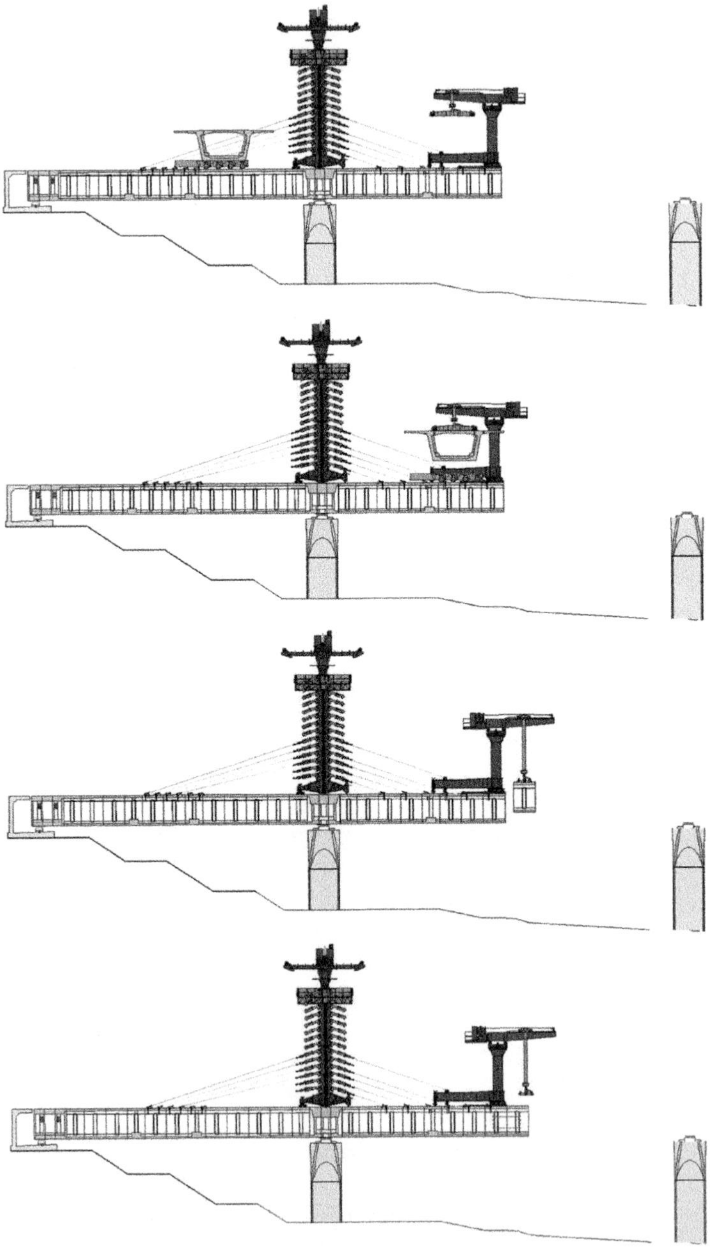

Figure 10.10 Swivel crane moving to the next position

Once the segment S is tied to the segment S − 1, the swivel crane is moved onto segment S. The moving can be either by skidding on a rail or by simple rolling.

The rear leg of the swivel crane is then tied to the deck by means of HT bars.

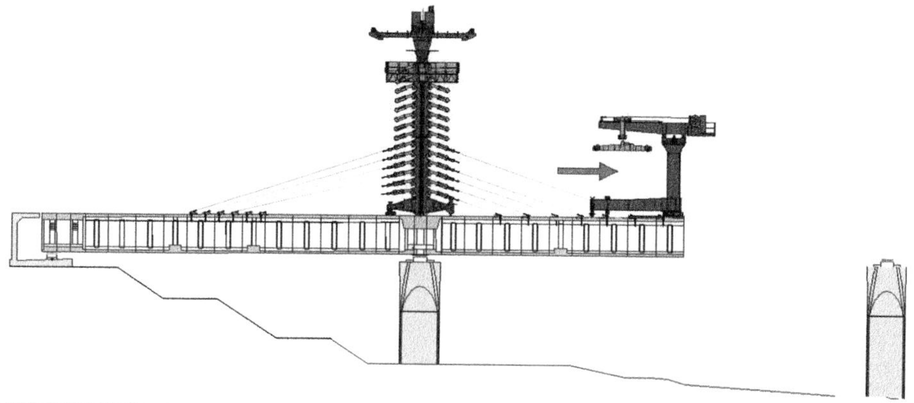

Figure 10.11 Stay-cable tensioning

The stay cables' deck anchorages are tied to the deck, at the front to segment S and at the rear to the back span. Subsequently the cables are connected at the mast to the tensioning jacks by means of a winch located at the very top of the mast.

The front cables are generally tensioned with a force that compensates the dead weight of the segment. The tensioning force in the rear cables is calculated to balance horizontally the front forces at the tower.

As stated previously, the tensioning of the two pairs of stays must be as synchronised as possible.

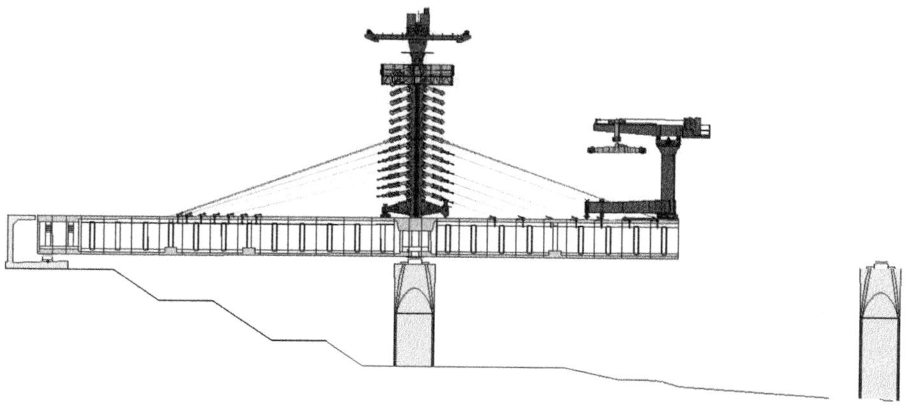

10.5. Impact on permanent work design

Using the PCM requires a very close interaction between design and detailing of the permanent works and the temporary works. The design of the deck (post-tensioning layout and stressing sequences, cross-section dimensions) is very much controlled by the erection method.

Figure 10.12 Pier segment and bearings

The bearings are usually set on the pier before the cantilever arrives and the pier segment is installed. Final levelling of the deck is carried out before the bearings are fixed and bearing plinths are grouted.

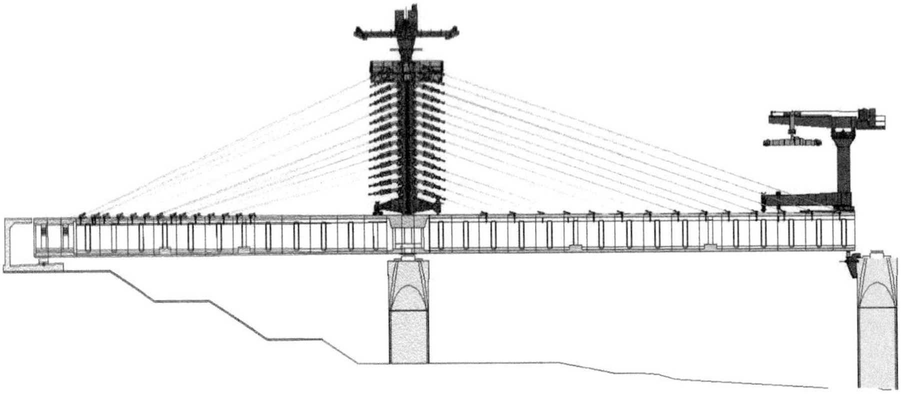

Figure 10.13 Post-tensioning

Once the span is supported on its permanent bearings, post-tensioning operations can start. Not all permanent PT in span N is installed at this point of time, but enough to make span N self-supporting without need for suspension by the stays.

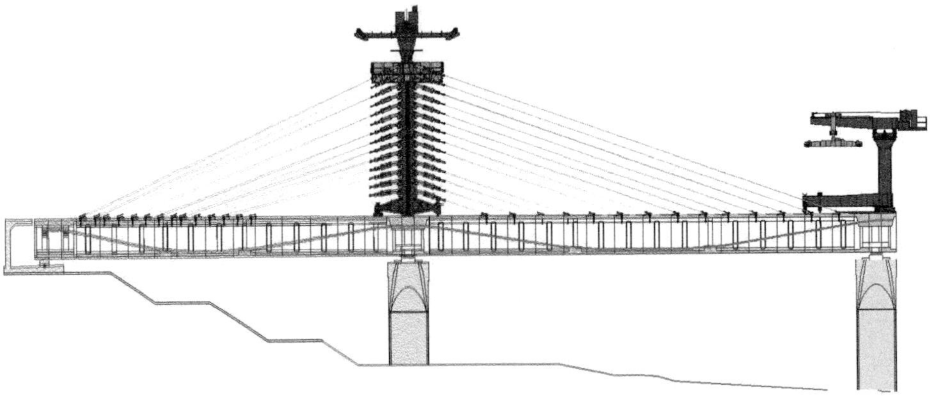

10.5.1 Local verifications

Local verifications are of three types.

- Stay cable anchoring: the anchoring of the stays generates high local loads. According to the weight of the segment, the tension in each stay cable ranges from 700 to 1100 kN. This force must be anchored to the deck through a relatively small device, whose dimensions have to be compatible with the reinforcement and the geometry of the segment. Properly detailed reinforcement has to be provided around the anchoring devices to avoid splitting or spalling of the concrete.
- Reactions from the construction equipment: the mast introduces very high loads into the pier segment. These loads are equivalent to twice the weight of the cantilevering span.

Figure 10.14 Starting a new span cycle

All cable stays are removed and the cycle can start anew with the next span.

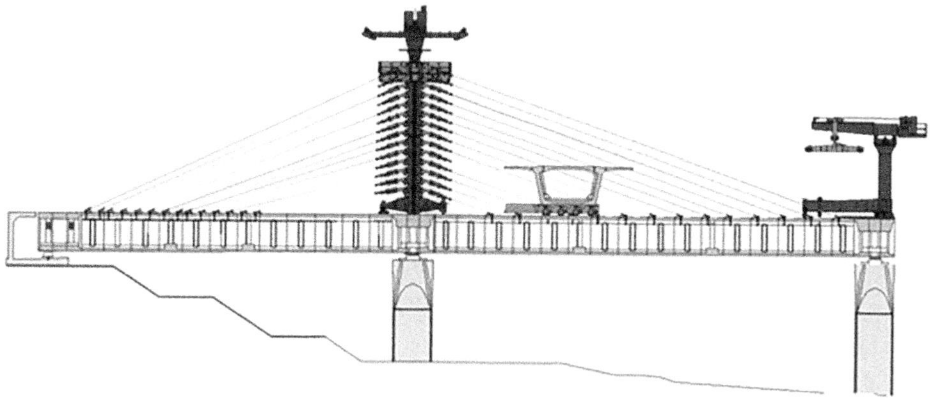

Figure 10.15 Stay forces

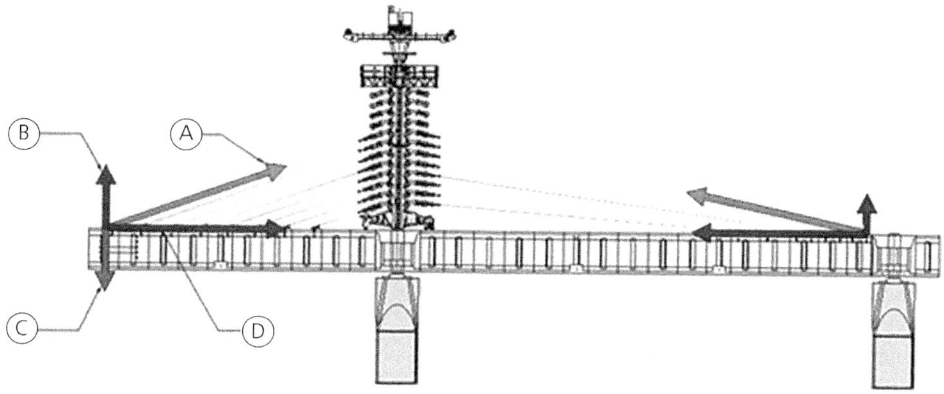

A. Stay force
B. Vertical component of stay force
C. Segment self-weight
D. Horizontal component of stay force

- Reactions and anchoring loads for the swivel crane: the swivel crane is not symmetrical. The swivel crane loads the deck over one of the two webs only. The self-weight of the crane plus the segment with spreader beam is generally 1000 kN or more. Hence in addition to bending about the deck's transverse axis the crane's reactions induce significant torsion into the box girder. The box is, however, a cross section able to cope with torsion. Additional reinforcement must generally be provided where loads are introduced by the swivel crane.

10.5.2 Accidental loads (segment fall, stay cable failure)

The usual accidental loads considered for construction with precast segments – such as segment fall – must be analysed for the PCM as well. In addition, one other accidental load specific to the method should be considered: the failure of one or more stay cables. This load case is very similar to what is done for permanent staying systems. During the construction, the failure has to be considered either during the tensioning of the stays, or at any time during the cantilevering process.

10.5.3 Pier head dimensions

As already stated, the geometry (deflection) of the cantilever is firstly controlled by the as-cast geometry of the segments, but also by field factors such as the tension introduced in the cable stays and climatic conditions, especially the thermal gradient in the deck. When reaching the next pier, the cantilever geometry must be stabilised with jacks, before fixing the bearings. Pier heads are best detailed to be large enough to allow installation of these jacks beside the permanent bearings, which are normally placed on the piers prior to the deck erection. Otherwise temporary brackets may be needed.

10.6. Cycle for the erection of a typical span

Assumptions made for the cycle shown in Table 10.1 are:

- span with 16 typical segments and 3 segments installed in front of the pier segment prior to launching of the mast
- segment weight ranging between 600 and 700 kN
- single shift per day (no night shift), 8 hours per shift
- temporary mast transfer system incorporated into the mast structure itself; the mast is moved with the segment carrier
- swivel crane equipped with double drum winch, movement by sliding or rolling on the deck
- pot bearings are installed before the installation of the segments
- maximum travel distance for the trailer is 1 km; a longer distance may require a second trailer.

Comments:

- The cycle presented in Table 10.1 is an optimum cycle, after the learning curve.
- Constructing the stitch between the pier segment and the first typical segment S1 is a time-consuming operation in the cycle. To reduce its impact on the cycle time, the stitch is done either at the end of a working day or at the end of a week. This removes the curing from the working time.
- The grouting and curing of the bearing plinths also require time. During this time, no movement is allowed on the span. This activity is therefore preferably performed at the end of a working day or at the end of a week.

10.7. Risk analysis

The list of risks given in Table 10.2 is by no means comprehensive, but addresses risks which are particular to BDE used for deck erection with the progressive cantilever method. For a specific project, a main contractor needs to carry out a full risk assessment of the operations performed with the BDE, considering its actual features, the erection sequence and the site constraints.

10.8. Performance criteria

The following issues have to be addressed when drafting the progressive cantilever erection method.

- Alignment: longitudinal slope, curvature in plan, cross fall variation.
- Span arrangement.
- Segment dimensions: segment size – weight – presence of starter bars.
- Number of segments per span.

Table 10.1 Cycle for a typical span erected by the PCM

TYPICAL CYCLE FOR PROGRESSIVE CANTILEVER METHOD USING CABLE STAYS										
ACTIVITY / STEPS	h	DURATION (shifts)								
		1	2	3	4	5	6	7	8	9
FIRST SEGMENTS INSTALLATION	**28hrs**									
- First segment S1 installation and setting - Stitching - Stitch concrete hardening - Segment S1 pre-tensioning - Segment S2 and S3 installation, fastening and pre-tensioning	4 hrs 5 hrs 10 hrs 3 hrs 6 hrs									
TEMPORARY MAST TRANSFER	**5 hrs**									
- Temporary mast transfer - First stay cable installation and tensioning	4 hrs 1 hr									
TYPICAL SEGMENTS INSTALLATION	**18 hrs**									
- Typical segment installation - Segment S4 to S16	1hr 20									
SEGMENTS ON PIER	**12 hrs**									
- Cantilever jacking - Segments on pier installation - Pot Bearings setting and grouting	2 hrs 1 hr 9 hrs									
POST-TENSIONING OPERATIONS	**22 hrs**									
- Post-tensioning operations - Temporary stay cable deactivation	18 hrs 4 hrs									

Table 10.2 List of risks specific to operation of BDE for progressive cantilever erection of precast segments

Activity	Risk	Mitigating measures
Installation of handrails on top and inside of segment Attachment of swivel crane's spreader beam to segment delivered on trailer	Vertical access and work at height: climbing up and down, work performed at height (at least 3 to 4 m above ground)	Access ladder fixed to stable base or access staircase, handrail or safety line fixed to segment and winch's segment lifting beam
Working on top of and inside of deck segment	Falling down from deck segment	Provision of temporary handrails around all exposed edges on top and inside of segment installed prior to lifting of segment from trailer
Stay cables tensioning	Unbalanced staying forces between rear and front cables	Step-by-step tensioning of the jacks
Stay cables installation and tensioning Staying of cantilevering deck	Failure of a stay cable (impact, fire) Failure of a stay anchoring device Improper fastening of stay anchoring device (pins, shear key)	Accidental design load for equipment and deck considering failure of one stay cable or anchoring device Locking system for upper anchoring device's pin Adequate procedure to check installation of lower anchoring device No fire source, such as power generators or diesel tank, near the stay cables Adequate speed control of any vehicle on the deck near the stay cables
Staying of cantilevering deck	Oil leakage in the tensioning jacks and corresponding hydraulic system	Back-up with fastened safety HT bars
Swivel crane tying down before lifting	Failure or improper fastening of rear tie-down	Back-up tying-down system (redundancy of HT bars) HT bars of tie-down prestressed before operation of the swivel crane Tie-down equipped with load sensor
Segment lifting and handling with the swivel crane	Swivel crane overturning or failure due to improper position of the swivel arm	Swivel crane system equipped with sensors and end stop
Segment erection over live traffic	Dropping of segment (unlikely), falling of debris, tools (possible)	No segment handling over live traffic
Epoxy gluing of segment joints	Falling down from deck segment	Front leg of the swivel crane equipped with gluing platform
Epoxy gluing of segment joints	Dropping epoxy	No traffic if over road, plastic sheets or similar below joints to catch epoxy if over water
Temporary staying mast – load transfer	Loss of control of transfer movement → loss of equilibrium → overturning of temporary mast	Presence of BDE engineer and supervisor; adequate speed of transfer; wind conditions checked before and during transfer

- Permanent work loading
  - permissible loading from segment transport by carrier
  - permissible temporary loading for permanent bearings
  - permissible temporary loading from swivel crane (rear anchorages and front supports) during segment installation or crane movement
  - load introduction of the staying forces into deck slab.
- Staying arrangement, for each span
  - stay angles
  - position of recess holes for stay anchors
  - interfaces between two consecutive stay positions for the back span
  - stay length: the length of each stay must be determined, considering the different span configurations; the number of length adjustments for each stay must be minimised.
- Tensioning jacks characteristics: the capacity of the jacks depends on the staying forces; the stroke has to be determined in regards to the cable length, its sagging effect and the upper winch capacity.
- Last typical segment installation and jacking
  - available space on the pier head
  - jacking forces
  - needs of additional support brackets.
- Access
  - into the segment for installation and prestressing of the temporary HT bars
  - on top of the segment being installed for connection/disconnection of the spreader beam
  - onto the temporary mast to connect the stays to the jacking system
  - at the extremity of the cantilever during gluing operations
  - access from ground to pier head, ahead of cantilever arrival
  - installation and stressing of the cantilever PT for the first three to four segments.
- Assembly and dismantling of equipment (temporary mast and swivel crane) – schedule, assembly area, method and lifting means.
- Closing of anchor and lifting recess holes.
- Removal and relocation of staying system (cables, anchorages) from one span to the other.
- Reusability, adaptability.

10.9. Photo gallery

Figure 10.16 First span on falsework, Boeme Viaduct, France (Vinci Construction)

Figure 10.17 Charente Viaduct, France (Vinci Construction)

Figure 10.18 Segment Carrier, Charente Viaduct, France (Vinci Construction)

Figure 10.19 Boeme Viaduct, France (Vinci Construction)

REFERENCES

Figures 10.1, 10.2, 10.16–10.19: Vinci Construction

Bridge Deck Erection Equipment
ISBN 978-0-7277-6193-4

https://doi.org/10.1680/bdee.61934.175

Chapter 11
In situ form travellers

11.1. Introduction

Form Travellers (FTs) are used to construct cast-in-situ post-tensioned box girders, cable-stayed bridge decks and arch bridges by the cantilever method. In general, the bridge deck is cantilevered symmetrically about the pier (Balanced Cantilever Method/BCM) and the segments are normally cast in one pour.

11.2. Application range

Form travellers can be used for construction of cast-in-situ bridge decks of theoretically almost any cross section. Form travellers can cope relatively easily with variation in segment length, height, web thickness, deck width and deck alignment (gradient, curvature, cross fall). The use of travellers to build cast-in-situ bridge decks by cantilevering is therefore popular all over the world.

However, it is advisable to observe some basic rules when sizing the cross section for a bridge deck that is being built cast-in-situ by travellers. The cross section's webs should remain parallel over the length of a cantilever built in one run by a traveller. The outer geometry should be kept constant except for the girder's depth and the span of the wing slabs. Variation in cross fall, if any, should be achieved by rotation of the entire cross section and not by the deck slab only.

Travellers are mainly classified according to their position in relation to the bridge deck: overhead and underslung. Modified underslung form travellers can be used for the construction of arches.

Table 11.1 indicates general application criteria of this type of BDE.

The flexibility of this construction method makes it very suitable for projects with high architectural demands and difficult topography. Cast-in-place construction with travellers is especially suitable for relatively small projects (for example, with only two main piers and one large main span) and/or when the set-up of a casting yard for the production of precast segments is not economically viable. It is also a good solution for projects with large spans where delivery of precast segments is not possible from the ground and the size of spans would require a huge overhead gantry to permit segment delivery over the deck.

Before starting deck construction with travellers, the pier tables have to be built by conventional means. For comments about pier table geometry, refer to Section III, Chapter 15. Different options are possible for the construction of the out-of-balance part of side spans. For details refer also to Section III, Chapter 15.

Table 11.1 Application criteria for form travellers

Item	Criteria
Number of spans	No limit
Span length	Box girders: 50 m to 180 m (occasionally up to 260 m) Stay-suspended decks: 100 to 500 m
Segment length and segmentation	■ Box girders: 2.5 m to 5 m (rarely 6 m) with overhead FT, 2.5 m to 4 m with underslung FT; variation in segment length along a cantilever should be minimised ■ Stay-suspended decks with 'open' section: 8 to 10 m ■ For general comments about segmentation refer to Section III, Chapter 15
Curvature radius in plan	No limit, but need for special launching kinematics for radius smaller than 200 m
Cross fall	0%–7%, variable, but preferably by rotation of the entire section
Gradient	0%–7%, variable
Pier geometry (height, width, thickness, shape)	No limit, but needs to be able to take the out-of-balance moments during cantilever construction unless taken by props (refer to Section III, Chapter 15)
Pier table geometry	Needs to be long enough to permit installation of at least one form traveller; has to be either fixed to the pier monolithically, stitched to the pier temporarily or supported by props to stabilise the deck during cantilever construction (refer to Section III, Chapter 15)
Cross-sectional area and variations	■ Box section: no limit for cross-sectional area; variable; but main webs should be parallel and variation of outer geometry should be limited to increase in depth of girder and span of wing slabs; preferably no internal top blisters ■ 'Open' section: no limit for cross-sectional area; preferably constant geometry; inner faces of main beams and cross beams with taper to permit stripping by lowering of form
Topography	No limit
Special load cases for bridge designer	Additional cantilever moment due to FT's self-weight: flexural capacity and impact on deflection control during construction Wind (transversal from traveller and cantilever into pier) Stability of balanced cantilever during construction: blocking of all 6 degrees of freedom between deck and pier
Mobilisation: from signature of contract to first pouring (excl. transport time)	3 to 6 months for simple cross sections; 6 to 12 months for complex cross sections and/or large segments

There are different ways to work with travellers.

- **Unbalanced cantilever construction:** Only one traveller is building, by cantilever method, half of a main span after the adjacent side span has been completed (e.g. cast in situ on stationary scaffolding).
- **Balanced cantilever construction (balanced cantilever method):** One pair of travellers is used to build part of a bridge deck from a pier table to mid-span on either side of the pier table by symmetrical cantilever construction on the up- and downstream side.

The basic objective when designing form travellers is to make them light. With regard to an overhead traveller's main frame, this is achieved by taking advantage of the following two principles.

- The stiffness of the traveller's main frame, which during concreting provides the front support for all formwork, is not controlling for the design, since the formwork is suspended from the main frame by rather flexible HT bars. These bars lengthen elastically considerably when loaded during concreting. Since the formwork is longitudinally line supported at its rear end, where it is fixed directly to the concrete of the previously cast segment, and it is good practice to start with the concreting of a segment from the tip progressing towards the formwork's rear support, the formwork will accommodate the front 'settlement' by rotation about its rear support line. This rotation can be compensated by setting out the main frame and the HT bars accordingly prior to casting. It is important to note, however, that the design of the formwork's walers and runners is controlled by deflection limits derived from permissible tolerances for flatness and thickness of concrete members as defined in the project's technical specifications.
- Applied loads are taken as much as possible by axial forces and not by bending.

Apart from the frame, the traveller consists of the outer and inner formwork (with walers, runners and form skin), both with weights proportional to the extent of formed area, and the access provisions. Access provisions – particularly for deep cross sections and long segments and in countries with stringent workplace safety regulations for work above ground – can contribute significantly to the overall weight of a traveller.

Standard modular overhead form travellers are typically designed for a maximum segment length of 5 m and load capacities (concrete and formwork) of 1000 kN to 3000 kN depending on the number of main frames being used.

11.3. General arrangement of travellers

There are three main types of form travellers:

- type 1: overhead form traveller
- type 2: underslung form traveller
- type 3: underslung traveller for cable-stayed bridges with 'open' section.

This subsection gives a description of the main elements of these three types of travellers. The kinematics are shown in the following subsection. At the end of this subsection some general explanations are given about the formwork.

11.3.1 *Overhead form traveller (type 1 traveller)*

A typical overhead form traveller is made up of the following main components (for illustration refer to Figure 11.1).

Figure 11.1 Isometric view of a typical overhead form traveller sitting on segment N – 1 and ready to cast segment N (formwork and access platforms not shown) (Doka GmbH)

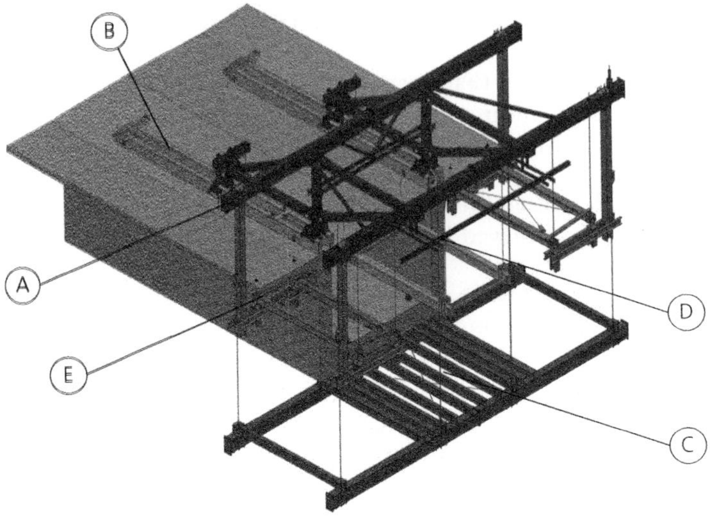

A. Pair of main frames with front and rear upper cross beams/trusses and hanger bars
B. Rail beam
C. Bottom platform with front and rear lower cross beams and longitudinal soffit beams; web restraining beams not shown for clarity
D. Internal tunnel form sitting on a pair of longitudinal roller beams (with roller bracket, not visible)
E. Wing slab and external side form

A. **Pair of main frames** with front and rear upper cross beams/trusses and hanger bars: The main frames are vertical plane frames. They sit on segment N – 1 (and sometimes also on segment N – 2). They cantilever over segment N to be cast. They are braced against each other by a front and rear upper cross beam/truss for lateral stabilisation of the frames but also provide the top fixity for the hanger bars, which support the formwork. During concreting and launching, the main frame is always the front support for the formwork. During launching it also provides, partially or fully, the rear support for the external formwork. For launching it is equipped with carriages which roll or slide over the rail beams and with longitudinal launching devices which connect the carriages with the rail beams.
There are two support points per frame, with downward reaction at the front support and uplift at the rear support. The front support is usually a hydraulic jack to allow fine adjustment of the level and load transfer (but with the load on the safety ring during concreting). For casting of segment N, this jack is positioned as close as is structurally feasible to the front end of the segment. The frame needs to be tied down to the deck at its rear support. This is achieved by installing and stressing at least two HT bars arranged symmetrically about the web. They are mostly detailed to be reusable from segment to segment.
For reasons of load introduction, the traveller's main frame is always aligned over a web. Hence there are normally as many frames as the cross section has webs. Single-cell box girders, for example, are fabricated by travellers with two main frames.

The front upper cross beam/truss and, if possible, also the rear upper cross beam are wider than the width of the deck slab. As well as stabilising the main frame, the rear upper cross beam/truss (located in the axis of the frame's front support) provides the suspension points for the hanger bars which support the external formwork at its rear, fully or partially, during launching. The front upper cross beam installed at the tip of the main frame supports all formwork at the front during concreting and launching. If the front upper cross beam/truss is too high above the deck (as is the case for a diamond-shaped main frame) a special platform has to be attached to the front upper cross beam/truss for the length adjustment of the hanger bars.

The main frame is set out longitudinally parallel to the deck's gradient and horizontally in transverse direction.

B. **Rail beams:** Each main frame is equipped with a rail beam. Rail beams consist of a pair of I-beams with a length of more than twice the maximum segment length. They are exclusively used for the launching operation of the traveller. The main frame's front support jack, which is used during concreting (as explained above), is generally placed between these I-beams. At the rear, the main frame is stressed down to a vertical prop, which is also located between these I-beams. During launching, the main frame's support points are transferred to a front and rear articulated carriage. The front carriage, which takes the traveller's downwards reaction and is sitting on top of the rail beam, is often a bogie with a pair of crane wheels or sliding pads. The rear carriage is generally made up of sliding pads, which bear against the top flange soffit of the rail beam's I-beams to provide stability against overturning. For the launching operation, the rail beam itself has to be tied down with HT bars to the concrete deck. The distance between the rail beam's support and tie-downs will control the minimum flexural capacity of the I-beams.

Launching of both the rail and the traveller is normally carried out by a double-acting hydraulic jack per main frame. This jack has a stroke of about 0.8 m. It is pinned against the front carriage and fixed against the rail beam – for example by a pin inserted through transverse horizontal holes arranged along the front half of the rail beam's webs at a spacing of a bit less than the jack's stroke. During resetting of the long stroke jack, the front carriage is secured against the rail beam by its own pin. Of course it is possible to achieve the movement of the traveller from segment to segment by more basic devices, such as centre hole jacks with HT bars or even with chain blocks. However, it is important that the movement of a traveller on all rail beams is synchronised to avoid jamming. If the devices used for movement work either in tension or compression only, a brake system has to be installed.

C. **Bottom platform:** The bottom platform supports the soffit form, provides the transverse fixity for the external web form (through the restraining beams) and carries the access platforms around the segment to be cast near or below its soffit level. For some travellers, during launching the outer side form and wing slab form are also supported on the bottom platform. Usually the main elements of a bottom platform are the front and rear lower cross beams arranged in the axes of the upper cross beams/trusses. The front lower cross beam is always suspended from the upper cross beam/truss by HT bars. The rear lower cross beam is only suspended from the upper cross beam/truss during launching, but is directly fixed to the bottom slab of segment N – 1 with stressed HT bars during concreting. The pair of lower cross beams are the support lines for a series of longitudinal soffit beams, which are basically the soffit form's waler beams. The soffit form, the outer web form's restraining beams and the longitudinal access platforms are directly mounted onto these longitudinal soffit beams. The transverse access platforms at the front and rear are attached to the lower cross beams.

If the girder has variable depth, the web's outer form passes through longitudinal slots in the bottom platform. If the webs are not vertical, the width of the bottom slab increases as the segment moves away from the pier. To cope with this situation, the longitudinal soffit beams, which take the concrete weight under the webs, and some additional longitudinal soffit beams to cater for the increased width of the bottom slab, have to be detailed to be movable along the lower cross beams. In addition, the restraining beams have to be movable as well. Longitudinally the bottom platform is always aligned parallel to the segment's soffit. For haunched bridge decks, the inclination of the bottom slab, particularly steep at the start of a cantilever, will create considerable biaxial bending in the lower cross beams. They are therefore flexurally connected to the longitudinal beams. In addition, the hanger bars need to have articulated connections at both ends whenever the deck has a gradient and/or the segment's deck and soffit are not parallel (as is the case for haunched girders). The cast-in ducts for fixity of the rear lower cross beam to the bottom slab of the previously cast segment need to be installed perpendicular to the soffit and not truly vertical.

D. **Internal tunnel form:** The internal formwork is a tunnel formwork which is sitting on and is movable against a pair of longitudinal roller beams. They are of similar length to the rail beams. These longitudinal roller beams are at the front, always suspended from the front upper cross beam/truss and have different support at the rear during launching and concreting. During launching from casting position for segment N – 1 to casting position for segment N they are rolling through brackets, which are fixed to the inner deck slab of segment N – 2. For concreting of segment N they are stressed with HT bars against the soffit of the inner deck slab of segment N – 1. The inner side forms are often individually suspended from the longitudinal roller beams, which permits the launch of the soffit form of the inner deck slab together with the main frames. The inner side form can be moved into its casting position after the reinforcement for the bottom slab and webs have been placed. At the starter segment next to the pier table's diaphragm special arrangements have to be made to permit the installation of the longitudinal roller beams. The longitudinal roller beams normally also support the access platform in front of the inner deck slab.

During concreting, hydrostatic pressure acting onto the inner and outer side form is balanced by form ties. For 'parabolically' haunched girders, the inner side form needs to be shortened for each new segment to be cast.

The tunnel form's walers for the inner deck slab's soffit form and the inner side form are arranged transversely to the bridge axis. That means that whenever the deck slab has a gradient, the inner side form's walers are not vertical, and when the gradient is variable, they are constantly changing their orientation against the vertically placed reinforcement stirrups. This makes the installation of the form ties, which are placed between the waler's double beams, a challenge. Furthermore, the overlap of the side form with the webs of segment N – 1 is not of constant length but variable over the girder's depth. This variation can be very significant for big-gradient and deep segments.

E. **Wing slab and external side form:** The wing slab and the external side form are supported by a pair of longitudinal beams which are connected by transverse beams at the front and rear. These beams are always suspended from hanger bars at the front. At the rear they are fixed by stressed HT bars to the wing slab of segment N – 1 during concreting. During launching they may be fully supported from the rear upper cross beam/truss (as shown in Figure 11.1), sit on the bottom platform or have a mixed support system. In the case of the mixed support system, the rear transverse beam is at its outer end suspended from the rear upper cross beam and the inner of the two longitudinal beams is extended by a rear nose so that it can be used during launching as a roller beam, together with an outer roller bracket.

As mentioned above, the hydrostatic pressure acting during concreting onto the inner and outer side form is balanced by form ties. For 'parabolically' haunched girders, the outer side form passes through longitudinal slots in the bottom form. It has therefore, unlike the inner side form, not to be shortened, unless vertical clearance becomes critical during construction. There are similar issues with the installation of form ties and the overlap of the side form with segment N – 1 for decks with gradient to those described in the paragraph on internal tunnel form, above.

F. **End form** (not shown in Figure 11.1): The end forms are always vertical and manually installed. They are generally made from timber and plywood, since they need to permit the protrusion of the longitudinal starter bars
   - bottom and top slab: the end form is propped against the soffit form and the deck form respectively
   - webs: the end form is fixed against the web's side forms; the reactions from the hydrostatic pressure are taken either by the side forms and transferred back to segment N – 1 or by longitudinal rebars coupled to starter bars protruding from segment N – 1.

G. **Access** (not shown in Figure 11.1): Each point of the traveller has to be accessed in every phase of construction. Figure 11.2 shows an example for the layout of these access platforms for an overhead traveller with diamond-shaped main frames.
   - Access around the wing slab (A): Used for the installation of reinforcement, to adjust the rear suspension of the bottom platform and to reach the bottom platform, usually over a ladder, but preferably over a stair.
   - Access along the upper cross beam (B): Needed to adjust all the front hanger bars.

Figure 11.2 Platforms on an overhead traveller (Doka GmbH)

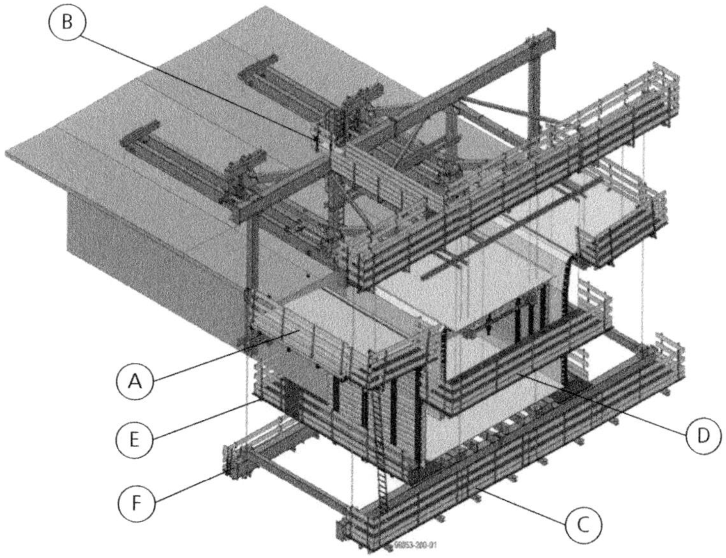

A. Access around the wing slab
B. Access along the upper cross beam
C. Front bottom access
D. Front top access
E. Lower longitudinal access
F. Rear bottom access

- Front bottom access (C): The access point to get onto the bottom platform and into the bridge deck. It is used for the adjustment of the soffit form for bridge decks with inclined webs of variable depth, for the installation of the reinforcement and the bottom slab's and web's end forms.
- Front top access (D): In this example, this access is a vertically adjustable platform, suspended from the upper front cross beam. It can be reached over the deck slab form. If people are permitted to be on this platform during vertical adjustment, a back-up suspension system has to be installed which ensures redundancy during movement. It is used for installation of the reinforcement, the deck slab's and web's end forms and the stressing of the cantilever PT.
- Lower longitudinal access (E): Usually protrudes by 2–3 m behind the bottom platform to provide passage into the rear bottom access. It is used for adjustment of the external web form, the web restraining beam and for installation and removal of the form ties.
- Rear bottom access (F): Used for the installation of HT bars, which connect the bottom slab form to segment N − 1, and for the adjustment of the soffit form's longitudinal beams for bridge decks with inclined webs of variable depth.

11.3.1.1 Weight

The weight of an overhead traveller depends on the cross section of a particular bridge and the access provisions but is normally between 25 to 45% of the deck's heaviest segment.

11.3.1.2 First pair of segments

As a form traveller overlaps with the adjacent upstream segment, there is often insufficient space on the pier table to assemble both travellers at the same time to start the deck construction in balanced mode. There are different options to cope with this issue.

- Option I: Only one traveller (A) is initially installed. After casting the first segment on one side of the pier table and launch of traveller A, the second traveller (B) is installed. The length of the pier table has to be between 6 to 7.5 m, depending on whether customised or modular travellers are being used. The regular cycle of the bridge deck production begins only with the casting of the second pair of segments. The impact on the construction programme is negligible, since upstream and downstream travellers A and B are rarely assembled simultaneously. Furthermore, if a pair of travellers is used more than once on a project, traveller B has to stay longer on the just-completed balanced cantilever since it is often used to cast the mid-span closure.
- Option II – Fork-solution (see Figure 11.3): The rear parts of modular travellers are often designed and detailed differently, so that the pair can be installed on an 8.5-m-long pier table simultaneously. The first launch of the travellers is non-typical, will require more time and needs special attention.
- Option III – The travellers are designed and detailed so that only the front parts of both travellers can be initially assembled simultaneously on the pier table linked by special connection pieces (see Figure 11.3). The advantage of requiring only a 6-m-long pier table is, however, hardly justified since considerable time, cranage and tight supervision are required after casting of the first pair of segments to separate the travellers and launch them forward for casting of the second pair of segments.

11.3.1.3 Mid-span closure

The mid-span closure is carried out by one traveller. The length of the closure pour is typically only 2 m, which is significantly shorter than the segment length the traveller has been set up to cast. The

Figure 11.3 Options to cast the first pair of segments on 'short' pier tables (Option II VSL International; option III NRS)

Option II: overlapping

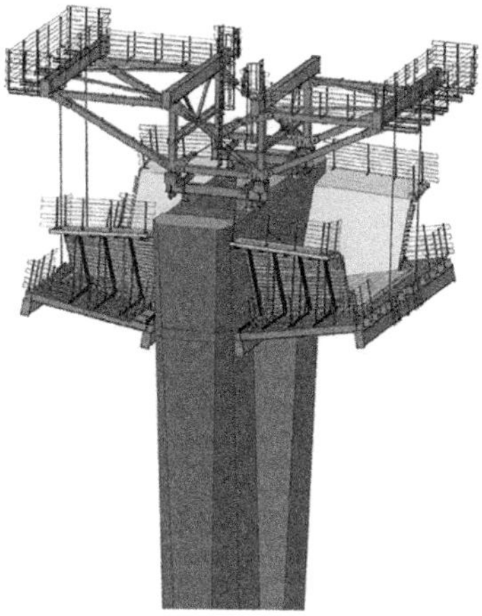

Option III: simultaneous

inner form is therefore often taken out before casting of the last segment of the cantilever. Light modular timber formwork is then used to cast this segment and the closure pour segment.

Special provisions are required to permit the 'overlapping' of the traveller with the tip of the adjacent cantilever, such as leaving longitudinal slots in the wing slabs for the passage of the inner hanger bars of the external formwork.

The traveller's main frames may be used to adjust the vertical tips of the cantilevers against each other and to transfer shear forces during casting of the mid-span closure. The travellers need, of course, to be checked for the assumed loading, particularly for loading acting upwards.

11.3.2 *Underslung form travellers (type 2 traveller)*

The development of travellers located not on top of the deck but below, for casting of haunched box girders, was driven by a desire to:

- remove any obstruction over segment N, so that a big part of the reinforcement cage can be prefabricated, with the objective to shorten erection cycles
- simplify installation and removal of the traveller by heavy-lifting the entire frame and outer form in one operation: from the pier table for installation and from anywhere along the deck for removal
- simplify launching back to the pier table for dismantling if required.

Particularly with regard to the first objective, the cycle time cannot be improved significantly with an underslung traveller when compared to a well-organised operation with an overhead traveller. The main reason for this is that about 25% of the reinforcement of a segment cannot be fully preassembled or fixed against each other even when using an underslung traveller, since the longitudinal starter bars protruding from segment N – 1 are about 1 m long. On the other hand, staggered installation of the reinforcement in the bottom slab and particularly in the deck slab helps to speed up the time-consuming coupling of the many tendon ducts. Another point is that considerable additional provisions have

Figure 11.4 Isometric view of a typical underslung form traveller used to cast haunched box girder segments: the traveller is supported on segment N – 1 for casting of segment N; access platforms not shown (VSL International)

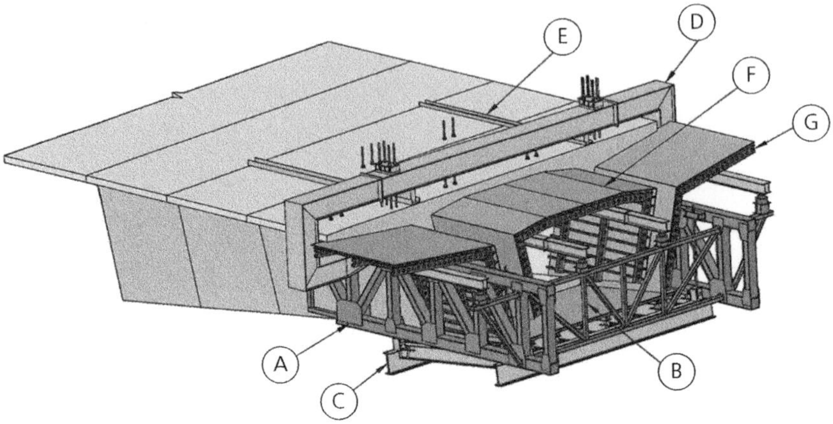

A. Main truss
B. Front transverse truss
C. Bottom platform with front and rear lower cross truss/beam and longitudinal soffit beams; web-restraining beams not shown
D. C-frame
E. Rail beam
F. Internal tunnel formwork sitting on a pair of longitudinal roller beams (with roller bracket, not visible)
G. Wing slab and external side form

to be made to enable prefabrication and handling of the big rebar cages, with constantly changing geometry. It is therefore no surprise that underslung travellers do not enjoy the same popularity as overhead travellers for construction of box girders with variable depth.

Underslung travellers can only be used for casting of single-cell box girders, since such travellers can, for reasons of load introduction (as explained later), have only one pair of main trusses.

A typical underslung form traveller used for casting haunched box girders is made up of the following main components (for illustration, refer to Figure 11.4).

A. **Pair of Main Trusses (MTs):** The main trusses are vertical rectangular space trusses. They are supported on segment N – 1 (and sometimes also on N – 2). They cantilever longitudinally over segment N to be cast. They are connected at the front by the front transverse truss (B). They are located below the wing slabs and as close as possible to the outer face of the webs, but still permitting the installation of the outer side forms. During launching, the main trusses provide the support of the entire outer form and the front support for the inner form. During concreting they are the supports for the wing slab and outer side forms and the front support for the soffit form and the inner form.
There are two support points per main truss, with downward reaction at the front support and upward reaction at the rear support. At the front support the pair of main trusses is sitting at the tips of the C-frame (D) during launching and is suspended for concreting by HT bars from the C-frame's main beam. The MT bears at its rear end directly against the soffit of the wing slab. During concreting the contact is through hydraulic jacks (with safety ring engaged during concreting). During launching the contact at the rear is through a roller. During launching the C-frame is slid along the rail beam on top of the deck.

B. Front transverse truss: The front transverse truss provides the structural connection between the main frames, the front support for the formwork and the support for the front access platforms.

C. Bottom platform: The bottom platform supports the soffit form, provides transverse fixity for the external web forms (through the restraining beams) and carries the access platforms around the segment to be cast near or below its soffit level. Usually its main elements are the front and rear lower cross beams. The front lower cross beam is always suspended from the main frame. During launching, the rear lower cross beam is suspended from the main frame, but is directly fixed to the bottom slab of segment N – 1 with stressed HT bars during concreting. The pair of lower cross beams are the support lines for a series of longitudinal soffit beams, which are basically the soffit form's waler beams. The soffit form, the outer web form's restraining beams and the longitudinal access platforms are directly mounted onto these longitudinal soffit beams. The transverse access platforms at the front and rear are attached to the lower cross beams.

For explanations on how to arrange the longitudinal soffit beams when the webs are inclined and of variable depth, the design of the lower cross beams, the articulation of the hanger bars and the orientation of the cast-in ducts, refer to the details given for the overhead traveller (11.3.1).

D. C-frame: The C-frame is the front support for the traveller during concreting and launching. It is supported by a pair of solid ram jacks equipped with safety rings and sledges, which sit on the rail beams and are always centred over the deck's webs. During launching, the main frames are supported on the tips of the C-frame and during concreting they are directly suspended by HT bars from the C-frame's main beam. For casting of segment N, the C-frame is moved as close as structurally feasible to the front end of segment N – 1.

E. Rail beams: The C-frame is sitting on a pair of rail beams. The rail beams are generally channels placed flat onto the concrete deck. During launching the traveller is slid longitudinally over the rail beams by activation of long stroke jacks fixed between the C-frame's support jacks/sledges and the rail beams.

11.3.2.1 Weight

The weight of an underslung traveller is greater than the weight of an overhead traveller. Segments are usually no longer than 4 m.

11.3.2.2 First pair of segments

The first pair of segments is often built with the pair of underslung travellers' main frames being temporarily connected. The rear half of the main frames is partially or fully omitted, depending on the length of the pier table.

11.3.2.3 Mid-span closure

This type of traveller is not suitable to be used to cast the mid-span closure, unless the main trusses are positioned below the deck.

11.3.2.4 Application for stay-suspended decks

This type of traveller is very suited for construction of box girders with constant depth. Box girders are ideal for decks suspended by a single plane of vertical stays located at the centre of the deck due to the high torsional stiffness of this type of section (refer to Section III, Chapter 15). When the deck has constant depth, the main trusses can be positioned below the deck. They are then connected by transverse trusses. The main trusses' front and rear reactions are introduced directly into the girder's webs. The C frame is 'flipped over' which means that the C frame's cross member is not located above but below the deck. The vertical-stay plane is thus kept free of any obstructions. An example of such an underslung traveller is given in Figure 11.12.

11.3.3 *Travellers for cable-stayed bridges with 'open' section (traveller type 3)*

As explained in Section III, Chapter 15, cable-suspended concrete decks have either a box section or an 'open' cross section. At the travellers' concept design stage, the introduction of reactions during the segment casting and the launching needs to be carefully assessed in order to:

- be within the local shear and flexural capacity of the cross section on which the traveller is standing or to which it is attached
- permit the installation and stressing of the stays and the subsequent forward-launching of the traveller.

For further explanations about the overall structural impact of the traveller on the deck and the options available to mitigate this impact, refer to Section III, Chapter 15.

This subsection gives details about underslung travellers for casting segments of decks with an 'open' cross section of constant depth. These travellers cast the main beams of segment N and the cross beams and deck slab of segment N − 1, while the traveller is supported by the main beams of segment N − 1.

Such an underslung form traveller is made up of the following main components (for illustration, refer to Figure 11.5).

Figure 11.5 Isometric view of an underslung form traveller for the construction of a cable-stayed deck with an 'open' cross section, without the use of stays for suspension and with the casting of the segment longitudinally staggered; the traveller is supported on the main beams of segment N − 1 to cast the cross beams and deck slab of segment N − 1 and the main beams of segment N; access platforms not shown (VSL International)

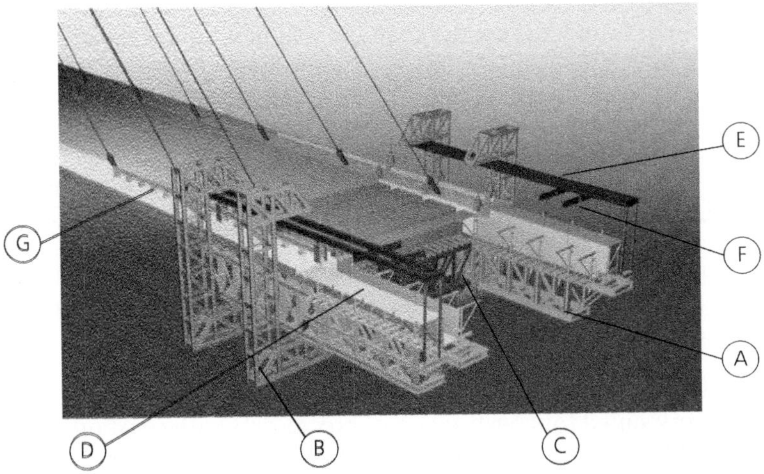

A. Main truss
B. Hanger frame
C. Cross trusses with the formwork for the cross beams and the deck slab
D. Formwork for the main beams and wing slabs of segment N
E. Formwork for the cross beams and the inner deck slab of segment N − 1
F. Auxiliary hoist
G. Rail beam

A. **Main truss:** The main trusses are vertical rectangular space trusses. They are supported on segment N – 1 (and sometimes also on N – 2). They cantilever longitudinally over segment N. They are connected at the rear by two to three cross trusses (C). They are located below the main beams. During launching and casting, the main trusses provide at the front the support for the form of segment N's main beams and wing slabs and at the rear the support for the form of the cross beams and the inner deck slab of segment N – 1. During launching, the main trusses are suspended from two pairs of hanger frames. For concreting, they are fixed to the deck by HT bars.
It is important to design and detail the main trusses so that they permit access for installation and stressing of the stays suspending segment N and for restressing, if required, of the stays suspending segment N – 1.

B. **Hanger frame:** The hanger frames support the traveller during launching. They also allow the vertical movement of the traveller by means of hydraulic jacks to:
   - break the bond between the formwork and the concrete and to
   - create enough vertical clearance between the deck slab's soffit form and the soffit of the cross beams to launch the traveller from casting to casting position without need for the partial removal of the inner formwork.

   This feature is a characteristic of this type of traveller, since it permits a simplification of the formwork and a reduction in the time required to strip, launch and set up the traveller.

C. **Cross trusses** with the formwork for the cross beams and the deck slab: The cross trusses connect the pair of main trusses structurally. They also provide the support for the formwork of the cross beams and the inner deck slab of segment N – 1.

D. **Formwork for the main beams and the wing slabs** of segment N: Whereas the general geometry of these elements is usually constant, the stay's blisters are often variable from segment to segment. The local adaptions required may be tailor-made with timber and plywood. Alternatively the stay blister can be prefabricated. However, this needs lifting devices on the deck or incorporated into the traveller which can handle the considerable weights of these blisters at rather big reach unless they are placed onto the main trusses during launching. Furthermore, the reinforcement layout has to be re-detailed and provisions on the main trusses have to be given to receive these blocks, fine-tune their positions and keep them firm against the traveller during concreting.

E. **Formwork for the cross beams and the inner deck slab** of segment N – 1: Cross beams are often deep and narrow. Hence it may be necessary to detail one side form per cross beam horizontally movable to give access for cleaning the forms and placing the reinforcement.

F. **Auxiliary hoist:** Auxiliary hoists are shown for this traveller. They are used to handle the stay guide pipes, bearing plates and anchorages.

G. **Rail beam:** The hanger beams are sliding on the rail beams during launching of the traveller. The longitudinal movement is generally by long stroke jacks installed between the hanger beam's sledges and the rail beams. These rail beams need to be supported by the deck's main beams but also have to clear the stay's guide pipes. If there is not enough space available, it may become necessary to provide a couple of transverse support beams. These transverse beams are fixed between the guide pipes to the deck's main beams and support the rail beams on their cantilevers.

11.3.3.1 Weight

In this kind of traveller the centre of gravity is closer to segment N – 1's stays than with a conventional overhead or underslung traveller. This significantly reduces not only the negative bending moment in the deck but also that in the traveller's main trusses. This has a positive effect on the traveller's overall weight.

11.3.3.2 First pair of segments

As with underslung travellers, the first pair of segments is often built with the pair of underslung travellers' main trusses temporarily connected. The rear half of the main trusses is partially or fully omitted, depending on the length of the pier table. If the deck is not connected to the pylon in the pylon's axis and the pylon is not detailed with a cross beam below the deck, the coupled travellers can even be used to cast the pier table.

11.3.3.3 Mid-span closure

This type of traveller is suitable to be used to cast the mid-span closure, since it can be easily overlapped with the tip of the adjacent cantilever.

11.3.4 *General comments about formwork*

Hydrostatic pressure during concreting in webs is generally equated across a member by horizontal form ties to minimise the weight of the travellers. Hence the form skin is full of holes. As well as holes for the horizontal form ties, there are also holes to be drilled for tie-downs and vertical suspension bars. Since drilling holes through plywood to cope with construction tolerances and unplanned obstruction by reinforcement and PT elements is, of course, much easier than drilling holes through steel plates, use of coated plywood is very common. Good-quality plywood, which is properly treated on site, can be recycled up to 20 times. This is generally enough to last for one cantilever run.

For the inner form of haunched box girders, the web's walers are often made partly from timber to permit easy shortening of the inner side form when moving from segment to segment.

11.4. Kinematics

This subsection gives the kinematics for a typical overhead traveller for casting of a haunched single-cell box girder (type 1 traveller) and for an underslung traveller to cast a stay-suspended deck, which has an open section and constant depth (type 3 traveller).

11.4.1 *Overhead traveller (type 1 traveller)*

Figure 11.6 (on pages 189 to 191) shows the kinematics for the overhead traveller being launched from segment N – 1 to segment N and the casting of segment N.

11.4.2 *Underslung traveller (type 3 traveller)*

Figure 11.7 shows the kinematics for the underslung traveller being used to cast main beams and wing slabs of segment N and cross beams and inner deck slab of segment N – 1, followed by forward-launching of the traveller onto the main beams of segment N.

11.5. Permanent work interfaces

11.5.1 Geometry

11.5.1.1 *Pier table*

For details about pier table geometry, refer to Section III, Chapter 15. Pier tables should always have kickers to permit the overlap of the traveller's inner and outer forms.

11.5.1.2 *Segmentation*

For explanations about symmetrical and asymmetrical segmentation and respective PT layout, refer to Section III, Chapter 15.

Figure 11.6 Kinematics for an overhead traveller (Doka GmbH)

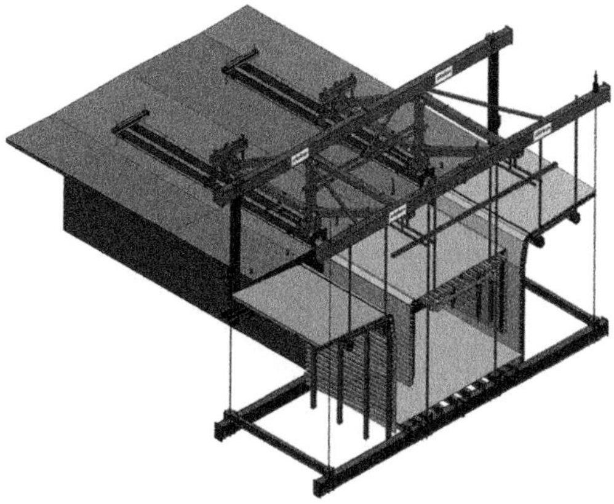

Stage 1: Segment N – 1 (on either side of pier) completed: concrete poured.

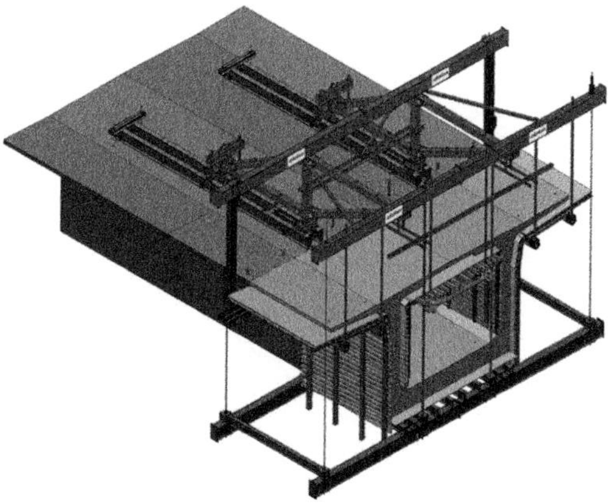

Stage 2: During curing of concrete, removal of end forms: installation of PT strands, launching of rail beams onto segment N – 1, relocation of roller brackets (for launching of tunnel form and, if required, for launching of external side and wing slab forms); once concrete has reached enough strength, stressing of cantilever PT (and if applicable transverse deck PT) and cutting of protruding strands; removal of all suspension bars and form ties and stripping of side forms; breaking bond between form and concrete by tilting/lowering of traveller at its front supports.

11.5.1.3 *Decks with box section*

External geometry: As explained in the subsection 'Application range' (11.2), over the length of a cantilever built in one run by a traveller, the cross section's webs should remain parallel, outer geometry should be kept constant except for depth of girder and span of wing slab, and variation in cross fall should be achieved by rotation of the entire cross section and not by the deck slab only.

Internal geometry: Top blisters should be avoided. The minimum inside vertical clearance at the centre of the span should not be less than 2 m. Thickness reduction of webs along a cantilever run are preferably achieved in steps at the front end of a segment. Thickness reduction of the bottom slab is done linearly.

Figure 11.6 Continued

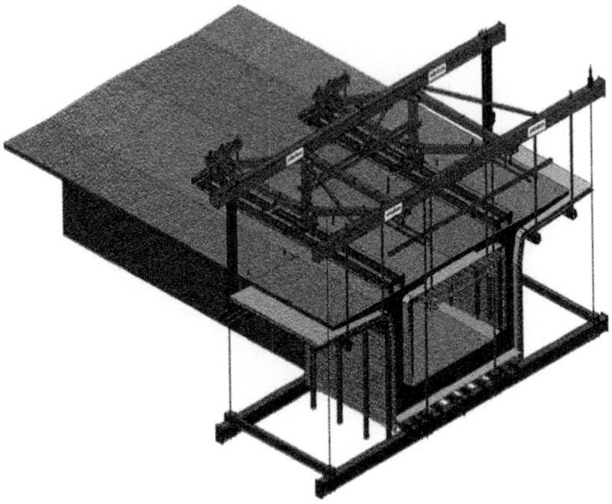

Stage 3: Load transfer main frame onto rail beams by de-tensioning rear tie-down bars and retraction of the front main jacks. Launch traveller without inner tunnel form forward onto segment N – 1 into casting position for segment N.

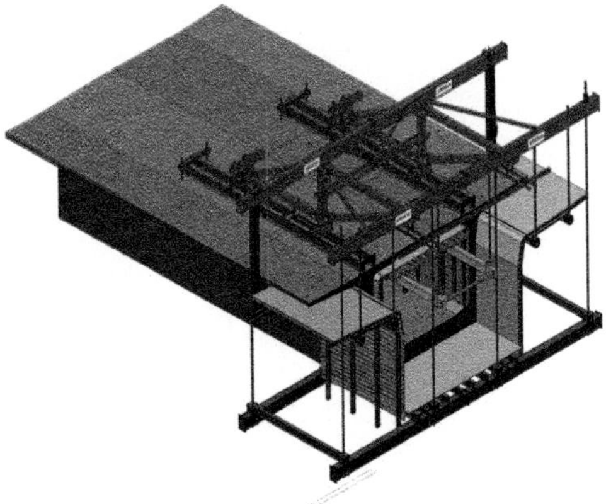

Stage 4: Load transfer main frame from rail beams to deck; set out external form to geometry of segment N and install form suspension bars at rear; clean outer form; place reinforcement in bottom slab and webs; place ducts for bottom PT.

11.5.1.4 *Decks with open section*

Vertical faces should have a taper of about 1 to 10 (width increasing with height) to permit stripping by lowering of the formwork. The width of deep prestressed cross beams is very often not controlled by structural requirements but by the need to give enough space for proper placing and vibration of the concrete.

11.5.2 High-tensile bars and form ties

Travellers require many HT bars (tie-down and form suspension bars) and form ties to be installed in each segment. Different travellers have different demands. Potential clashes between such bars and PT and reinforcement need to be sorted out well ahead of the commissioning of the traveller on site. The contractor and the consultant need to show some flexibility. The objective must be to arrive at solutions that are structurally acceptable but that also improve the constructability, which ensures that

Figure 11.6 Continued

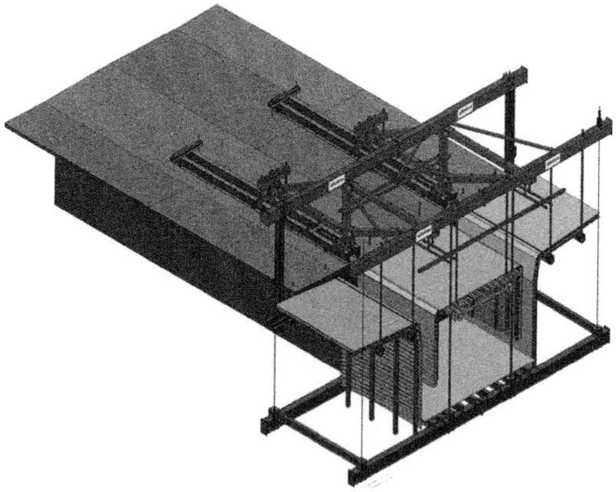

Stage 5: Launch tunnel form into casting position of segment N and set out; install form suspension bars at rear; place reinforcement of deck slab and ducts and anchorages for the cantilever PT; close inner web forms; install horizontal form ties; complete installation of end forms.

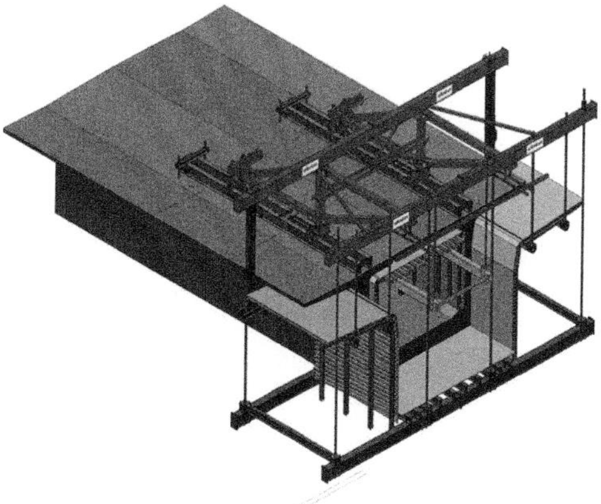

Stage 6: After final survey, pour concrete.

the specified quality can be realistically achieved in the field. For each cantilever to be built by travellers, it is vital to produce drawings that consolidate all relevant cast-in components and, the temporary tie-down and, suspension bars and form ties and define their exact locations. Curvature in plan adds another complication to this exercise.

11.5.3 Weight of the traveller

The weight of form travellers that are custom-made for a specific project is normally (including forms and access provisions) close to 25 to 45% of the weight of the heaviest segment. However, this ratio is sometimes exceeded, especially when modular form travellers are used for segments smaller than the maximum 'permissible' segment in terms of weight and length. Due to its position at the tip of the bridge cantilever under construction, the traveller's weight is of importance for verifying that the specified cantilever PT is sufficient to cope with the cantilever moment during construction. The traveller's weight also has to be considered to define when the cantilever PT needs to be grouted during cantilevering to increase the deck's flexural ultimate capacity. Furthermore, the weight of the traveller has

Figure 11.7 Kinematics for an underslung traveller (VSL International)

Stage 1: Traveller ready for construction of segment N's main beams and wing slabs and segment N – 1's cross beams and inner deck slab.

Stage 2: Placing of stay pipes/bearing plates/anchorages and reinforcement; closing of side forms of cross beams; installation of horizontal ties; installation of end forms.

Stage 3: Pouring and curing of concrete for cross beams/internal deck slab of segment N – 1 and main beams/wing slab of segment N; removal of form ties and end forms.

Figure 11.7 Continued

Stage 4: Installation of stays suspending segment N, including first stage stressing; opening of side forms; stressing of PT in cross beams; removal of form suspension bars.

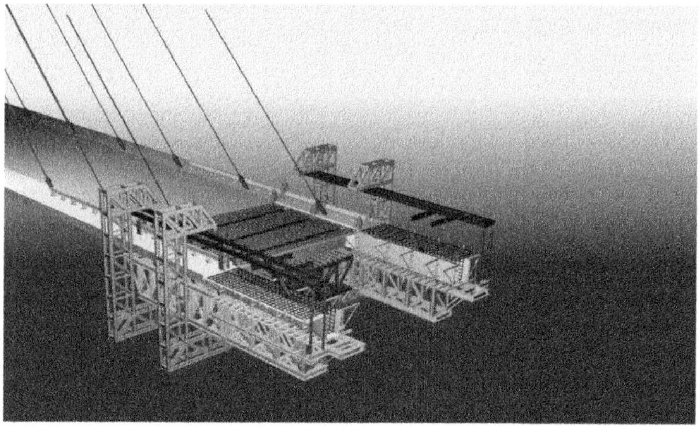

Stage 5: Launching of rail beams forward onto main beam of segment N.

Stage 6: Load transfer of traveller from deck to hanger frames and removal of traveller's direct deck fixity; breaking bond between deck and form by lowering traveller with hanger frames' jacks; lower traveller until deck slab form clears the cross beams' soffit.

Figure 11.7 Continued

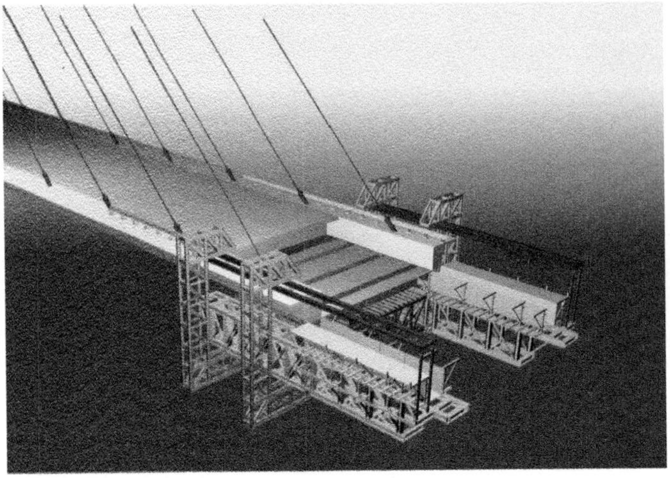

Stage 7: Launch the traveller forward onto main beam N, that is into casting position for main beam of wing slabs of segment N + 1 and cross beams and inner deck slab of segment N.

Stage 8: Lift traveller up into casting position; load transfer traveller from hanger frames to deck; fix main trusses directly to deck; clean forms; second-stage stressing of stays suspending segment N.

to be known accurately for the calculation of the segment's setting-out geometry, to enable an effective control of the deck's geometry during cantilevering.

11.5.4 Layout of tendons

For discussion of sensible PT layouts for BCM bridges, refer to Section III, Chapter 15. In any case, the consultant should always be open to revisiting the assumed tendon layout in plan, once it has been decided which traveller shall be used for construction. Specialist bridge deck contractors experienced with travellers have generally not only knowledge about the means and methods on how to build but also have the in-house skills to design and detail state-of-the-art concrete bridges. They usually propose sensitive compromises to the consulting engineer.

11.5.5 Load introduction

It is important to check the load introduction of the traveller's front reaction, since travellers are positioned as closely as possible to the tip of the deck. This check is best carried out using a strut-tie model. It may be necessary to add horizontal links below this support point to provide a properly anchored longitudinal tension tie. For underslung travellers whose main trusses are positioned below the wing slabs and not below the webs, the load introduction from the traveller's rear reaction is usually critical. To satisfy the structural demands, the deck and web may have to be locally strengthened (punching shear, shear and transverse bending).

As explained above and in Section III, Chapter 15, when building stay-suspended decks, the introduction of the traveller's reactions during the segment casting and during launching needs to be carefully assessed with regard to magnitude and location in order to:

- be within the local shear and flexural capacity of the cross section's element the traveller is standing on or is attached to and
- permit the installation and stressing of the stays and the subsequent forward-launching of the traveller.

11.5.6 Stability

Decks built by BCM need to be fixed against the pier for all 6 degrees of freedom. For more details on this subject, refer to Section III, Chapter 15.

11.5.7 End span construction

For options on how the out-of-balance part of end spans can be built, refer to Section III, Chapter 15.

11.5.8 Mid-span closure

Mid-span closures have to be properly engineered. In order to achieve a good concrete, the tips of the two adjacent cantilevers need to be fixed against each other vertically and horizontally so they do not move or rotate against each other differentially until the concrete has set and the initial post-tensioning has been activated. Whereas it may be possible to use the traveller to transfer the vertical shear through its main frames or main trusses, the transfer of the normal forces and moments generally needs to be achieved by other means. For medium to large spans, a well-proven way to fix the two cantilevers against each other for the transfer of vertical bending moments and the normal forces is to use longitudinal props at the top and bottom. One pair of steel props is located within the deck slab bearing against the cantilever tips' end faces and a second pair above the bottom slab acting against permanent concrete blisters. These props are pre-compressed by temporary longitudinal stress bars above the deck and by one pair of the permanent continuity PT. For additional comments about casting of mid-span closures, refer to Section III, Chapter 15.

11.5.9 Forming and closing of insert holes

There are hundreds of holes to be closed after a traveller has passed. It is important to define a proper method on how these holes are formed and how they are filled, so that they do not negatively affect the structure's long-term durability.

Table 11.2 Typical cycle for form traveller type 1

Activity	H	Day 1				Day 2				Day 3				Day 4				Day 5				Day 6				Day 7			
Stressing of PT	5		A																										
			B																										
Launching FT from N – 2 onto N – 1 and setting out of FT	5			A																									
							B																						
Placing of bottom slab and web reinforcement; placing of bottom PT ducts	20						A	A			A	A																	
								B			B	B		B															
Launching of tunnel form and setting out	5													A															
															B														
Placing of deck slab reinforcement; placing of cantilever PT ducts and anchorages	10															A			A										
																			B	B									
Pouring concrete	10																		A	A									
																							B	B					
Curing of concrete; removal of end forms; opening of side forms; threading of PT strands; launching of rails	36																					A	A	A	A		A	A	
		B																							B	B	B	B	B

11.6. Cycle times

In this chapter cycle times are shown for an overhead traveller (traveller type 1) and an underslung traveller (traveller type 3). It is important to note that the cycles shown are for illustration only and that actual cycles need to be worked out for individual projects, considering project-specific features and constraints.

Table 11.2 shows the cycle for a traveller type 1. Assumptions made are:

- segmentation: symmetrical; balanced cantilever built with pair of travellers (A and B)
- segment characteristics: 4 and 5 m long and 12 m wide, maximum girder depth 7 m, maximum concrete volume 70 m^3, maximum rebar tonnage: 13 ton; pair of 6–19 cantilever tendons to be stressed per pair of segments; no transverse PT
- starting point: segment N – 1 ready for post-tensioning
- use of standard concrete (36 hours to reach minimum strength required for load transfer)
- single 10-hour shift per day with one extended shift for pouring of concrete.

Table 11.3 shows the cycle for traveller type 3. Assumptions made are:

- segmentation: symmetrical; stay-suspended deck built by BCM with pair of travellers (A and B)
- segment characteristics: 9-m-long and 15-m-wide segments, open section having constant depth of 2 m; pair of main beams and three cross beams; stays anchored directly in main beams; concrete volume: 136 m^3; reinforcement tonnage: 20 ton; pair of parallel strand stays per segment with 61 strands per stay, stay length 100 m
- starting point: installation and stressing of stays suspending segment N – 1
- use of standard concrete (36 hours to reach minimum strength required for load transfer)
- stays stressed in two stages during cantilever construction: the stay suspending segment N is firstly stressed during installation, which takes place after curing of the main beams of segment N; second stage takes place after launching of the traveller onto the main beam of segment N
- double shift per day with 10 hours per shift.

11.7. Risks

The list of risks given in Table 11.4 is by no means comprehensive, but addresses risks which are particular to form travellers. For a specific project, a main contractor needs to carry out a full risk assessment of the operations performed with travellers, considering the actual features of the traveller, the erection sequence and the site constraints.

11.8. Performance criteria

The following items need to be addressed when drafting a performance specification for a traveller.

- Segmentation: variations, length of mid-span closure.
- Vertical and horizontal alignment, cross fall.
- Pier table geometry: length, with or without kickers.
- Segment characteristics
  - weight and location of centre of gravity per segment
  - geometrical variations
    - depth of girder, bottom slab and webs
    - blisters
    - cross fall.

Table 11.3 Typical cycle for form traveller type 3

Activity	H	Day 1				Day 2				Day 3				Day 4				Day 5				Day 6				Day 7			
Stressing of transverse and longitudinal PT; installation and stressing of stays suspending segment N – 1	15	A	A	A																									
		B	B	B																									
Launching of rails; launching FT forward onto main beams of segment N – 1 and setting out of FT	10				A	A																							
							B	B																					
Second-stage stressing of stays suspending segment N – 1	2										A																		
	2										B																		
Placing of guide pipes, bearing plates and stay anchorages in main beams of segment N; placing of reinforcement in main beams of segment N and in cross beams and deck slab of segment N – 1; placing and coupling of longitudinal PT; placing of transverse PT in crossbeams	40								A	A	A	A	A	A	A	A													
											B	B	B	B	B	B	B	B											
Pouring concrete	10																		A	A									
																							B	B					
Curing of concrete; removal of end forms; opening of side forms; preparation of construction joints	36																				A	A	A	A	A	A			
		B																							B	B	B	B	B

Table 11.4 List of risks specific to the operation of travellers

Activity	Risk	Mitigating measures
Working on top of and inside segment and traveller	Falling from height	Provision of handrails with kicker boards around all exposed edges on top and inside segment, installed prior to launching of traveller; provision of handrails with kicker boards along all access walkways of traveller; as far as possible, use stairs and not ladders for vertical access
	Injured people (from work accident) cannot be rescued within acceptable time	Creation of a second rescue route; repetitive training of site staff and local rescue forces on site
First launching of traveller after completion of first segment and change to typical FT configuration	Instabilities during exchange of parts	Detailed method statement; managed by experienced superintendent
Installation of tie-down bars (for rail beam and FT's rear support)	Wrong position of holes	■ Setting out position defined on drawings ■ Use of rigid void formers ■ Void formers effectively fixed against formwork and held in place during concreting
	Incompletely coupled or anchored bars or damaged bars	■ Visual inspection ■ Spray-marking ■ Proof-loading or redundancy in terms of number of bars
Installation of form suspension bars (to clamp form work at rear end to previously cast segment)	Wrong position and alignment of holes (generally need to be perpendicular to element's soffit and not truly vertical)	■ Setting out position defined on drawings ■ Use of rigid void formers ■ Void formers effectively fixed against formwork and held in place during concreting
Installation of rail beam	If rail beams are not correctly installed with regard to position (alignment and level, pair of rail beams not parallel) FT may get jammed during longitudinal movement and/or not arrive at correct position for casting of the next segment (FT's front and rear support are out of position)	■ Engineered and not site-improvised solution for shimming; required for each cantilever run ■ Provision of sufficient number of shims on the deck as defined in the operation manual ■ Marking with spray rail beam's position on the deck

Table 11.4 Continued

Activity	Risk	Mitigating measures
Launching of traveller	Loose parts may fall down from traveller when launching	Removal or fixing of all loose parts or tools before lowering traveller for stripping (storage racks); no launching over live traffic
	Loss of control during longitudinal movement (downhill launching, friction overestimated, etc.)	Provision of braking system: sledge/trolley pin engaged to rail beam when resetting long stroke jack; double-acting long stroke jack with brake valves and jack designed so that piston's end stopper can hold the load
	Change of alignment of hanger bars from which bottom platform is suspended during stripping and moving of FT: deforms flexurally, damages or even breaks hanger bars or their fixities, with subsequent tilting or even drop of the bottom platform	■ Provision of articulation at both ends of bars ■ Back-up suspension system (e.g. hanger bars and chain blocks) ■ Minimum number of people on platform during launching (only to check clearance between form and concrete); people with safety harness and sling connected to outer side form (not to platform's handrails) ■ Visual checks during launching of FT, so that form does not get stuck against concrete surface
	Outer roller bracket not relocated prior to launching: pulling of longitudinal roller beam off its rear support on the roller bracket and subsequent drop of the side form	■ Checklist ■ Provision of end stopper at end of longitudinal roller beams
	Wrong positioning and/or inadequate shimming of FT's main support on deck for concreting of next segment results in traveller losing its footing during concreting. Traveller 'settles', whereby already-placed, but still 'green', concrete is damaged. Traveller may even collapse	■ Mark correct position on deck ■ Visual inspection to ensure concrete is sound in this area (no honeycomb, etc.) ■ Use of engineered and not improvised shims if required ■ Centring of shims below jacks
	Lateral instability due to wind	Do not launch when wind is above permissible speed or likely to go above permissible speed during launching

Table 11.4 Continued

Activity	Risk	Mitigating measures
Coupling of HT bars	Incomplete engagement of bars in coupler, resulting in loss of tie-down or suspension	Provision of centre pin; spray-marking of bars
	HT bars can turn out of couplers and subsequently fall down due to wind-induced oscillation about bars' vertical axis	Provision of safety pins
Construction over live traffic	Objects, concrete laitance, etc. falling on public space frequented by cars, bikes, trains, pedestrians	■ Provision of trays below the outer web form if it protrudes through the bottom platform (normally the case for haunched girders); no traffic under traveller for a period starting with stripping of outer form and ending with closing of outer form ■ Alternatively, use of traveller with a closed bottom platform, whereby the outer web form has to be shortened like the inner web form for each segment

- Post-tensioning and rebar layout: need to superpose with layout of tie-downs, form suspension bars and form ties to determine clashes and work out solutions.
- Starter bars
  - longitudinal between segments
  - if applicable: for parapets, for stitches between wing slabs.
- Structural permanent work interfaces
  - upper weight limit for travellers
  - load introduction at deck contact points (downwards reaction at front, upwards reaction at rear; support for rails): limitations for magnitude, location and contact area.
- Clearances: vertical below and above deck (overhead power lines, overhead bridges, etc.); horizontal (adjacent decks {twin bridge or deck construction}, houses, power lines, etc.): important for detailing of rear suspension system.
- Formwork:
  - material for form skin (plywood uncoated/coated, steel)
  - surface finish requirements (flatness; surface finishes prohibiting use of form ties should not be specified), layout of form tie positions, architectural demands
  - inserts for forming circular holes for tie-downs and form suspension bars
    - material used
    - how to fix against form skin to avoid displacement during concreting
    - follow-up work.

- Inner formwork: main contractor needs to clearly specify what kind of features are required
  - vertical clearance: sufficient for use of tunnel form supported on longitudinal beams below inner deck slab?
  - inner deck form movable on top of longitudinal beams?
  - reduction in thickness of webs: by telescopic translation of inner side form?
  - inner side form: with top articulation to ease stripping? Independently movable against inner deck form?
  - removal of inner form after having cast second-last segment of a cantilever when adjacent cantilever has already been completed and use of light modular form panels for casting of last segment?
- End forms
  - concept
  - fixity
  - who supplies?
- Mid-span closure
  - use of form traveller for vertical adjustment between the tips of adjacent cantilevers?
  - use of form traveller for transfer of vertical shear during concreting and curing?
  - use of light modular form panels for casting of mid-span closures?
- Rebar placing
  - rebar logistic: lowering of rebars for bottom slab and webs delivered over deck (e.g. launching of inner deck soffit form before or after placing of this reinforcement)
  - vertical access for placing, if reinforcement is partially prefabricated (e.g. web reinforcement).
- PT
  - access for stressing of cantilever and transverse PT (level and width of platforms)
  - handling of stressing jacks (suspension points)
  - need to have transverse PT's protruding strands cut prior to launching of traveller (lateral clearances against rear suspension)?
- Concreting
  - need for access holes in inner web forms and inner deck soffit form?
  - passage for pipes for pumping of concrete.
- Target cycles
  - double or single shift (lighting)
  - time restrictions for noisy type of work (e.g. strand pushing)
  - environmental factors like rain and snow: need to stop work, need to delay concreting, longer curing time (cover, insulations, heating)
  - labour costs, skill level
  - degree of mechanisation (auxiliary devices like hoists, etc.); provision of lighting, cover, insulation and heating.
- Vertical tie-downs for FT and rail beams
  - proof-loading or redundancies
  - recovery of bottom anchor plates and nuts if below wing slab or bottom slab.
- Assembling weight: the maximum weight of components to be lifted must be specified by the supplier both for assembling and for dismantling operations.
- Launching of FT
  - braking system
  - suspension of bottom platform: need for back-up suspension system (e.g. HT bars and chain blocks).

- Access
  - into segment (from deck, from pier)
  - on traveller during different stages of operation: from deck slab to traveller and along traveller to the front top transverse platform; from top of traveller to bottom of traveller; access for placing of reinforcement, concreting and stressing (cantilever and transverse PT); access for closing/opening forms and installation/removal of form suspension bars and horizontal form ties (working at height); access for retrieving bottom anchor plates and nuts of vertical tie-down if below wing slab or bottom slab
  - access for follow-up work on holes used for form ties, form suspension bars and tie-down bars and for touch-up of external concrete surfaces.
- Shipment costs (initial, if reused).
- Reusability.
- Method, schedule and crane definition for assembly, relocation, dismantling.

The buyer needs to be given a time frame for fine-tuning of the traveller concept/support reactions, insert layout/PT layout rationalisation and delivery/commissioning of the traveller.

11.9. Photo gallery

Figure 11.8 Overhead traveller type 1, Viaducto de Teror, Spain (Doka GmbH)

Figure 11.9 Underslung traveller type 2, La Sioule valley, central France (SYSTRA)

Figure 11.10 Overhead traveller type 1 used to construct the composite deck of an extradosed bridge; deck and bottom slab in concrete, corrugated steel webs (M43 Highway, Hídépítö, Hungary) (Doka GmbH)

Figure 11.11 Overhead traveller with a pair of main frames rotated in plan to provide vertical clearance above the webs for placing of prefabricated rebar cages (Second Gateway Bridge, Brisbane, Australia) (VSL International)

Figure 11.12 Underslung traveller with main trusses below deck for construction of a stay-suspended box girder deck of constant depth and with single vertical plane of stays: pair of travellers coupled for construction of the pier table (Panama Canal bridge, Panama) (Bilfinger Berger)

Figure 11.13 Modified version of an underslung traveller type 3 for construction of a stay-suspended deck with 'open' section: full cross section cast in front, use of permanent stays to support traveller at its tip (Dolsan Bridge, South Korea) (VSL International)

Figure 11.14 Pouring of bottom slab (Litomerice Bridge, Czech Republic (Doka GmbH))

Figure 11.15 Pouring of deck slab (Doka GmbH)

REFERENCES

Figure 11.1, 11.2, 11.6, 11.8, 11.10, 11.14, 11.15: Doka GmbH
Figure 11.3 (option II), 11.4, 11.5, 11.7, 11.11, 11.13: VSL International
Figure 11.3 (option III): NRS
Figure 11.9: SYSTRA
Figure 11.12: Bilfinger Berger

Bridge Deck Erection Equipment
ISBN 978-0-7277-6193-4

https://doi.org/10.1680/bdee.61934.209

Chapter 12
Movable scaffolding systems

12.1. Introduction

Bridge construction using Movable Scaffolding Systems (MSS) is a full-span method for multiple-span cast-in-situ concrete bridges. The bridge superstructure is cast in situ by using an external formwork system which is mounted onto a launching girder. After one bridge span is built at its final position, the MSS is shifted into the next span.

The flexibility of this construction method and its level of mechanisation suggest use on sites with high architectural demands, difficult topography or within existing infrastructural facilities where access to the bridge is limited. The use of the MSS construction method is widespread and is a well-accepted technology.

Since 2011, MSS has been used for demolition of bridges within areas deserving protection or where access is poor, predominantly as an alternative to explosives.

12.2. Application range

This construction technology can be used for different deck cross sections (refer to Section III, Chapter 15). Typical bridge deck cross sections are T-beams, TT-beams, trough sections and single- or multiple-cell box girders. Box girders are usually produced in two casting phases. After partial pre-stressing of the trough, the deck slab is poured. The trough takes some of the deck slab's fresh concrete load. The most efficient use of MSS can be achieved by pouring the entire cross section in one pour, which, however, requires a heavier MSS.

Table 12.1 indicates common limits of optimal economic use of this construction method.

For continuous beam bridges, the construction joint is mostly placed at L/5 of the span length (specifically at the zero-moment point for dead-load bending). Spans are either cast with the MSS being suspended from the cantilever of the previously constructed span or, in special cases – such as simply supported spans – supported only on piers. Supporting the MSS for continuous bridge decks at the cantilever joint is easier for building the permanent works as designed because there is no need for jacking at the rear pier support during concreting to achieve the locked-in bending stresses. Furthermore, loading and deflection of the MSS can be reduced, which helps to optimise the weight of the MSS.

12.3. General arrangement of the BDE

Movable scaffolding systems are divided into two groups according to the position of the main girders in relation to the bridge deck: overhead (type 1) and underslung systems (type 2). As a subtype of underslung systems, 'alongside systems' (type 2B) are used in locations where available space under

Table 12.1 Application criteria for MSS

	Typical application	Technically possible
Number of spans	>8	>1
Span length	25 m–70 m; constant	90; variable
Ratio curvature radius/span length	>10	5
Cross fall	0%–5%; variable	No limit; variable
Gradient	0%–5%; construction direction uphill	No limit; construction direction downhill
Cross section with 30 m spans	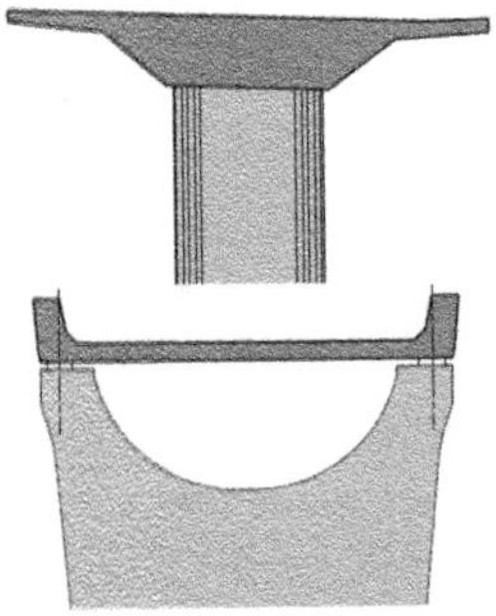	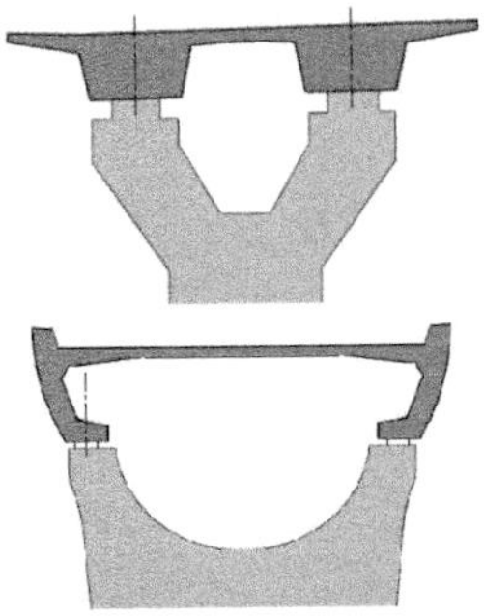
Cross section with 45 m spans	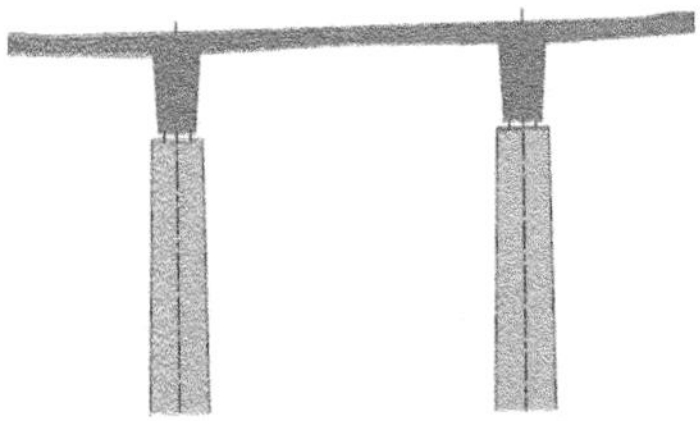	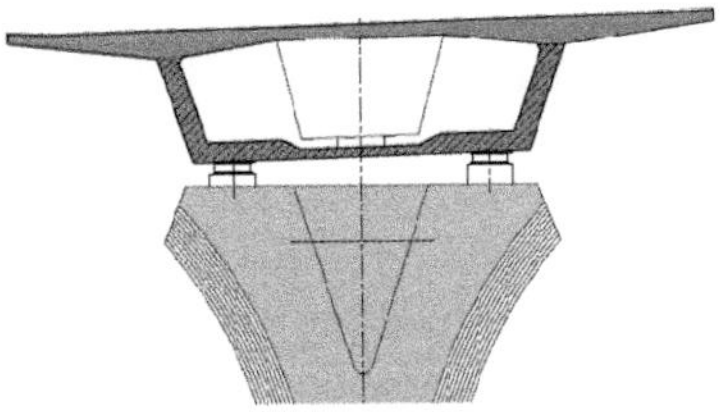
Cross section with 60 m spans	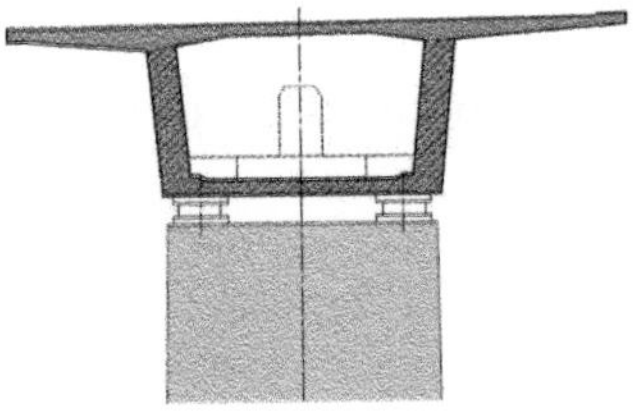	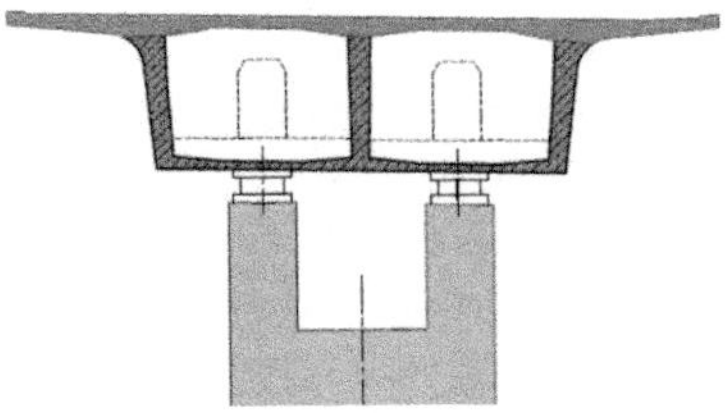
Topography	No limit	
Special load cases for bridge designer	Launching (longitudinal forces on piers) Wind (transverse forces from MSS into pier) Cantilever suspension loads in pouring stage	
Mobilisation: from signature of contract to first pouring (excl. transport time)	4 months for simple bridges with small spans	12 months for complex bridges with long spans

Figure 12.1 Types of movable scaffolding systems (formwork and access provisions not shown) (thyssenkrupp Infrastructure, saul ingenieure)

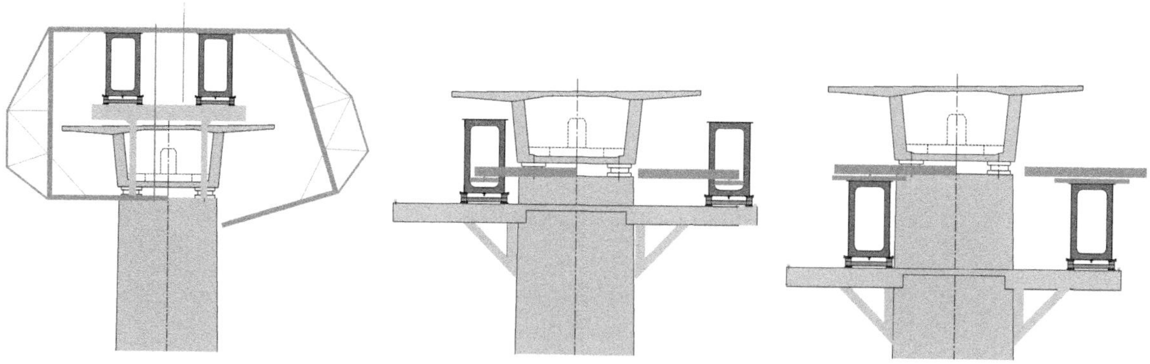

Type 1: Overhead | Type 2A: Underslung (typical) | Type 2B: Underslung (alongside)

Formwork frames/cross beams (for support of the external formwork)
Anchor bars
Main girders
Support and moving devices
Pier support (front support of MSS)

the bridge deck is limited. For details, refer to Figure 12.1. Indications of typical weights of parts of the MSS, which are self-launched ('moving weights'), are shown in Table 12.2.

Table 12.2 Indications of weights of parts of a typical MSS which are self-launched (excluding supports, including formwork)

	Deck weight	Overhead (type 1)	Underslung (type 2)
30 m T-section single pour	180 kN/m	3000 kN	2200 kN
45 m TT-section single pour	285 kN/m	7500 kN	5700 kN
45 m box section two pouring phases	260 kN/m	7000 kN	4900 kN
65 m box section two pouring phases	310 kN/m	9000 kN	7000 kN

Note: weights in Table 12.2 are for reference only and should not be used to determine the weight of an MSS for specific projects

The individual components of the two types of MSS are defined in Table 12.3.

A comparison of the various types of MSS is given in Table 12.4.

12.3.1 Description of MSS elements

The following items are now described in detail.

- Formwork detailing
- Formwork cross beams
- Precamber adjustment
- Main girders
- Support and moving devices

Table 12.3 Definition of MSS main components

		Overhead (type 1)	Underslung (type 2)
1	Formwork (not shown in Figure 12.1)	Steel or timber	Steel or timber
2	Formwork cross beams	Frame with hangers, hinged or transversely movable	Fixed on main girders or transversely movable
3	Pre-cambering adjustment	By adjustment of hangers or formwork	By means of spindles or shimming plates or by adjustment of formwork
4	Main girders	Box or truss girders, numbers depend on cross-sectional shape, truss girders with underpinning and integrated compensation of deformations	
5	Support and moving devices	Hydraulic equipment for launching, transversal shifting and vertical alignment	
6	Pier support (front support of MSS)	Struts on top of piers through box-outs of bridge deck or on prefabricated pier heads	Brackets, longitudinal insert beams in piers, heavy load towers
7	Rear support (not shown in Figure 12.1)	Rear support of main girders on bridge deck	Suspension of main girders through bridge deck

- Pier supports
- MSS rear support

12.3.1.1 Formwork detailing

The choice of formwork construction has an impact on the weight of the MSS and on the operation method. Whereas the external formwork is mounted onto the MSS's formwork cross beams, the internal formwork is made of independent modules with their own moving and handling system.

External formwork: The external formwork is spliced longitudinally along the bridge's centre line, so that the soffit form can be moved transversely or rotated to clear the pier during launching of the MSS. To enable launching of an MSS at two parallel bridge superstructures with small distance between the decks, the side-wing formwork of a type 2 MSS can be hinged (see Figure 12.2) or the whole section can be lowered if the deck's webs are tapered. In the case of a type 1 MSS, the complete soffit form will be rotated at the free side of the bridge deck.

Internal formwork: While bridge decks with a solid cross section (T-section) are poured in one stage, box girders are mainly poured in two stages to minimise the weight of the MSS. For the production of the trough, wall formwork panels are used as internal formwork for the webs, which are removed after the first pour and either stored on the deck of the previous span or on the ground until they are reused for the next span.

The inner deck slab's formwork can be fixed onto tables, which can be pulled into the trough from the previous span if diaphragms (see Figure 12.15) or, for overhead systems, the temporary front supports of the MSS do not obstruct that movement. Casting as much of the pier diaphragm as is structurally feasible after departure of the MSS is preferable for efficient production. If it is not possible to provide

Table 12.4 Comparison of MSS types: components and related project aspects

	Overhead (type 1)	Underslung (type 2A)	Alongside (type 2B)
Weight of system on superstructure	Heavy	Light	
Introduction of launching load	Above deck	Low point of load introduction	
Wind loads	Large area	Medium area	Small area
Weather protection	Can be easily integrated (but higher wind loads)	No	No
Application at narrow curvature radius (ratio radius: span length <10)	Recommended	Applicable	Difficult
Application at wide decks with wide single piers	Light formwork cross beams with intermediate suspension	Heavy formwork cross beams, large transverse shifting in launching stage	
Assembling time	Many components	Smaller number of components; type 2B is more complex	
Required vertical clearance under bridge	If small, transversally movable formwork If high, hinged formwork	Medium	Small
Handling in end spans	Easy Only formwork and cross beams need to be changed	Time-consuming Remove launching noses or use stationary shoring systems (if abutments are not built in stages)	
Application for parallel superstructures	Formwork's frames may collide with adjacent deck, especially at curves in plan	Rotation of side formwork, additional transverse moving of formwork and cross beams	
First assembling and dismantling	80% behind abutment	Shortened MSS; to be completed after first span if it cannot pass abutment	
Follow-up work: closing of superstructure holes and box outs used for:	Suspension of formwork's cross beams, support legs	Insertion of needle beams through pier shafts or fixity of pier brackets or towers	
Additional works: provision of auxiliary supports	Behind abutments	Under end spans	
Difficult access to piers	Self-relocated supports, to be shifted during launching		
High gradient	Self-locking launching devices, direction of construction: uphill		
Variable span length	Mechanically or hydraulically adjustable pre-cambering provisions; removable internal formwork modules		

Table 12.4 Continued

	Overhead (type 1)	Underslung (type 2A)	Alongside (type 2B)
Variable cross fall	Rotation of formwork's cross beams around centre line of bottom slab	Rotation of cross beams about a main girder	Rotation of formwork cross beams around centre line of bottom slab
Technical level of operation	High	Range of complexity: Low: with fixed formwork cross beams on pier towers High: movable formwork cross beams and self-relocated brackets	
Average number of launching crew per shift for 45 m spans (regular launching cycle)	6	4 for box girder section with simple type 2	6 for TT-section with 3 main girders

a large enough opening through the pier diaphragm or the temporary front support for the launching of the internal deck slab form, this form has to be made up of light panels and props, which can be handled manually.

Single-stage pouring of a box-girder section requires as inner form a heavy tunnel-type formwork. This tunnel formwork is usually a complex piece of machinery as it has to be foldable and launchable. The internal geometry of the box girder has to be fine-tuned to minimise obstacles for launching and to keep the folding mechanism as simple as possible.

Steel formwork is stronger and heavier than timber formwork. It is more appropriate for multiple uses without geometrical changes (long viaducts). Steel formwork can contribute significantly to the global rigidity of the MSS. Modular timber formwork can be adapted relatively easily to geometrical variations. A great advantage of wooden formwork is that the majority of the parts are rentable. For typical weights of formwork refer to Table 12.5.

Table 12.5 Average weight of formwork types related to the area of bridge deck for a typical 45 m span bridge. Weights include skin, runners and walers. Weights are for reference only.

	TT-type cross section	Single-cell box girder section	
		Trough pouring	Deck pouring
Steel	1.6 kN/m^2	1.4 kN/m^2	1.4 kN/m^2
Modular timber	1.4 kN/m^2	1.2 kN/m^2	1.4 kN/m^2
Tailor-made timber	1.7 kN/m^2	1.6 kN/m^2	1.6 kN/m^2

12.3.1.2 Formwork cross beams

Formwork cross beams transfer loads from the concrete and the formwork into the main girders of the MSS. They have to be split in order to let the MSS pass the pier during the launch stage.

Working in tight curvature in plan may require adjustments within the formwork to create a polygonally curved alignment. This adjustment is done on the cross beams by side shifting sections of the formwork, sliding on the cross beams.

The longitudinal distance between cross beams is kept within 1.5 to 4 m, controlled by the capacity of the cross beams and the formwork system and the weight of the bridge deck. The length of the cross beams depends on the formwork system and on the distance from any adjacent superstructure.

Overhead systems: Formwork cross beams are usually suspended from the main girders for the concreting. These suspension elements are removed before the launching operation begins. Cross beams are mostly rotated for launching stage, rarely shifted sideways. The point of rotation defines the required clearance under and next to the bridge deck during the launching stage.

Weights: 0.9 kN/m^2 of bridge deck for a simple frame construction with mechanical opening; 4 kN/m^2 of bridge deck for heavy cross beams with complex hydraulic opening mechanism, inclusive of walkways.

Formwork's cross beams are designed as frames enveloping both main girders and the bridge deck. For the construction of parallel and curved bridges, these frames need to be detailed carefully in order to avoid collisions during the launching. Safe access to install and remove the lower suspension connections and to open and close the cross beams' splices has to be part of the MSS design.

Underslung and alongside systems:

The construction of formwork cross beams is simpler and lighter. They can be shifted sideways or fixed on the main girder with or without hinges. For an example refer to Figure 12.2. The use of hinges reduces the need to shift the main girders sideways for the launch. The ability to move the cross beams sideways also reduces the need to shift the main girders. This can be advantageous for bridges with tight radius in plan and/or when there are other structures next to the bridge being built.

Weight: 0.65 kN/m^2 of bridge deck for a light formwork cross beam construction without centring beams that distribute loads equally on the webs of the main girders; 3 kN/m^2 of bridge deck for heavy cross beams that can shift sideways, including walkways.

12.3.1.3 Pre-cambering adjustment

At bridges with construction joints at zero-point of bending moment, the back end of the concrete will be poured last. Virtually all the deflections in the main girder occur before the concrete is cast against the end face of the previous pour.

There are different ways to compensate for deformations of the MSS during concreting (refer to Figure 12.3). The basic solution is to pre-camber the formwork. This pre-camber can be set to a fixed value when installing the formwork by using shimming plates under the formwork cross beams (at overhead systems under the frames). Alternatively, when spans are variable, normally jacks (mechanical or even hydraulic; with a weight of about 50 kg to 150 kg per main girder and cross beam axis) are provided, which allow the pre-camber to be adjusted prior to any pour.

Figure 12.2 Hinged formwork's cross beams, handled with hydraulic cylinders, on underslung MSS (thyssenkrupp Infrastructure, saul ingenieure)

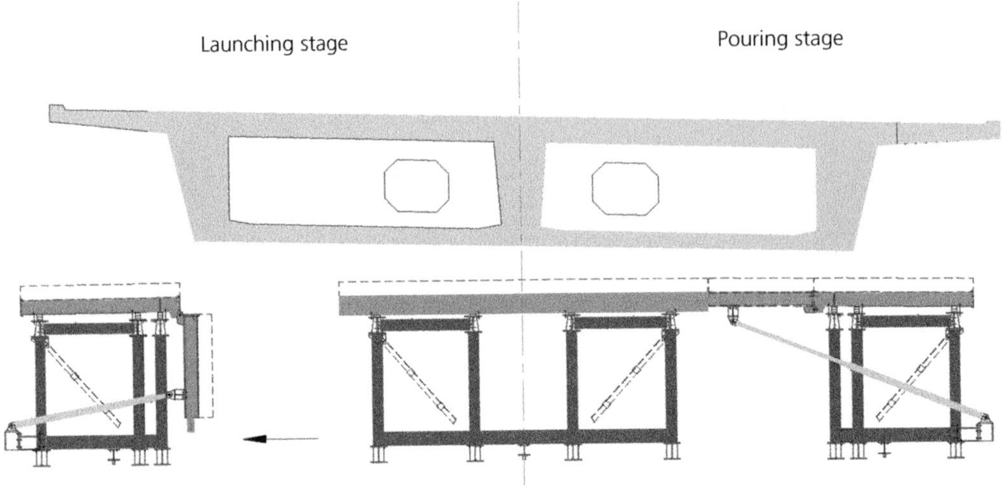

The area between and around the formwork cross beams has to be accessed for adjustment of the pre-cambering, if required, and for splicing/un-splicing and moving the formwork's cross beams sideways at an underslung MSS. Walkways have to be detailed to provide access to these places both from the pier supports and from the tip of the cantilever of the previously cast bridge span.

It is difficult to adjust the pre-camber for hinged formwork systems.

Another option is the use of a computer-controlled progressive underpinning system (a patent-protected solution), where deflection of the main girders is compensated during pouring (refer to Figure 12.4). This reduces the need for pre-cambering against weight-induced deformations. It allows comparatively lighter main girders while the complexity of the MSS structure is significantly raised by the underpinning system and its associated components. The need for vertical clearance either under or over the bridge deck will be much bigger than for conventional MSS. Commissioning time and cost of

Figure 12.3 Different solutions for pre-cambering adjustment; access platforms not shown (thyssenkrupp Infrastructure, saul ingenieure)

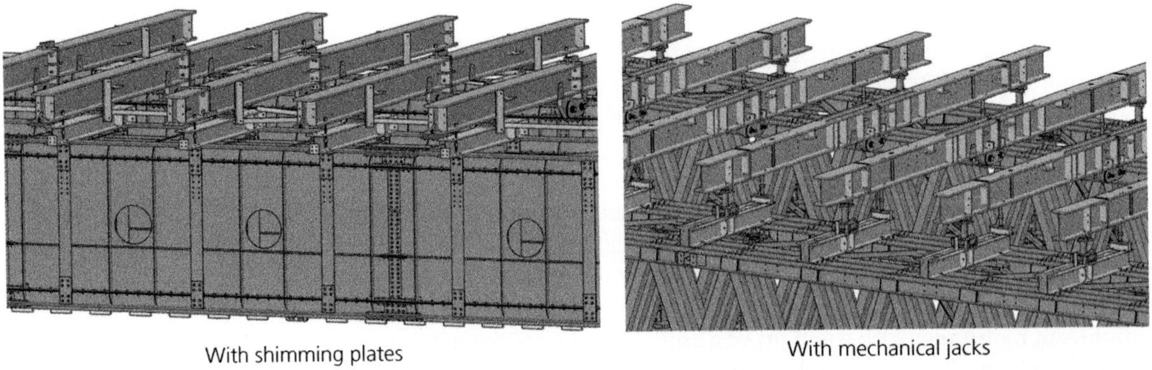

Figure 12.4 Comparison of conventional and underpinned MSS (patented solution) (thyssenkrupp Infrastructure, saul ingenieure)

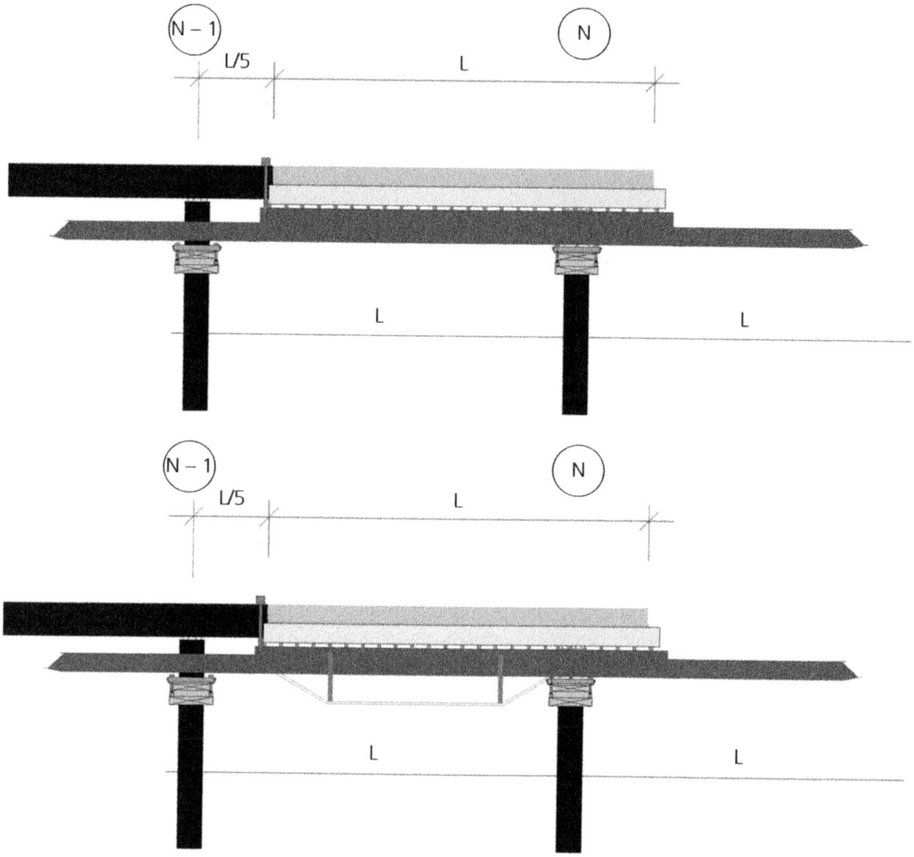

equipment are high. If the span length varies during the construction of a deck, the underpinning system needs to be adapted. This requires the changing of several major parts of the MSS. However there are benefits. Spans of up to 90m are possible. No post-tensioning of the trough section is required for two-stage pouring. It is possible to monitor the development of the bridge deck's deformation during fabrication.

More information about the impact of casting a box girder in one or two pours on the design and detailing of an MSS can be found in Section III, Chapter 16.

Soffit formwork may have to be pre-cambered transversely, as well. This will depend on the width of the bridge and the cross beam's support system during pouring.

12.3.1.4 Main girders

Main girders transfer the concrete load to the supports. Their length is defined by the maximum pour length. Usually MSS have a heavier centre part, which takes the concrete weight, and front and rear launching noses, which are only used for the launching. The total length of the main girders is generally slightly longer than the two longest successive spans.

Figure 12.5 Different types of main girders (thyssenkrupp Infrastructure, saul ingenieure)

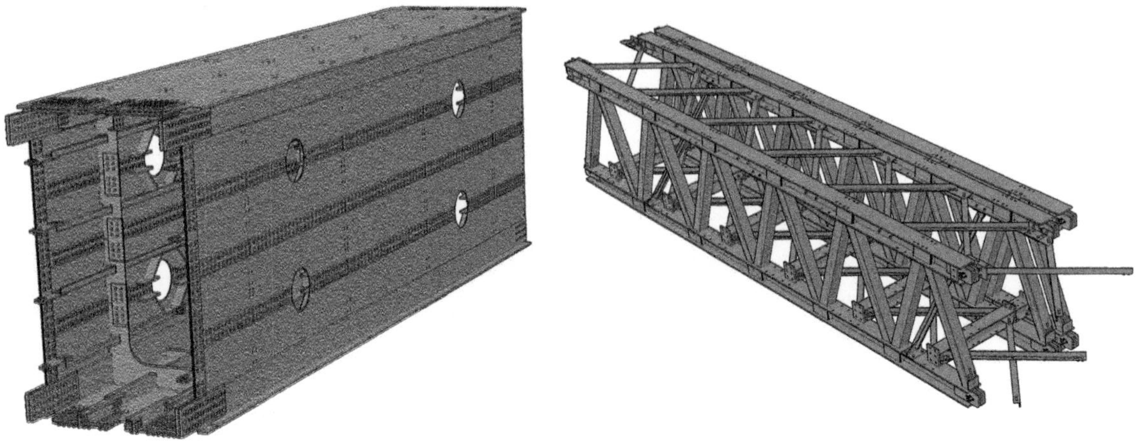

Main girders can either be lattice truss girders or box girders (refer to Figure 12.5). Truss girder elements are often connected by pins, which enable a quick assembly on site, while box girders are mostly bolted. The deflection of the main girders due to self-weight during launching has to be calculated. Provisions (sliding noses, hydraulic rams) have to be engineered which allow the recovery of the cantilever deflection when approaching the next pier or leaving the last one.

Support points of truss girders should be carefully observed. If not verified by calculation, truss girders should always be supported at truss nodes. Webs of box girders may need additional stiffeners to resist web buckling if the vertical stress from bearing rollers or pads is high.

The transverse stability of the main girders during launch has to be considered. The formwork needs to be opened to enable transverse clearance for passing the pier in launching stage. Very wide piers and/or a lack of torsional stiffness of main girders may require stabilising counterweights prior to opening the formwork's cross beams for the launching stage. Counterweights have been used to enhance stability during launch, although it is a slow operation if the weights have to be added for each launch. The extra load affects the launch and support systems and is therefore not recommended.

Special consideration must also be given to the situation where wind acts on the 'front' end of the web's formwork as the MSS is launched into the next span while the rear portion is still sheltered by the previous span. Even if 'launch wind' speed is restricted for operational purposes, the torsional loading from the girder onto the supporting structure can cause difficulty, particularly in the mechanical supporting elements.

Main girders of MSS are often tailor-made for a project. However, some companies offer modular and rentable solutions (for details refer to Table 12.6). These can be:

- transversely modular truss girders where load capacity can be raised by connecting more trusses to the basic truss box, consisting of two vertical truss planes with a steel bracing 'X', or
- vertically modular box girders.

Table 12.6 Examples of two types of modular and rentable main girders

	Element height: m	Max. V_{Rd}: kN	Max. M_{Rd}: kNm	Weight: kN/m
Truss girder simple box IXI	2.260	2700	14 400	9
Truss girder with 3 trusses IIXI	2.260	4050	21 600	13
Truss girder with 4 trusses IIIXI or IIXII	2.260	5400	26 400	17
Box girder	2.300	6480	57 800	22
Box girder	3.200	10 500	81 800	27
Box girder	4.100	14 530	105 700	30
Box girder	5.000	18 550	129 700	32

12.3.1.5 Support and moving devices

Movements of the MSS are mainly carried out with three moving devices. Figure 12.6 shows an example of support and moving devices, which are often compact units that include:

- hydraulic moving devices for launching operations
- hydraulic jacks for vertical loads (recommended with locknut)
- hydraulic cylinders or winches for transverse shifting of the MSS.

MSS are mostly pushed forward. They are either moved on rollers, on PTFE-based sliding pads or metallic sliding planes. For values of friction coefficients, refer to Section III, Chapter 17. Depending on which method is selected, the horizontal moving force during the launching operation, which is calculated by multiplying the weight of the moved part of the MSS with the sum of the applicable friction coefficient and the longitudinal inclination in per cent, varies significantly. However, a reduction in friction does not come free of charge. The additional costs may be compensated by the smaller

Figure 12.6 Moving devices at MSS supports (thyssenkrupp Infrastructure, saul ingenieure)

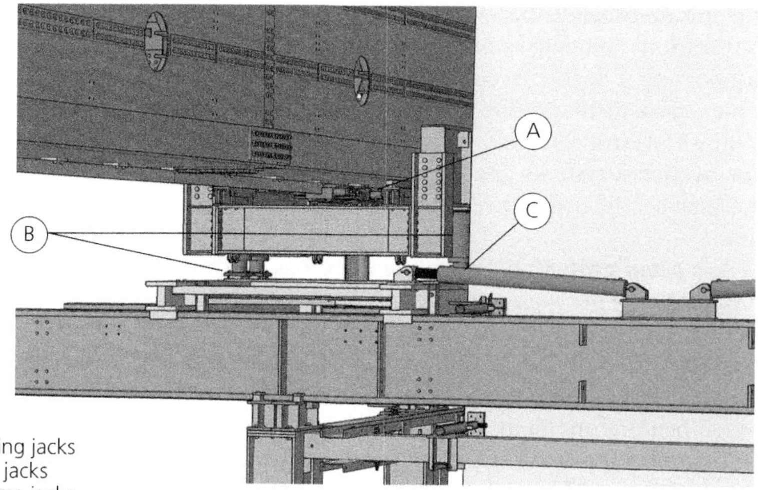

A. Launching jacks
B. Vertical jacks
C. Transverse jacks

investment in launching equipment and, potentially, by savings in the reinforcement of the piers which can have smaller resistance against the launching forces.

Usually, MSS are launched from the front pier of the previous span. Forces from launching are introduced in a structure already stabilised by the weight of the bridge deck. From this launch position there is no need to change the position of the launching devices during the launch process and the launch forces are kept 'internal' as far as possible.

It is vitally important to keep the longitudinal restraint location and method clearly defined. Main girders on roller supports will probably need to be held in place while the launch system is retracted for resetting because the bearing friction can be lower than the longitudinal slope of the bridge. In any case, an MSS should never be left unrestrained on its sliding/rolling supports.

MSS often use load-balancing support systems to equally distribute vertical loads on the piers and vertical reactions into the MSS in the longitudinal direction. The weight of these balancing support systems varies between 10 to 150 kN per main girder support, depending on the length of the pier support cross beams and the magnitude of loads which have to be transferred. For simple systems, hydraulic jacks for vertical alignment can be positioned under the pier support cross beams.

12.3.1.6 Pier supports

All loads will be transferred to the piers and foundations through the pier supports. In modern systems, horizontal loads will be introduced into the pier walls by mechanical jacks. Usually three sets of pier supports are supplied. While two sets are located under the MSS, the third one is shifted. With self-launching supports, only two sets are required. Various types of pier supports exist (refer to Figure 12.7).

- Pier Towers consist of a pair of cross beams placed in the case of an underslung MSS on vertical props standing on the pier's pile cap or in the case of an overhead MSS on the pier itself. Towers are suitable for underslung systems if pier heights are short. Support on the pile cap is the favoured solution for an MSS on demolition projects as the condition of the old piers and the increased loads for this task are an unacceptable risk compared to new construction. In contrast to stationary shoring systems, these towers have to be designed also for the launching forces. Appropriate bracings between the towers and the pier have to be provided. To avoid on-site modifications, bracing connections must be carefully planned and detailed, particularly for piers supporting a deck curved in plan. For overhead systems, pier towers can be placed on top of the pier, next to the bridge bearings. The bridge deck will be cast around these towers. Vertical nailing to restrain launch forces has to be provided. As another option, the pier supports for overhead systems can be positioned on pier heads, cast ahead of the arrival of the MSS. These pier heads, if they are sitting on bearings, need to be temporarily stabilised during construction.
  Weight of cross beam grid: 80 kN (light, small cross section with slim piers) to 450 kN.
  Weight of tower shafts per metre of height: 10 kN (4 legs of 2000 kN capacity, light MSS) to 24 kN (4 legs of 6000 kN capacity, heavy MSS).
  Brackets consist of cross beams cantilevering on either side of the pier with vertical reactions either introduced in box-outs in the pier walls or by vertical props into the pier foundations. The transverse bending moment is taken by a horizontal force couple, which is balanced through a horizontal frame or HT bars around the pier at the top and is pressing against the pier shaft at the bottom. The weight of bracket elements can be higher when compared with pier tower elements. In order to reduce the weight of brackets, the range of the main girder's

transverse movement is minimised by provision of transverse movement of the formwork's cross beams.
Weight per bracket without vertical props: 100 kN to 300 kN.

- Insert beams (needle beams): Cross beams are supported by longitudinal beams which are positioned through openings in the pier shaft. For safe operation of this system, attention should be paid to the interface between steel and concrete at these openings as loads are high. A safe assembling and dismantling procedure is also important. This is an efficient method for wide, high piers.
  Approximate size of each opening: width 0.3 m to 0.45 m × height 1 m to 1.30 m.
  Weight: 60 kN (two pieces of small insert beams) to 120 kN (two pieces of strong insert beams plus load distribution plates).
- A-type: If the cross-sectional shape of piers permits the positioning of a suspension beam between the bridge bearings, this type of support can be an alternative for high piers. Apart from the suspension beams, assembling and dismantling of the support structure can be carried out on the ground. The pier support can be lifted safely as one unit. The dismantling procedure of the suspension beams after fabrication of the bridge deck has to be examined carefully.
  Weight: 100 kN to 200 kN.
- Self-relocated supports: Support systems of overhead systems and brackets for underslung or alongside systems can be designed and detailed to be self-relocated. Supports are shifted during launching operation, a relevant load case. The advantage of self-relocated supports is a ground-independent construction of the bridge deck. However, they require more time for the launching operation, a higher level of mechanisation of the MSS and are more demanding for the

Figure 12.7 Pier support types (thyssenkrupp Infrastructure, saul ingenieure)

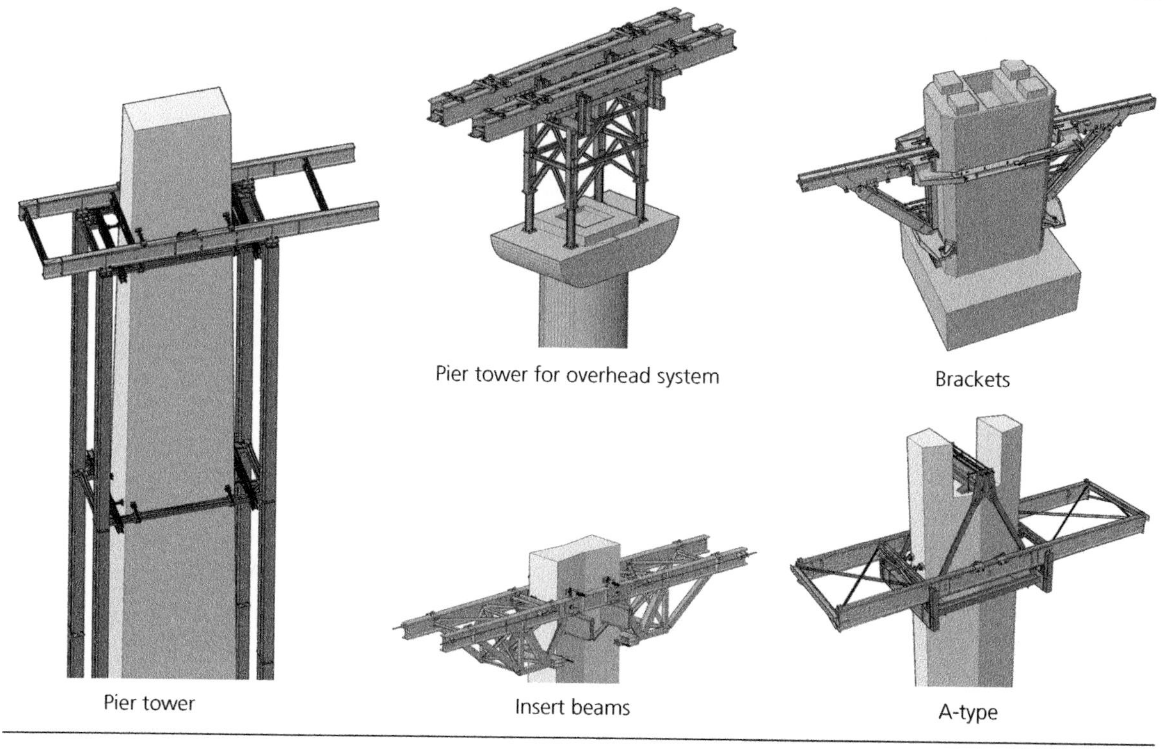

operating staff compared to a conventional MSS. Another disadvantage is that vehicular access over the completed deck to the cantilever support is not possible until all supports have been launched to the position of the next pouring stage. Many systems of self-relocated brackets are not suitable for curved bridges. Refer to Figure 12.10 for the kinematics of self-relocated brackets.

- Intermediate temporary supports: Intermediate temporary supports are steel towers with a similar shape as pier supports. They have to be stabilised against launching and wind loads. Intermediate towers require additional foundations. They may need to be used when there are irregularities in the span length along the bridge deck.

12.3.1.7 MSS rear support

The MSS rear support can be located on the rear cantilever or centred over the pier. The following explanations refer to an MSS rear support at the tip of the cantilever.

The MSS rear support grants a smooth transition between the existing bridge deck and the fresh concrete because the support is close to the construction joint and does not allow visible deformations of the MSS at this point. The load case 'construction stage', whereby the MSS is loading the previously cast deck at its cantilevering tip, is relevant for the design of the bridge. This case may potentially even be critical for the stability when during the casting of the second span of a bridge deck the MSS is supported on the cantilever of a short end span. The use of this support type creates a shortening of span length for the pouring stage of the MSS main girders, which helps to make the MSS lighter.

The cantilever support consists of hydraulic jacks, a cross beam which is supported over the webs of the bridge deck and, in the case of underslung systems, of tension bars which are anchored to the main girders (refer to Figure 12.8). With underslung MSS with box-shaped main girders, the position of the

Figure 12.8 Cantilever suspension at underslung MSS (thyssenkrupp Infrastructure, saul ingenieure)

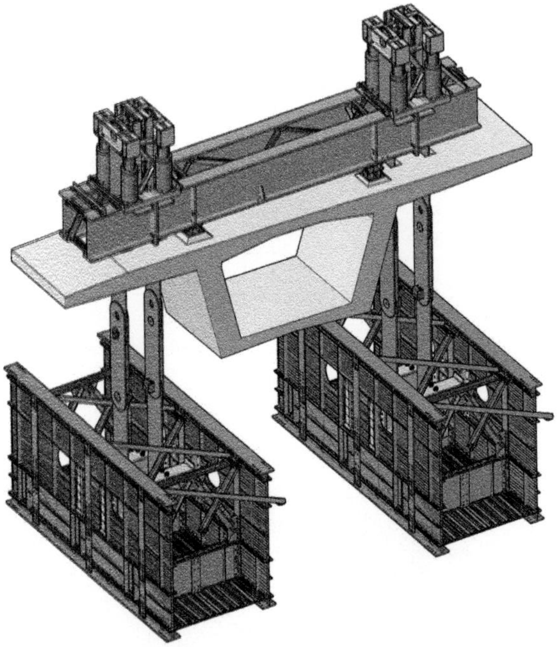

suspension is often fixed within the main girders, which poses some design and detailing challenges for variable span lengths.

Weight: 100 kN (light bridge deck with two main girders); 300 kN (wide and heavy bridge deck with three main girders).

12.4. Kinematics

Kinematics for three different type 2A MSS are presented. The launching procedure for all of them is described for the same bridge type – a single-cell or twin-cell box girder bridge with 50 m spans. One launching operation is described starting with the launching of the MSS and ending with casting the next span.

Figure 12.9 shows the kinematics for an MSS with crane-shifted supports.

Figure 12.9 Kinematics of an underslung MSS with crane shifted supports (thyssenkrupp Infrastructure, saul ingenieure)

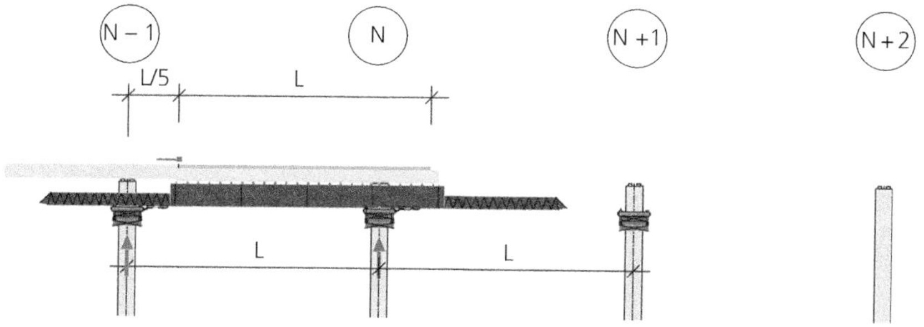

1. Lowering down MSS after prestressing of span N – 1/N: deactivation of cantilever suspension

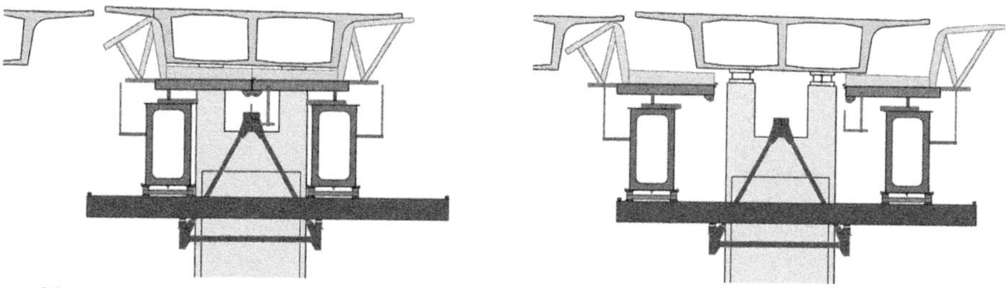

2. Opening of formwork, transverse moving of MSS (and formwork)

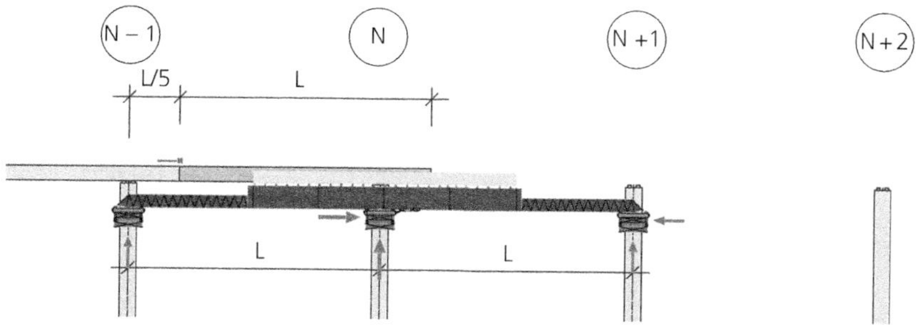

3. Longitudinal launching to next pier

Figure 12.9 Continued

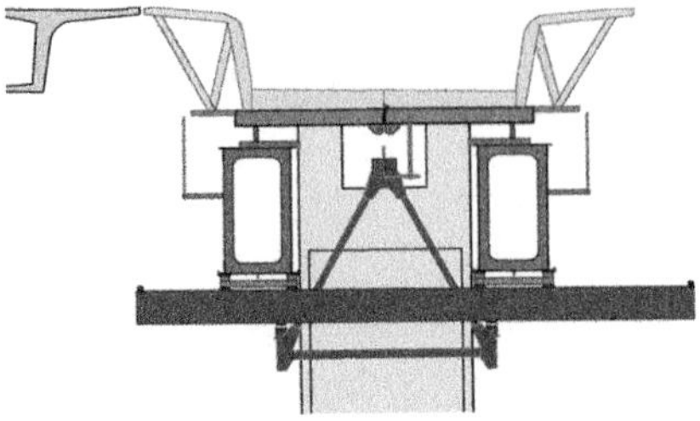

4. Closing of outer formwork, levelling and activation of cantilever suspension

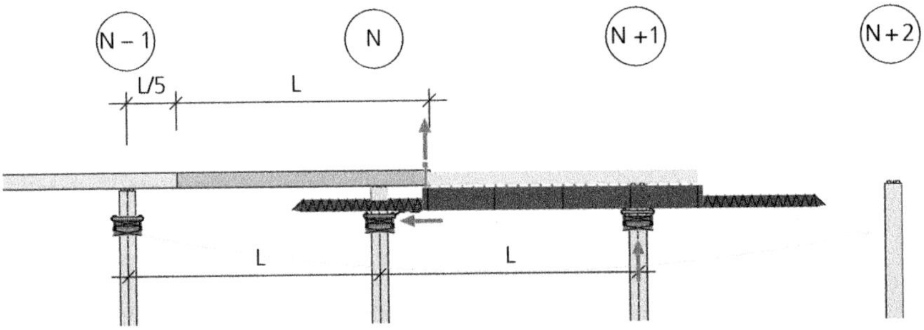

5. Placing of reinforcement and PT tendons; installation of inner web forms; first pouring phase: casting of trough; at this point the longitudinal restraint of the MSS must be installed directly at the previously cast span

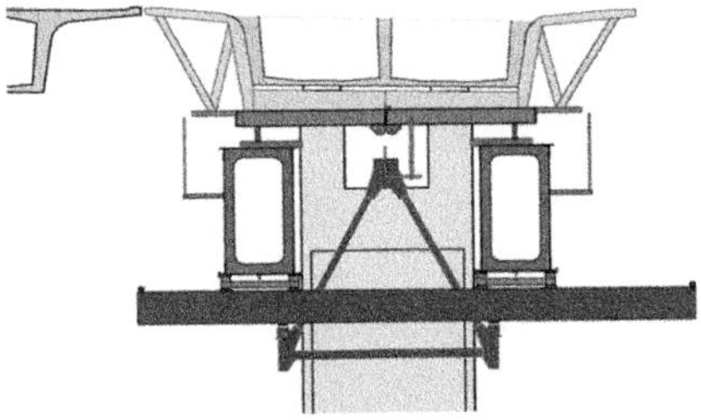

6. Concrete curing; removal of inner web forms; partial prestressing

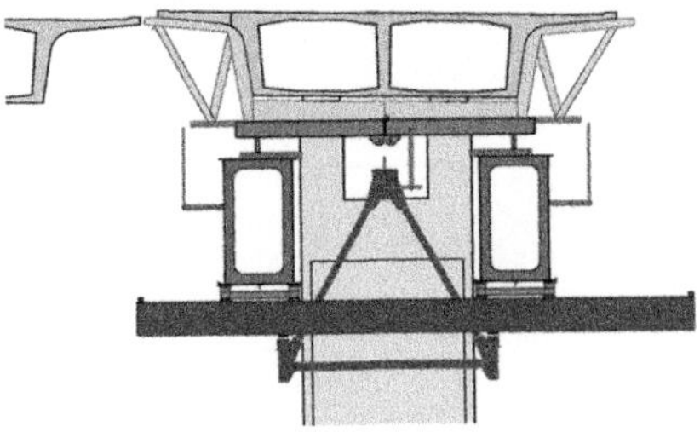

7. Installation of inner deck form; placing of deck slab reinforcement; second pouring phase: deck slab; note inner form not shown for clarity

Figure 12.10 Kinematics for an MSS with self-relocated supports (thyssenkrupp Infrastructure, saul ingenieure)

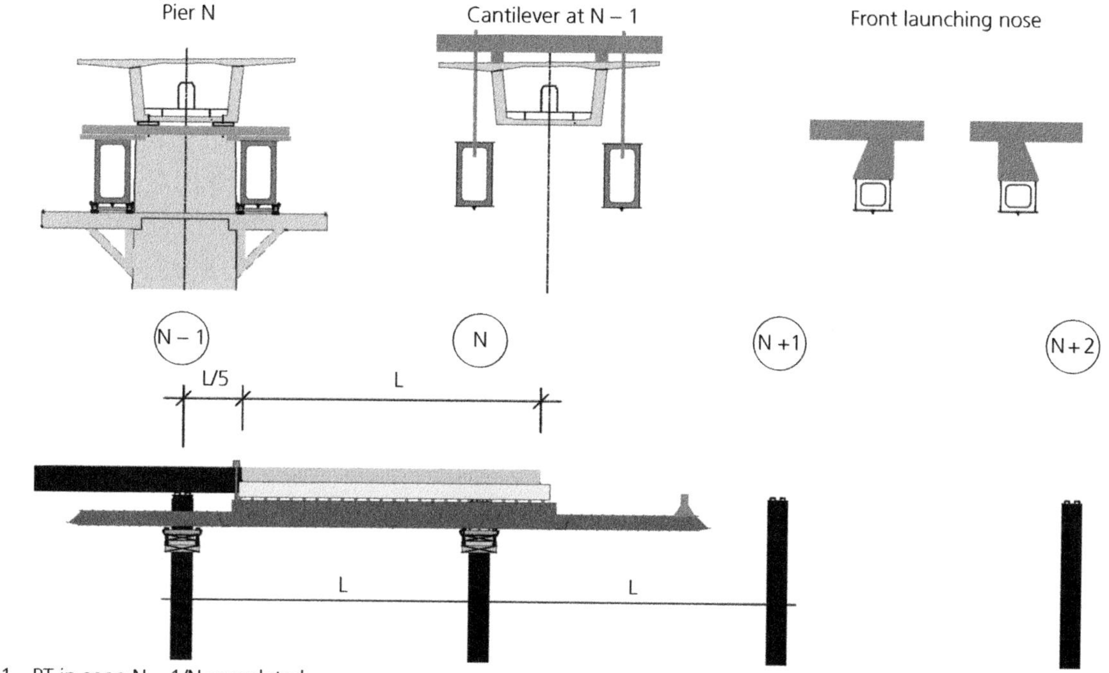

1. PT in span N – 1/N completed

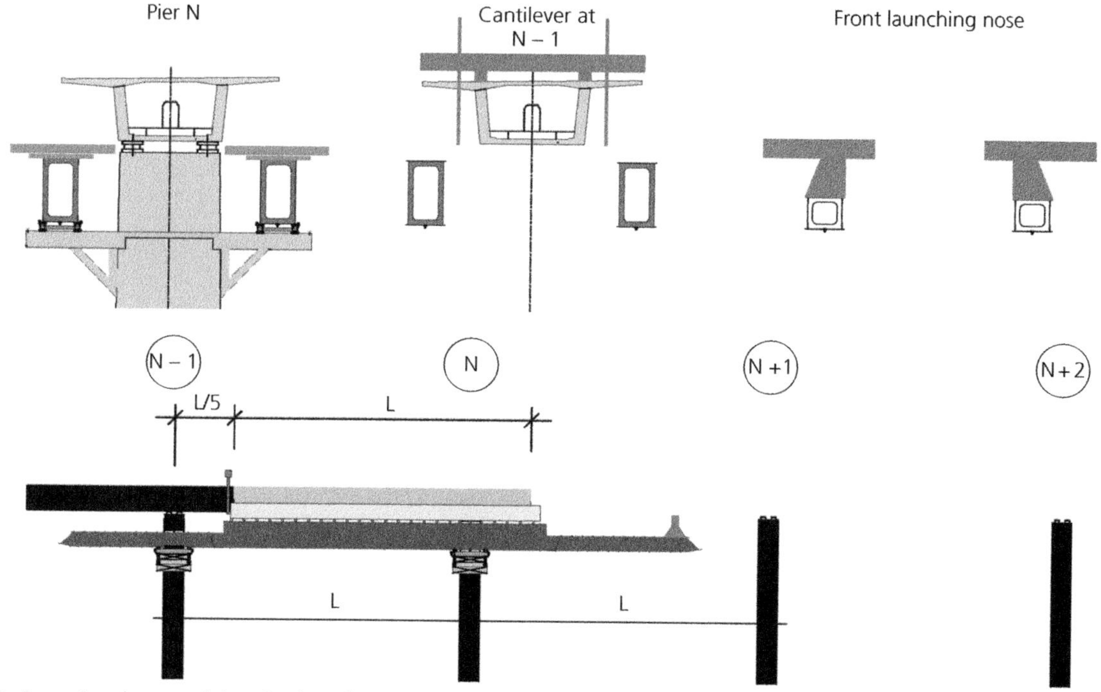

2. Lowering down and deactivation of suspension, opening of formwork, transverse shifting of formwork's cross beams and main girders

Figure 12.10 Continued

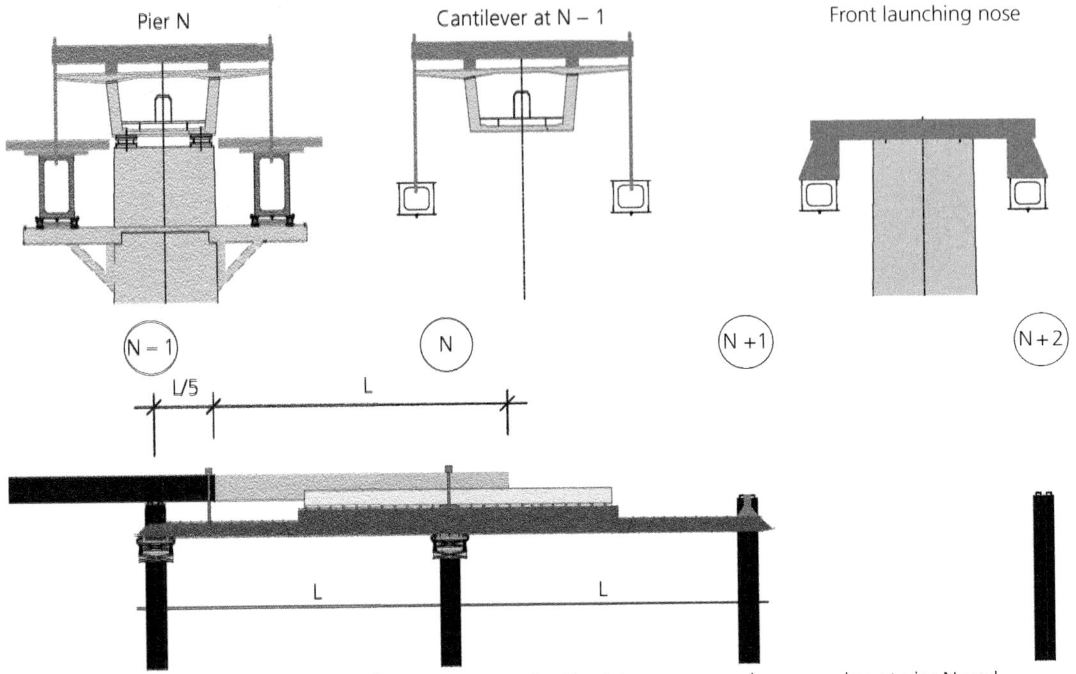

3. Launch to next pier, activate temporary front support at pier N + 1 temporary main suspension at pier N and auxiliary suspension at construction joint N – 1

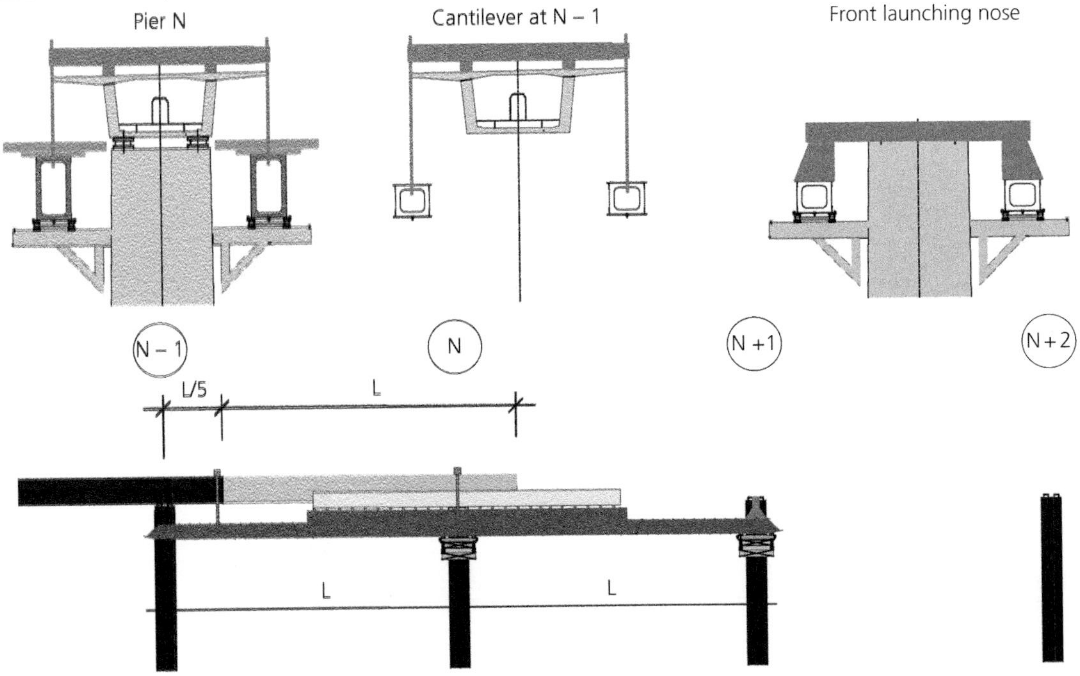

4. Shift pier brackets one after the other

Figure 12.10 Continued

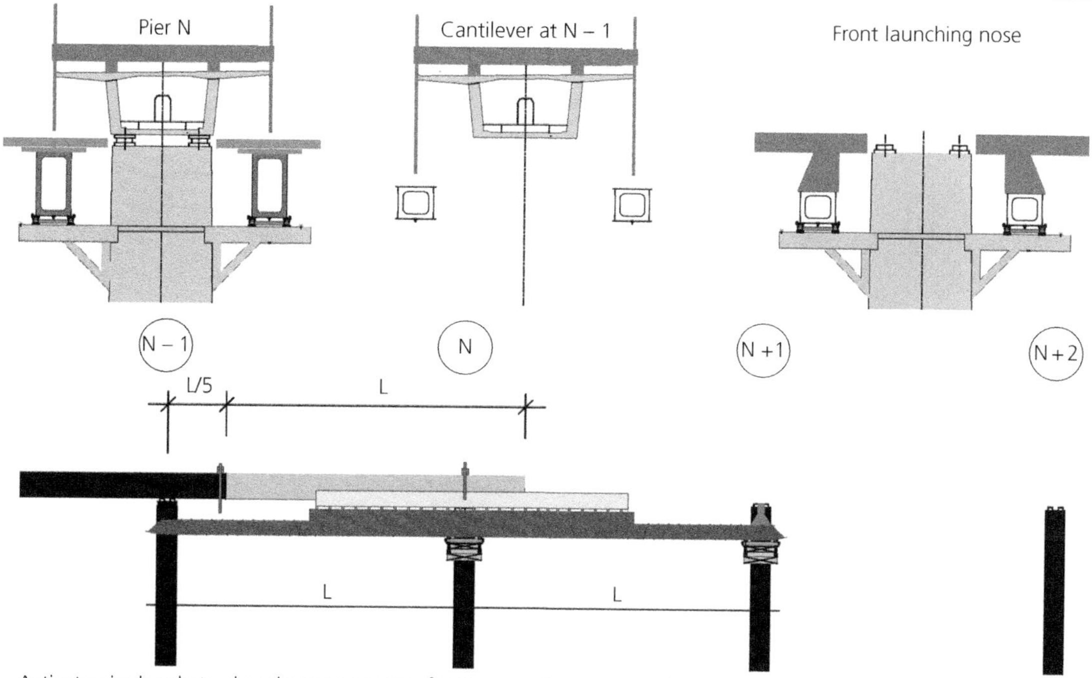

5. Activate pier brackets, deactivate temporary front support and suspensions

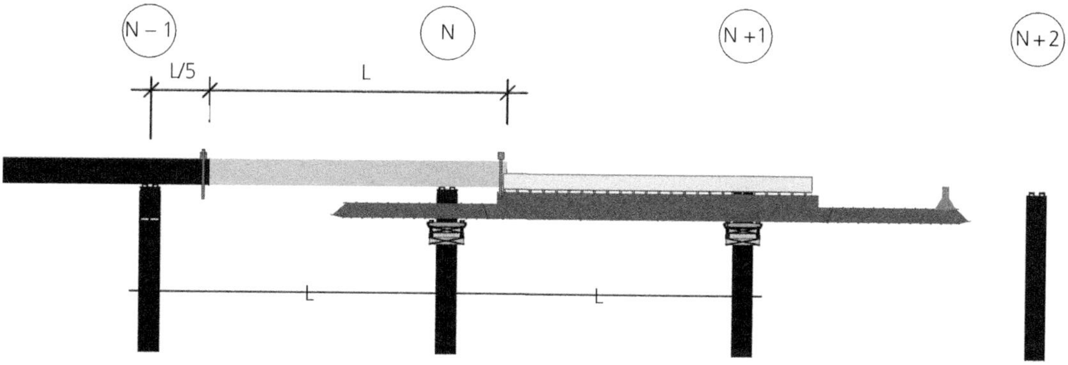

6. Launch to next pouring position, close external formwork, activation of cantilever suspension; preparation (rebar, PT, inner form) and casting, curing and stressing of first and second pour

Figure 12.10 shows the kinematics for an MSS with self-relocated supports.

Figure 12.11 shows the kinematics for a 'C-frame'-type MSS with self-relocated supports (main girders with front nose only).

Figure 12.11 Kinematics for a 'C-frame'-type MSS with self-relocated supports (thyssenkrupp Infrastructure, saul ingenieure)

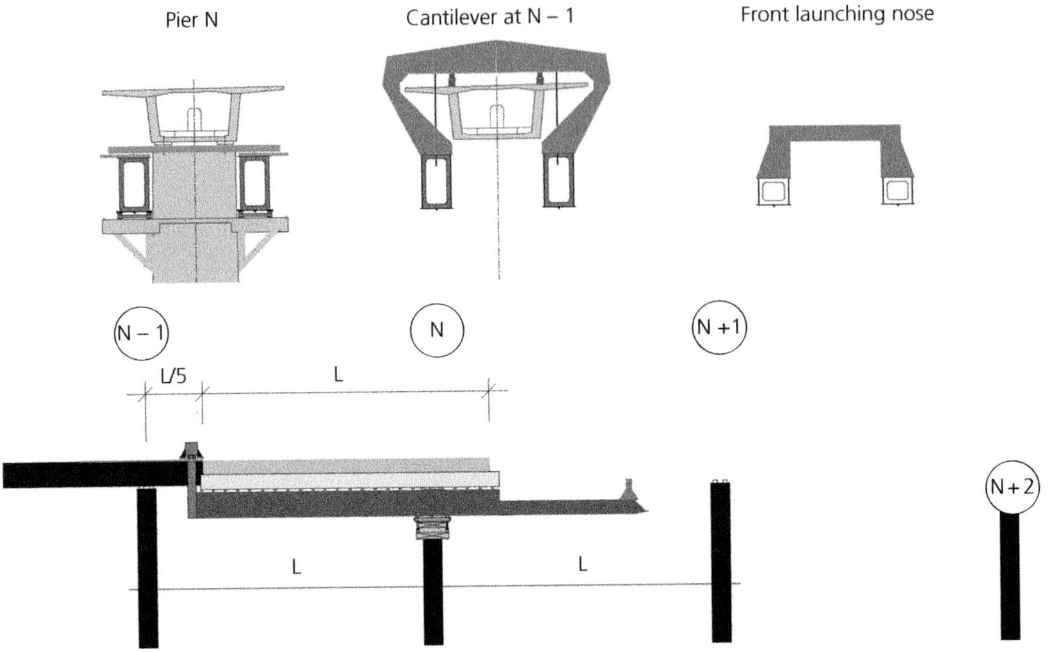

1. PT in span N – 1/N completed; installation of rails over webs of new span

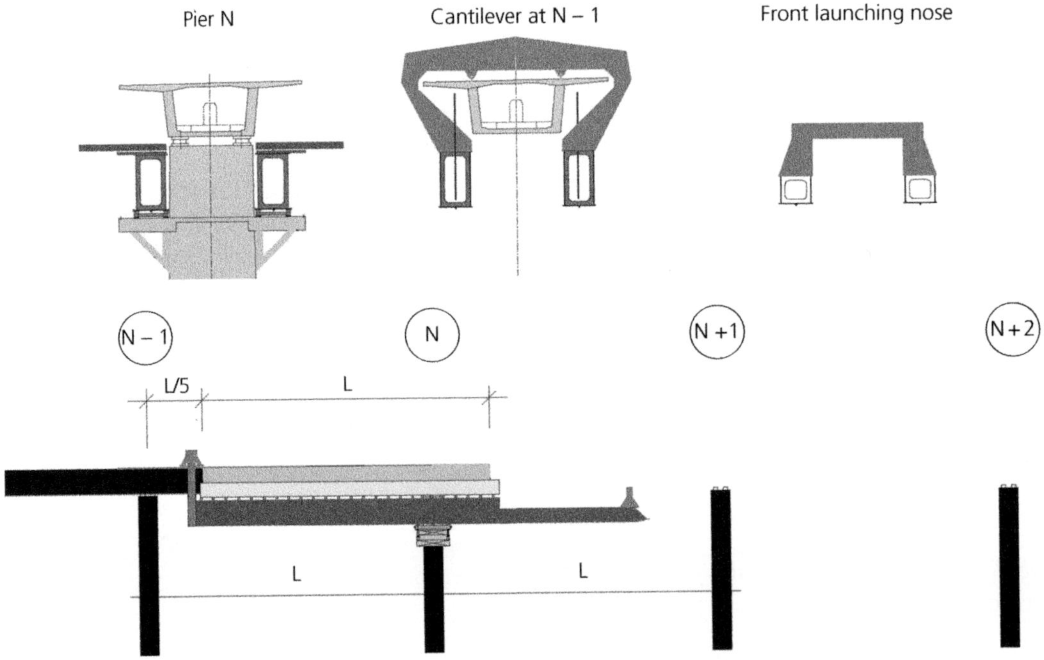

2. Lowering down, deactivation of suspension

Figure 12.11 Continued

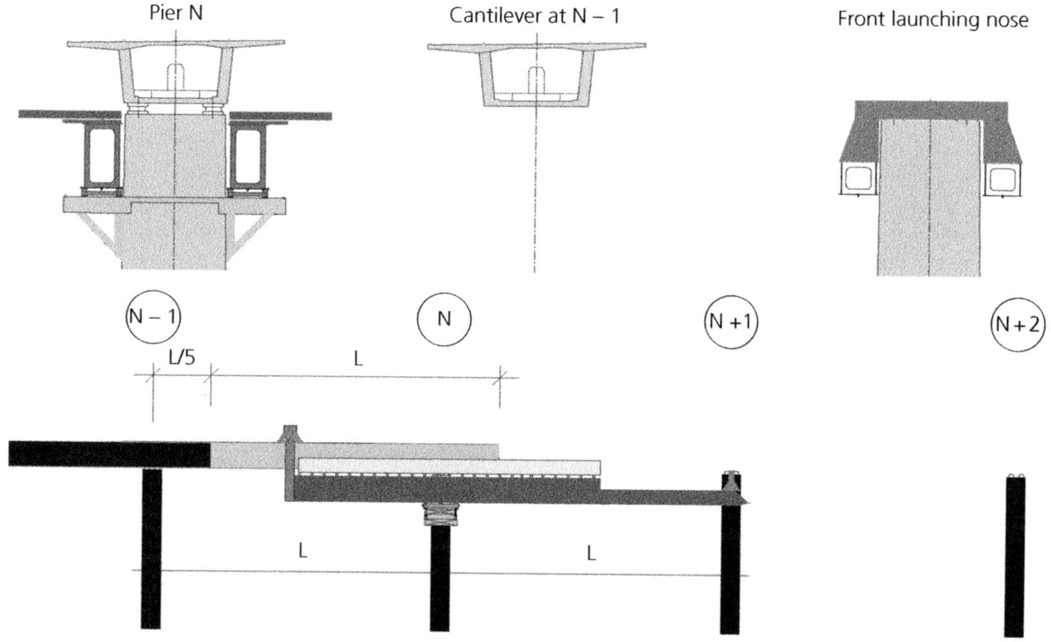

3. Launch of MSS to next pier by rolling/sliding 'C-frame' at rear over span N – 1/N and in front over bracket at pier N; during launching open and close formwork segments (by transverse shifting of formwork and crossbeams) to pass pier N, activate temporary front support once tip of front nose is over pier N + 1

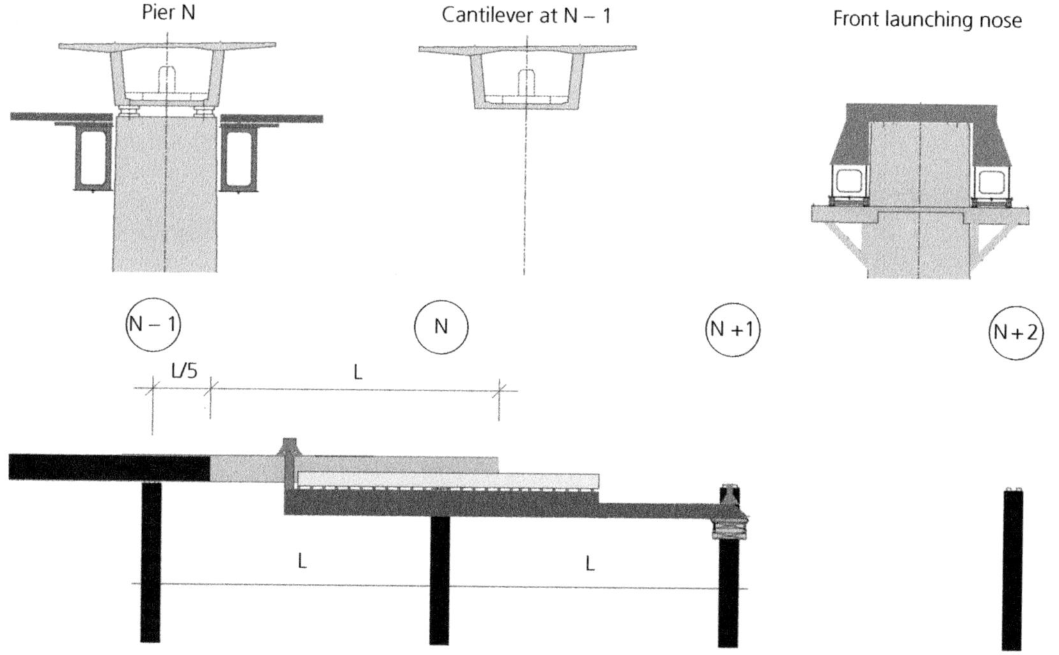

4. Lock brake at front suspension, deactivate pier bracket, launch bracket to pier N + 1

Figure 12.11 Continued

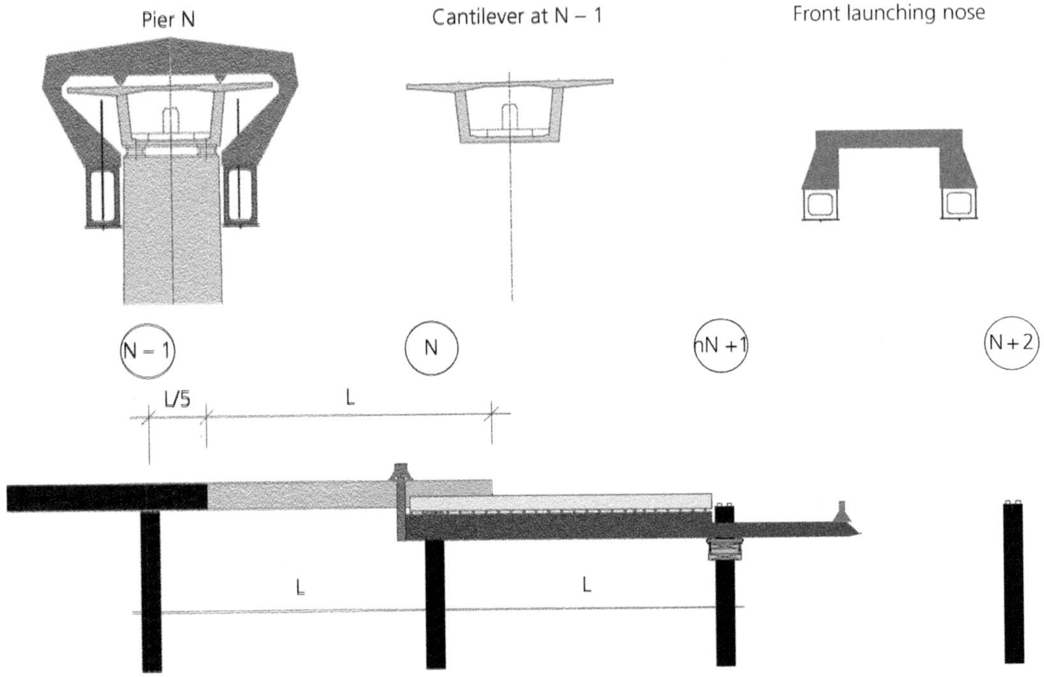

5. Release brake at front suspension, launch MSS forward, during launching open and close formwork segments to pass pier N

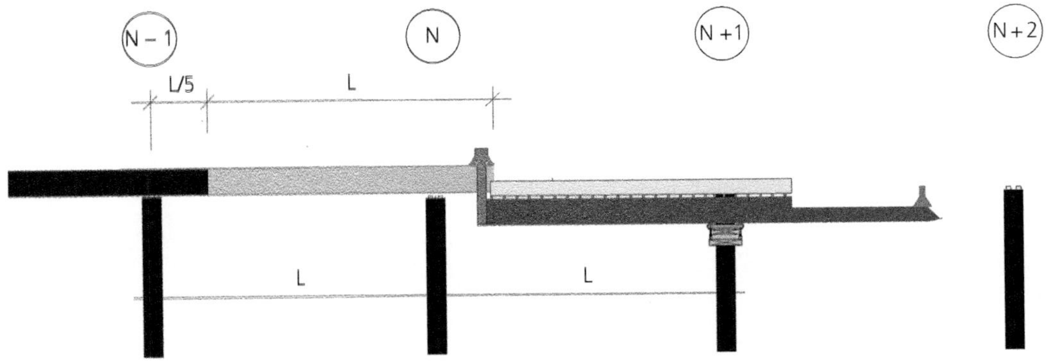

6. Launch MSS forward to next pouring position, close formwork, activate hydraulic supports of cantilever suspension; preparation (rebar, PT, inner form) and casting, curing and stressing of first and second pour

12.5. Permanent work interfaces

- The pier design has to allow the MSS and the formwork to pass. Piers should not be wider than the bottom edges of the bridge superstructure.
- Box girder sections are often cast in two phases. After the pour, the composite system of the MSS and the trough is loaded by additional reinforcement and fresh concrete for the deck slab. Consequently, the trough will have to take over a share of the load according to the rigidity

Table 12.7 Guiding values for distribution of support loads in construction stage (assuming a ratio of bending stiffness between MSS and concrete trough of 0.5)

	Pier support		Cantilever suspension at L/5	
	Weight of movable part of MSS	Concrete weight	Weight of movable part of MSS	Concrete weight
Single-phase pouring	60–65%	60%	35–40%	40%
Two-phase-pouring		60% of trough 30% of top deck*		40% of trough 20% of top deck*

*Remaining weight is taken over by trough

ratio between MSS and trough (for indicative distribution factors, refer to Table 12.7). Different measures can be taken to grant a proper production result without cracks. Each measure has a different impact on the MSS loading, the construction cycle and the outer appearance of the permanent structure. The most important measure is generally to partially stress the deck's longitudinal PT (normally internal, located in the webs with parabolic profile; for details refer to Section III, Chapter 15).

- Single-stage pouring requires complex inside formwork. The design and detailing of the deck need to provide big enough temporary or permanent openings through the pier diaphragms to permit the passage of a tunnel form. This tunnel form is left behind in the just-completed span and launched forward into the next span only after placing of the reinforcement and PT in the bottom slab and the webs.
- Cantilever support at point of zero bending moment due to dead load gives a gradual loading of the cantilevering deck during concreting compared with the sudden jack-down of an MSS supported at the previous pier. High suspension loads during concreting have to be considered in the post-tensioning layout. For an indication on magnitude of loading to be considered, refer to Table 12.7. With rear support at the cantilever tip of the previously cast span the MSS develops a negative moment over the front pier which serves to reduce the critical positive moment in the back-span. At smaller span lengths, a cantilever support is not always required when using rigid main girder systems. Refer to Section III, Chapter 16 for comments about the design of MSS with regard to pouring sequence.
- The deck has to be carefully checked for all construction stages considering the actual weight, weight distribution and load introduction of the MSS.
- Direction of construction: for safety reasons, uphill is always recommended. At the cantilever joint, fresh concrete flows against the existing structure.
- For long semi-integral bridges, pre-deformations of piers can be introduced with suitable launching devices. A longitudinal fixation between the MSS and the already built bridge deck might be required. Post-tension losses, and creep and shrinkage of concrete will be greater with this form of bridge construction than for pre-cast bridges as the concrete is much younger when the stressing is activated.
- If pier diaphragms of girders with TT section can be fabricated after the MSS has been positioned in the next span, the inside formwork of underslung systems can be launched to the next span without any need of dismantling. The fabrication of diaphragms on the rear launching noses of the MSS saves labour costs for the project. Similarly the casting of the internal pier diaphragm for a box girder can also be done when casting the following span to

ease the transport of the inside deck formwork or tunnel form. The effect on the load introduction between the deck and the pier has to be examined if the pier diaphragm is not yet available, particularly with regard to the reactions from the MSS's rear support if placed at the tip of the cantilever. Generally the temporary omission of the diaphragms is only possible when the webs can temporarily be supported directly on the piers.

- Pier diaphragms of box girders which are built by overhead MSS have to be designed and detailed with temporary box-outs for the vertical props of the support tower.
- For bridges curved in plan the following aspects have to be considered:
  - asymmetric position of the cantilever support loads
  - breakdown of the formwork into longitudinal polygonal segments to enable a smoother shape of the bridge and prestressing layout
  - unequal cross-sectional load distribution on the straight main girders of the MSS.

12.6. Cycle times

In Tables 12.8 to 12.10, typical cycles are shown for the construction of a single-cell box girder with standard spans of 60 m with different pouring sequences and different methods for relocation of the pier supports.

- Table 12.8: poured in one stage; three sets of pier supports to be shifted by crane
- Table 12.9: poured in two stages; three sets of pier supports to be shifted by crane
- Table 12.10: poured in two stages; two sets of self-relocated pier supports.

Assumptions and comments regarding Tables 12.8 to 12.10 are as follows.

- Cycle times depend on the number and quality of the workforce and therefore are for guidance only. They assume 'typical spans', which includes the improvement in performance that comes from repetition. The first two or three cycles will be significantly slower than shown. Changes to sequences caused by variations in the permanent works will reduce work speed.
- Use of standard concrete (36 hours to reach minimum strength required for stressing of PT and load transfer).
- One shift per day (10 hours per shift).
- Cycle times can be significantly reduced by preassembling rebar cages. The preassembled rebar cage can be transported over the bridge deck and the wing slab formwork (by means of special straddle carriers). Several parts of the MSS have to be adjusted for use of this method. The straddle carrier's reactions will be a relevant load case for the formwork design.

12.6.1 Project duration

Due to their simplicity, underslung systems have remarkable advantages in comparison with overhead MSS regarding assembly time. As the number of bridge spans increases, this advantage will lose its dominance. However, as most components of an overhead MSS are located above the bridge deck, the dismantling operation is easier.

12.7. Risks

The list of risks given in Table 12.11 is by no means comprehensive, but addresses risks which are particular to MSS used for full-span in-situ deck erection. For a specific project, a main contractor needs to carry out a full risk assessment of the operations performed with the MSS, considering the actual features of the MSS, the erection sequence and the site constraints.

Table 12.8 Regular cycle for MSS for construction of a single-cell box girder, 60 m span; poured in one stage; three sets of pier supports to be shifted by crane

TYPICAL CYCLE FOR MSS: Single-cell box girder section with 60m spans																		
ACTIVITY / STEPS		DURATION (working hours)																
		8	16	24	32	40	48	56	64	72	80	88	96	104	112	120	128	136
Launching to next pier		■	■															
- lowering of MSS - opening of formwork cross beams - launching - closing of formwork cross beams																		
Shifting of pier supports				■	■	■	■											
- dismantling in previous position - transport to nex pier - reconfiguration and assembling																		
Installation of reinforcement at full section				■	■	■												
- cleaning of formwork - installation bottom formwork at piers - reinforcement not pre-fabricated																		
Pouring and curing of trough and deck slab							■	■	■									
- full post-tension																		

Table 12.9 Regular cycle for MSS for construction of a single-cell box girder, 60 m span; poured in two stages; three sets of pier supports to be shifted by crane

TYPICAL CYCLE FOR MSS: Single-cell box girder section with 60m spans																		
ACTIVITY / STEPS		DURATION (working hours)																
		8	16	24	32	40	48	56	64	72	80	88	96	104	112	120	128	136
Launching to next pier																		
- lowering of MSS - opening of formwork cross beams - launching - closing of formwork cross beams																		
Shifting of cantilever support																		
- relocation - activation																		
Shifting of pier supports																		
- dismantling in previous position - transport to nex pier - reconfiguration and assembling																		
Installation of reinforcement at trough																		
- cleaning of formwork - installation bottom formwork at piers - reinforcement not pre-fabricated																		
Pouring and curing of trough																		
- stripping of inner wall formwork - partial stressing																		
Installation of reinforcement at deck slab																		
- installation of deck formwork - reinforcement not prefabricated																		
Pouring and curing of deck slab																		
- full post-tension																		

Table 12.10 Regular cycle for MSS for construction of a single-cell box girder, 60 m span; poured in two stages; self-relocated pier supports

TYPICAL CYCLE FOR MSS: Single-cell box girder section with 60m spans and self-launching supports																		
ACTIVITY / STEPS		DURATION (working hours)																
		8	16	24	32	40	48	56	64	72	80	88	96	104	112	120	128	136
Launching to next pier																		
- lowering of MSS - opening of formwork cross beams - launching - closing of formwork cross beams																		
Shifting of cantilever support																		
- relocation - activation																		
Shifting of pier supports																		
- activation of launching suspension - launching of supports - reconfiguration and de-activation of suspension																		
Installation of reinforcement at trough																		
- cleaning of formwork - installation bottom formwork at piers - reinforcement not pre-fabricated																		
Pouring and curing of trough																		
- stripping of inner wall formwork - partial stressing																		
Installation of reinforcement at deck slab																		
- installation of deck formwork - reinforcement not prefabricated																		
Pouring and curing of deck slab																		
- full post-tension - relocation of launching suspension																		

Table 12.11 General risks and mitigating measures when working with MSS

Activity	Risk	Mitigating measures
First assembling or dismantling Launching Self-relocation of supports	Overturning of MSS due to loss of control of longitudinal movement → loss of equilibrium → overturning of MSS's supports and loss of entire gantry	■ Indicate CoG (centre of gravity) in all adaptation/moving phases in longitudinal and transversal direction ■ Periodically counter-check CoG ■ Back-up fixity, nominal tie-down of supports, presence of MSS engineer and gantry supervisor
Launching on pier	Overturning of permanent structure	■ Check resistance of pier and strengthen it if necessary against horizontal loads with real friction and launching forces ■ Limit maximum launching force to match pier's capacity ■ Control or prevent displacement of pier in launching stage
Pouring stage	Overturning of MSS due to failure of MSS support	■ Third-party check after each assembling of support
	Overturning of MSS due to failure of permanent structure	■ Match load dimensions and introduction points with pier and deck design within geometrical tolerances given by the engineer in charge of structural checks of the construction stages ■ Stress tie-down bars to at least SLS level (as defined by engineer); verify locked-in forces (use of calibrated pressure gauges) ■ Adequate edge distances, so that support's contact area is within concrete confined by reinforcement; shimming with similar material over entire footprint of support's contact area
Opening/closing of formwork's cross beams for launching stage	Longitudinal and vertical access and work at height, falling objects	Provision of stairs and access platforms with handrails and kicker boards No traffic underneath MSS
Shifting of pier supports	Falling down of people or materials from >3 m of height	Man-lift for all assembling operations, access staircase, safety line fixed to pier
Pouring stage	Damage of formwork in pier and cantilever joint area due to overloading	Permanent survey of support forces during pouring
Working on top of and inside bridge deck Working on top of and inside MSS	Danger to workers when at height	Provision of handrails (or at least safety lines) around all exposed edges on top and inside of MSS/bridge deck installed prior to launching of MSS
Free-hanging hanger bars exposed to wind	Bars may turn out of couplers/nuts and subsequently fall down due to wind-induced oscillation about bars' vertical axis	Do not leave single hanger bars suspended unloaded, or provision of locking pins/nuts. Mark the bars

12.8. Performance criteria

The following items need to be addressed when drafting a performance specification for an MSS.

- Characteristics of spans to be built by MSS
  - cross section and its variation along the span
  - weight and weight distribution
  - post-tensioning layout (longitudinal and, if applicable, transverse PT in deck slab), anchorage details, stressing sequence (including for two-pour construction if applicable), stressing access (lateral clearance against web forms), etc.
  - rebar layout (starter bars).
- Alignment: gradient, curvature in plan and cross fall variations.
- Permanent work interfaces
  - permissible deck loading (standing on or suspended from deck during casting, rolling/sliding over deck during launching)
  - permissible deck loading (cantilever support)
  - permissible deflection in pouring stage
  - permissible eccentric loading of piers
  - deck geometry at bifurcations and ramps
  - support conditions: bearings/monolithic connection between pier and deck, need for temporary deck support during construction, longitudinal fixation during construction stage
  - positions for MSS supports: permissible positions and dimensions of box-outs in deck, pier diaphragms, piers and permissible positions on pile caps.
- Pouring sequence: deck cast in one or two pours?
- Formwork
  - hydrostatic pressure and curing time
  - skin (plywood or steel, position of tie-bars, architectural demands)
  - material for runners and walers (steel or timber)
  - inner formwork
    - deck cast in one pour: tunnel form?
    - deck cast in two pours: crane-handled wall forms as internal forms for webs? Deck slab form on tables?
- Material supply (rebar, PT, concrete): with tower crane from back/from side, with on-board equipment.
- Reinforcement: degree of prefabrication and, if required, weight of loaded transporter and position of its tyres.
- Follow-up work (closing of holes and box-outs in piers, deck, pier diaphragms, etc.).
- Clearances: vertical below and above deck (overhead power lines, overhead bridges, etc.); horizontal (houses, power lines, etc.).
- Cycles (double or single shift, restrictions for delivery, restrictions for noisy type of work such as strand pushing, etc.).
- Mechanisation (labour costs, skill level, etc.).
- Maximum weight of assembled pieces and weight of pieces of pier supports to be shifted by cranes.
- Environmental conditions: need for cover, need for insulation of formwork.
- Limits for location or loading of auxiliary towers for assembling/dismantling.
- Access: on MSS during different stages of operation; access from ground to piers (during deck construction, for follow-up work) and onto and into deck.
- Shipment costs (initial, if reused).

- Reusability.
- Method, schedule and crane definition for assembly, relocation, dismantling.

In addition, the buyer needs to give a time frame for fine-tuning of the MSS concept, support reactions/inserts and delivery and commissioning of the MSS.

12.9. Photo gallery

Figure 12.12 Underslung MSS with self-launching brackets at a semi-integral railway bridge with a haunched T-beam section of variable depth (Germany, Scherkondetalbrücke, Adam Hörnig) (thyssenkrupp Infrastructure, saul ingenieure)

Figure 12.13 Underslung MSS for demolition of single-cell box girder highway viaduct (Germany, A45 Lützelbachtalbrücke, Adam Hörnig) (thyssenkrupp Infrastructure, saul ingenieure)

Figure 12.14 Formwork types (from left to right side): steel formwork, modular timber formwork, customised timber formwork (thyssenkrupp Infrastructure, saul ingenieure)

Figure 12.15 Inner deck slab formwork at single-cell box girder with two-stage pouring, launchable to next span (12.15a thyssenkrupp Infrastructure, saul ingenieure; 12.15b Xervon Austria; 12.15c thyssenkrupp Infrastructure)

Figure 12.16 Formwork cross beam types (from left to right side): hinged frames for overhead, transversely fixed on underslung MSS, transversely movable on underslung MSS (thyssenkrupp Infrastructure)

Figure 12.17 Pier support types (from left to right side): pier tower, pier tower for overhead system, brackets (12.17a and 12.17b thyssenkrupp Infrastructure; 12.17c VSL International)

Figure 12.18 Pier support types (from left to right side): insert beams, 2× A-type (thyssenkrupp Infrastructure)

Figure 12.19 Self-launching pier support types: left side for overhead system, right side for underslung system (thyssenkrupp Infrastructure)

Figure 12.20 Modular main girder types: left side truss girders, right side box girders (12.20 thyssenkrupp Infrastructure)

Figure 12.21 Dismantling of underslung MSS between existing infrastructure after removal of pier supports by lowering with HT bars (thyssenkrupp Infrastructure)

Figure 12.22 'C-frame'-type underslung MSS with front nose only and fixed temporary front support: configured for construction of simply supported spans (Bilfinger Berger Bauaktiengesellschaft)

REFERENCES

General references

Aubel P (1989) Die Anwendung von maßgeschneiderten Vorschubrüstungen in Stahlbauweise auf Neubaustrecken der Deutschen Bundesbahn. *VDI-Berichte*, **788**: 203–222.

Daebritz M (2011) Movable Scaffolding Systems. *Structural Engineering International*, **21(4)**: 413–418.

Daebritz M (2016) Vorschubgerüste im Massivbrückenbau. *Bautechnik*, **93(7)**: 452–461.

Maas R-B (1989) Anwendungsbeispiele für Vorschubrüstungen. *VDI-Berichte*, **788**: 223–242.

Marx S and Seidl G (2011) Integral Railway Bridges in Germany. *Structural Engineering International*, **21(3)**: 332–340.

Figures and tables

Figures 12.1–12.15a: thyssenkrupp Infrastructure, saul ingenieure
Figure 12.15b: Xervon Austria
Figures 12.15c, 12.16b–17b, 12.18–12.21: thyssenkrupp Infrastructure
Figure 12.16a, 12.17c: VSL International
Figure 12.22: Bilfinger Berger Bauaktiengesellschaft
Tables 12.8, 12.9, 12.10: thyssenkrupp Infrastructure, saul ingenieure

Bridge Deck Erection Equipment
ISBN 978-0-7277-6193-4

https://doi.org/10.1680/bdee.61934.247

Chapter 13
Incremental launching method

13.1. Introduction

The Incremental Launching Method (ILM) is used for the construction of concrete, steel and composite decks, whenever a bridge of short to medium length has to cross an obstacle that hinders access for ground-based cranes, scaffolding girders and formwork. Another reason to use ILM is lack of vertical clearance to install falsework girders and formwork over an existing traffic corridor.

The basic principle of ILM is described in Section III, Chapter 15 and is shown in Figure 13.1. ILM is used for construction of concrete, steel and composite bridge decks. The deck is broken down for construction by transverse cuts into elements of similar length, which are added to the already completed deck behind one of the abutments at a fixed location, either concreted in a formwork or assembled in an assembly area. The deck is pushed or pulled out of the formwork or assembly area and over the abutment onto the piers into its final position.

Provision of a launching nose stressed against the first deck segment can reduce the big bending moments in the deck near the front during launching. For more explanations of by how much these bending moments can be reduced, refer to Section III, Chapter 15. In viaducts with only a few spans that are larger than typical, the deck bending moments can be reduced by provision of temporary towers.

Special temporary bearings are required on the launching abutment and every pier to move the deck over the piers towards its final position during the launching operation.

The ILM has the following characteristics, shown in Table 13.1.

13.2. Application range

Theoretically bridge decks with all kinds of cross sections can be incrementally launched if the bridge has a steel or a composite deck. For concrete decks, the optimum cross section is a section whose bottom core value $k_{bottom} = Z^{top}/A$ is about two times its top core value $k_{top} = Z^{bottom}/A$ in order to minimise the amount of the centrical PT. For detailing aspects of decks built by ILM, refer to Section III, Chapter 15.

Approximate span ranges for the ILM method are as follows.

- Post-tensioned concrete girders: 35 m to 65 long spans, where the most cost-effective span is around 40 m. For longer spans (more than 60 to 80 m) temporary intermediate supports can be used. If the pulling/pushing is only done from the abutment and with minor slope, the bridge length can generally not exceed 1200 m per launching side (limitation in launching forces).

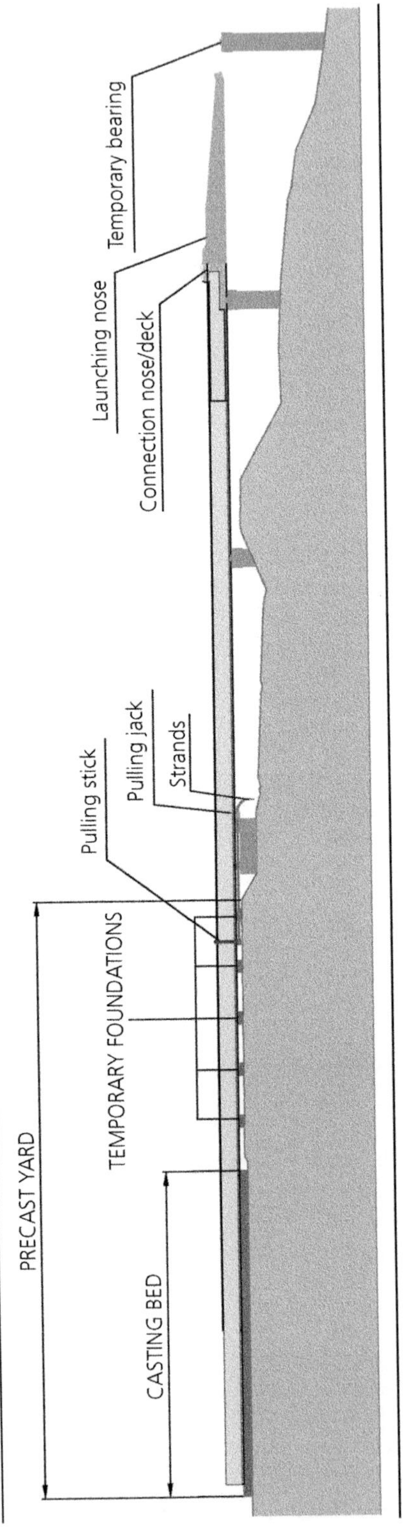

Figure 13.1 ILM – basic arrangement; if the deck is in steel, the the casting yard (CY) is replaced by the assembly yard (AY) (Freyssinet)

Table 13.1 Characteristics of the incremental launching method

	Concrete deck	Steel deck	Composite deck
Scaffolding	Not required	N/A	Travellers required when deck slab is cast in situ after main girders have been launched into final position (unless precast planks are used)
Quality	It is easier to control quality during the deck construction or assembly, since the majority of work is carried out under 'factory' conditions in the casting yard or the assembly area (which can even be covered) and not along the bridge alignment. On the other hand, tighter tolerances need to be achieved with regard to geometry of the soffit of concrete girders and steel main girders		
Launching equipment (jacks, temporary bearings, nose)	Components are modular and can therefore easily be recycled on other projects		
Permanent work material quantities	More post-tensioning, since deck needs to be prestressed centrically for the launch	Since steel is as good in tension as it is in compression, provided stability is not controlling, increase in quantities could be limited to thickening of bottom flanges/lower part of webs to cope with load introduction during launch (patch loading forces)	
Supervision	Tighter supervision is required during segment construction or assembly to achieve specified geometrical tolerances of the deck. Additional workforce is required during launch for long viaducts, since often up to four people are needed on each pier, who need to be well trained (correct insertion of sliding plates etc)		

- Composite and steel girders: 35 m to up to 150 m. As for concrete girders, launching span can be reduced by means of temporary supports or by the use of a temporary stayed mast.

13.3. Alignment in plan and elevation

For concrete bridges, the girder's soffit geometry should allow an 'extrusion' of the structure from the casting cell. For casting cell concepts for concrete decks built by ILM (type 1 and 2), refer to Section III, Chapter 19.

With regard to the launching, the alignment (in plan and elevation) of that part of the deck's or steel girder's soffit which slides or rolls over the temporary bearings needs to fall within one of the following six categories.

1. Straight line on a horizontal plane
2. Straight line on an inclined plane
3. Curved line (R = constant) on a horizontal plane
4. Curved line (R = constant) on an inclined plane (Figure 13.2)
5. Helix (Figure 13.3)
6. Curved both in plan and elevation; bottom flange lying on a cone (Figure 13.4)

Figure 13.2 Curved line on an inclined plane (Placidi 1998)

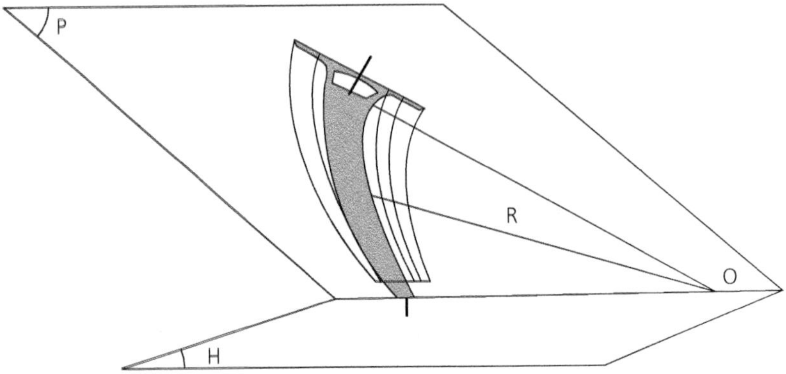

Figure 13.3 Helix configuration (Placidi 1998)

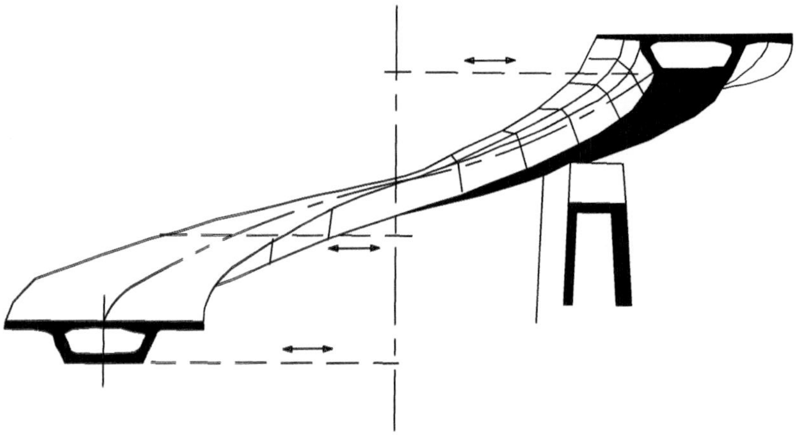

Figure 13.4 Launching on a cone (Placidi 1998)

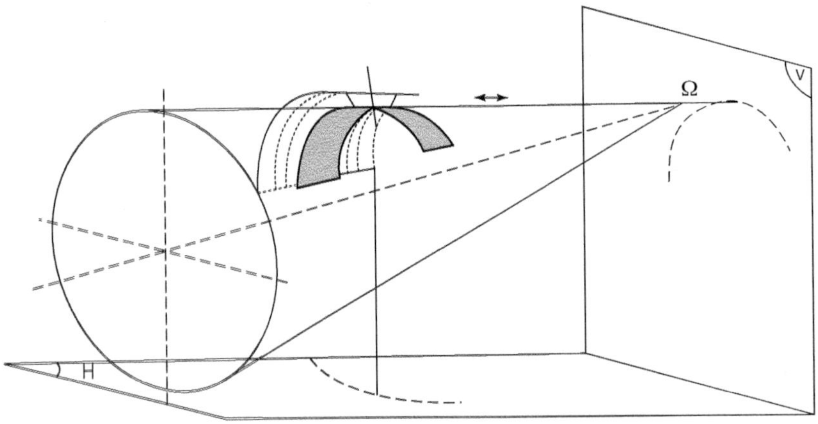

Table 13.2 Possible solutions for non-standard bridge decks

Problem	Solution	Comments, disadvantages
Vertical alignment (locally)	Increase locally the girder depth and/or the upper slab thickness and/or the ballast thickness (if present)	Increase of the deck weight
Horizontal alignment (locally)	Increase locally the wing slabs of the top deck	Increase of deck weight and introduction of torsion in the deck because of the eccentricity
Double curve (horizontal or vertical plan)	Launching from both abutments	Increase of the launching costs (two casting yards)
The combination of vertical and horizontal alignment is not compatible with the six spatial geometries listed above (example: constant vertical radius + constant horizontal radius)	A similar alignment can be proposed or if discrepancies are small and for steel main girders an investigation can be done to check if the stresses induced by launching with the conforming alignment are acceptable	
Steel deck with variable depth	Depending on the extent of the girder's variation in depth and the girder's flexibility, the deck can be launched as it is but with need for a launching system with an automatic control of the levels and reactions on each pier	In general, steel girders can accommodate more complex geometry (non-circular alignment, variable depth, smaller radius...) but with direct impact on the number of mitigation measures (numerous launching steps, jacking operations, special adaptable bearing devices, all resulting in increased global duration and complexity)

ILM can be applied even for decks whose soffit does not fall into the above classifications by means of some design modifications. Table 13.2 gives some examples of bridge decks that do not fulfil the main requirements for ILM but can be incrementally launched after some modifications.

As already mentioned, the previous requirements are only needed for that part of the deck or the girder's soffit which slides or rolls over the temporary bearings. Even so, for concrete decks generally the entire cross section complies to simplify the casting moulds. The girder's depth as well as its width may vary as long as the geometry of the sliding surface falls within one of the above categories.

It is fundamental to assess these geometrical conditions at the very beginning of the design or tender stage, to verify whether the deck geometry falls within one of the six categories or whether some modifications to the deck geometry are required to make the bridge buildable by ILM.

The horizontal radius is generally bigger than 250 m, even though prestressed concrete decks have been launched with a horizontal radius of 150 m. A tight radius increases dramatically the lateral guiding forces necessary to direct the bridge during the launching (and therefore the transverse force acting on the piers). It also creates significant bending in the horizontal plane of the deck and complicates the formwork arrangement and the geometry control.

13.4. General arrangement of the BDE

In this subsection the general arrangement of the launching equipment is described, both for post-tensioned concrete bridge decks and for steel/composite bridge decks. The following components are looked at.

- Movement device
- Temporary bearings
- Assembly yard (AY) and casting yard (CY)
- Launching nose.

13.4.1 The movement device

When using ILM, the bridge deck is moved by one of the four movement devices shown in Table 13.3.

For more details about hydraulic jacks, strand moving units and winches refer to Section III, Chapter 17.

Systems 1A, 2 and 3 will now be explained in more detail. System 1B is similar to system 1A, however the lifting-pushing jacks are installed not just at the abutment but on all support points (on all piers).

Table 13.3 Horizontal movement methods

	Method	Need to lift deck	Positioning of the equipment (standard case)	Launching 'plane'	
				Launching along gradient	Launching horizontally (level correction after every longitudinal movement step)
1A	Lifting-pushing with single set of hydraulic solid ram jacks	Yes, at abutment only	At abutment	X	
1B	Lifting-pushing with multiple sets of hydraulic solid ram jacks	Yes, at all support points simultaneously	At every support		X
2	Pushing with hydraulic long stroke jacks	N/A	At abutment	X	
3	Pulling with strand moving units, HT bars or winches	N/A	At abutment	X	

The advantage of this system is that minimum horizontal longitudinal forces are introduced into the abutment. The resistance against moving the deck at every support point is directly equilibrated within the launching system at this point. Hence this method is very suitable for long viaducts and/or viaducts with considerable gradient. On the other hand, it requires more launching equipment.

13.4.1.1 Lifting-pushing system

The equipment for this system generally consists of a set of jacks – a lifting jack ('vertical') and one or two pushing jacks ('horizontal'), normally placed on the abutment and always located under the webs. Launching can be upwards or downwards. Refer to Figure 13.5.

Each lifting jack has a hardened, chequered top bearing plate which permits the transfer of a 'horizontal' launching force H from the jack to the deck. This force can go up to a maximum value which is equivalent to 50% of the force built up in the 'vertical' jack when lifting up the deck for a longitudinal launching step. At the bottom, the lifting jack has a PTFE bearing plate allowing the lifting jack and the deck to move at a friction of about 1% over a stainless steel coated sliding plate by the stroke length of the pushing jack.

Figure 13.5 Typical jacking assembly on the launching abutment for the lifting-pushing system (Freyssinet)

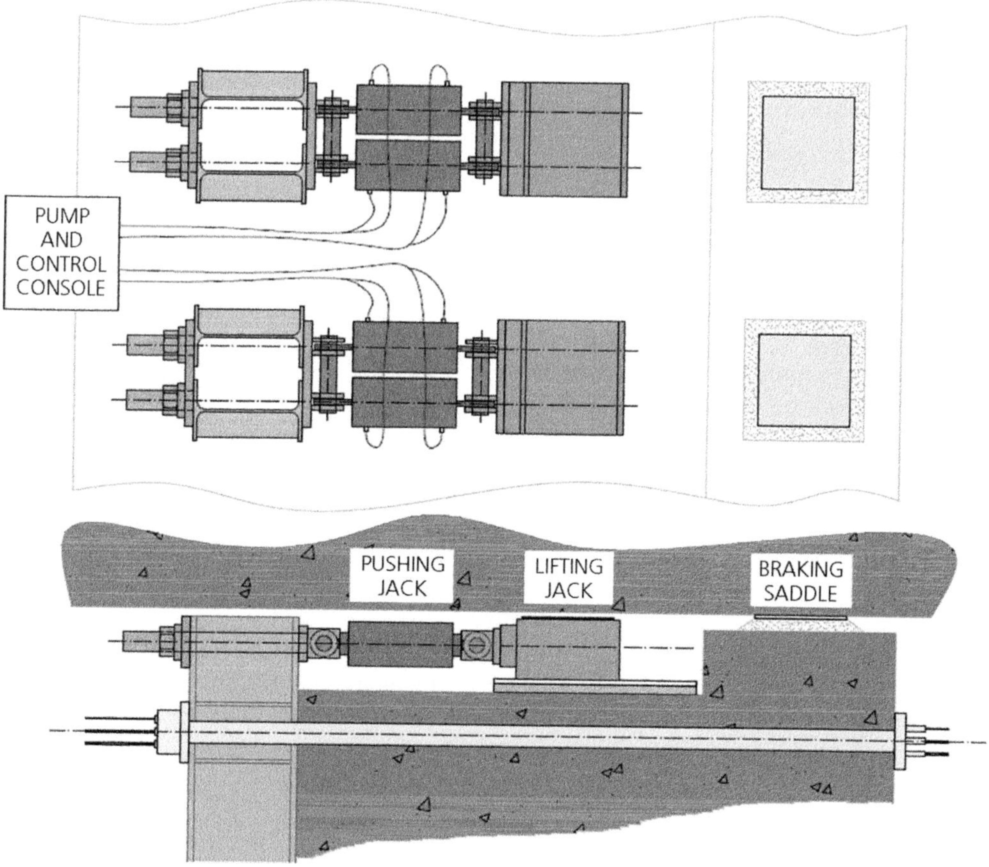

Figure 13.6 Typical cycle with the lifting-pushing system (Freyssinet)

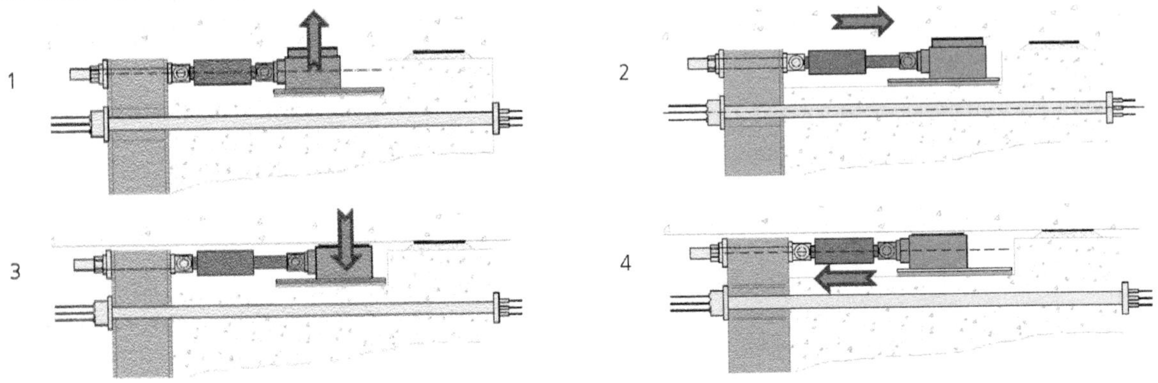

The first step is to lift the bridge (at the abutment) with the lifting jacks by about 5 to 10 mm and then to move the lifting jacks together with the deck over the sliding plate forward by the pushing jacks' stroke, which is typically about 250 mm. The bridge is afterwards lowered on the so-called brake saddle and the lifting jacks are retracted. Finally the pushing jacks are retracted. This action is repeated step by step, and provides, under standard conditions, a speed of about 6 m/hours. Refer to Figure 13.6.

The lifting load of the system can reach up to 38 000 kN, while the pushing force can reach 12 000 kN.

The following points should be noted.

- Pushing of the first span is generally not possible without an additional system (pulling) because the vertical reaction on the bearing plates is at this point of time not sufficient yet to mobilise enough friction to transfer the longitudinal pushing force to the deck (the launching nose is a lightweight element). An additional system has to be used so that the deck can be pulled at the very beginning (over the first 15 to 30 m).
- End span: the same applies as for the first span (see above). The additional system has to provide a big launch force, as the horizontal force is at its maximum.
- For long bridges (for example, a deck with more than nine spans and zero gradient) the vertical reaction is not enough to mobilise the launching force.

The additional system may be to connect the pushing jacks through special brackets to longitudinal HT bars, which are either anchored in brackets attached to the soffit of the nose or the deck (for the initial launch) or to vertical beams attached to the back of the completed deck (for the last launch).

One advantage of the lifting-pushing system is that it allows the bridge deck to be pushed backwards. This may be beneficial when there is a particular problem arising during the pushing operation. However, problems during launch are generally related to wrongly inserted sliding pads. Such problems can only be solved by jacking up over the affected pier.

13.4.1.2 Pushing system

With the pushing system, the launching force is transmitted to the deck by installing the pushing jacks behind the last cast or attached segment. For a typical pushing jack, refer to Figure 13.7.

Figure 13.7 Main components of a 2000 kN pushing system (Freyssinet)

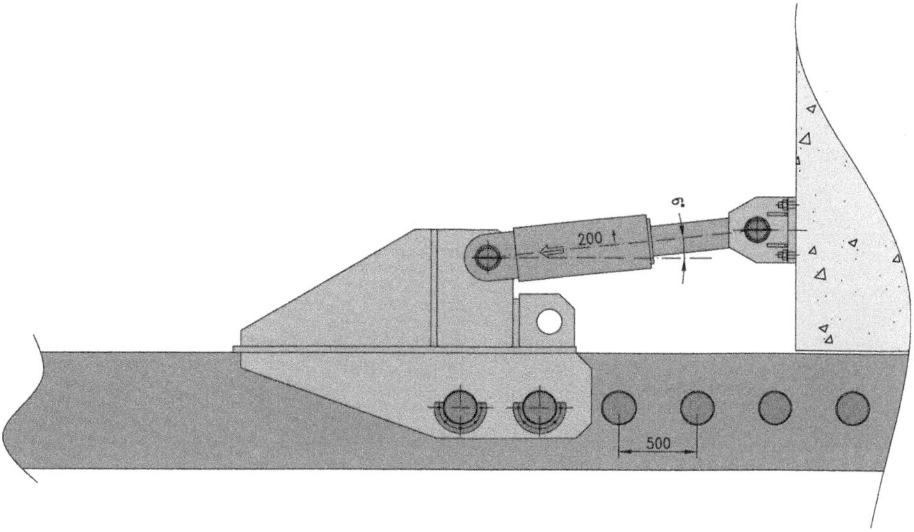

The pushing jacks are reacting against the launching track beams, which are, in the case of concrete decks, the fixed soffit form's support beams (type 1 formwork, refer to Section III, Chapter 19). The launching forces are transferred from the jacks to the launching track by one of the following three methods.

A. Direct bearing against steps in the launching track (see Figure 13.8)
B. Bracket with shear pin (see Figure 13.9)
C. Friction brackets with transverse jacks (see Figure 13.9)

When a concrete deck is pushed directly from the back, the outer formwork needs to be opened enough to give space for the pushing jacks and, if applicable, the brackets and the transverse jacks. Therefore the most popular method for launching of concrete decks by pushing is method A.

Figure 13.8 Pushing system A: fixed soffit form and sliding track (left); jack detail (right) (VSL International)

Figure 13.9 Typical pushing equipment fixed to the launching track by bracket with shear pin (left) or by friction brackets with transverse jacks (right) (Freyssinet)

Standard systems have a pushing force/jack of 2500 kN and a pulling force/jack of 1200 kN with 1 m stroke.

13.4.1.3 Pulling system

The most common system for launching is to pull the deck. Depending on the magnitude of the pulling force, the pulling device can be

- a winch for lightweight structures, such as steel girders (winches have relatively small capacity but can operate at quite high speed)
- Strand Moving Units (SMU) for heavy structures (mainly prestressed concrete decks); there is almost no limit to the capacity of SMUs; they operate at relatively low speed
- centre hole jacks with HT bars as pulling elements (for small projects with a few launching steps only).

Winches, SMUs and HT bars with centre hole jacks are generally fixed at the abutment. Winches and SMUs are explained in more detail below.

13.4.1.4 Winches

The most common system used to launch steel girders (steel and composite bridges) is to pull the deck by an electrical winch, whose ropes are reeved through a pulley block system to achieve the launch force required.

In general, single-rope winches with a pulling capacity of between 100 and 300 kN are used. The rope is reeved 8 to 12 times, which results in an overall winch capacity between 1000 and 2000 kN.

Compared to the other launching equipment, winches have the advantage of being able to move the bridge deck over a segment length without intermediate stops. This increases the operation speed significantly. However in reeved rope winch-driven launches, particularly if the ropes have significant catenary, considerable strain energy builds up, which can result in a stop-go type of movement and potential overlaunch. With the provision of a back winch such overlaunch, which could be catastrophic on some projects, can be avoided.

Steel and composite bridges are generally launched upwards, as the launching force is not an issue and proceeding this way there are no risks of uncontrolled deck sliding. If a deck needs to be launched downwards, a back winch has to be installed in order to restrain the deck from uncontrolled sliding or to reverse the launching direction if necessary. The back winch is generally installed behind the launching area and anchored to a dead weight or to a permanent structure.

If the deck has a variable depth and is moved along its soffit, the back winch is active each time the deck has to 'descend'.

Both front and back winches are generally attached to a lightweight steel structure, which is fixed to the rear part of the deck's steel girders. A typical set-up of a winch for launching of steel girders is shown in Figure 13.10.

13.4.1.5 Strand Moving Units (SMUs)

With this system, the deck is pulled by strand moving units. The jacks are generally fixed at the abutment using vertical steel beams and HT bars and the strand bundles are fixed to the deck either by brackets for steel girders and concrete decks or by pulling sticks for concrete decks (refer to Figure 13.11).

The strand bundles pass below the girder's or the deck's soffit at approximately 40 cm.

Another option to fix the strand bundles to the deck is to fix them to vertical beams attached to the rear end of the last segment cast or assembled. Such an arrangement works if the deck is made from concrete and it is best with a type 2 formwork system (refer to Section III, Chapter 19). This is because it requires the lowering of the soffit by at least 500 mm and the pulling through of the strand bundle over the length of the formwork (anything between 15 to 30 m).

SMUs are available with a wide range of capacity. The force per strand is usually limited to 50% of the breaking load. The stroke of the jacks is generally between 150 to 500 mm. The minimum number of jacks is a pair. More jacks can be used for bigger launching forces. SMUs up to 55 strands are available on the market.

For long bridge decks, load introduction of the launch force into the deck may become controlling.

A local check of deck slabs needs to be carried out and eventually an increase of the slab's depth could be necessary for sections where launching forces exceed a certain value (see Figure 13.11).

For illustration of various attachment details, refer to Figure 13.12.

No special operations are required for the initial and the last launch, since the deck can always be launched by pulling with the same equipment.

13.4.2 Temporary bearings

In order to move a deck over abutments, piers and temporary supports, temporary bearings are required on all support points during the launch. The available options are

- sliding
  - over the final bearings, which need to be specifically designed for this operation (concrete decks)
  - over temporary sliding bearings (for all kind of decks)
- rolling over temporary roller saddles (for steel or composite bridges).

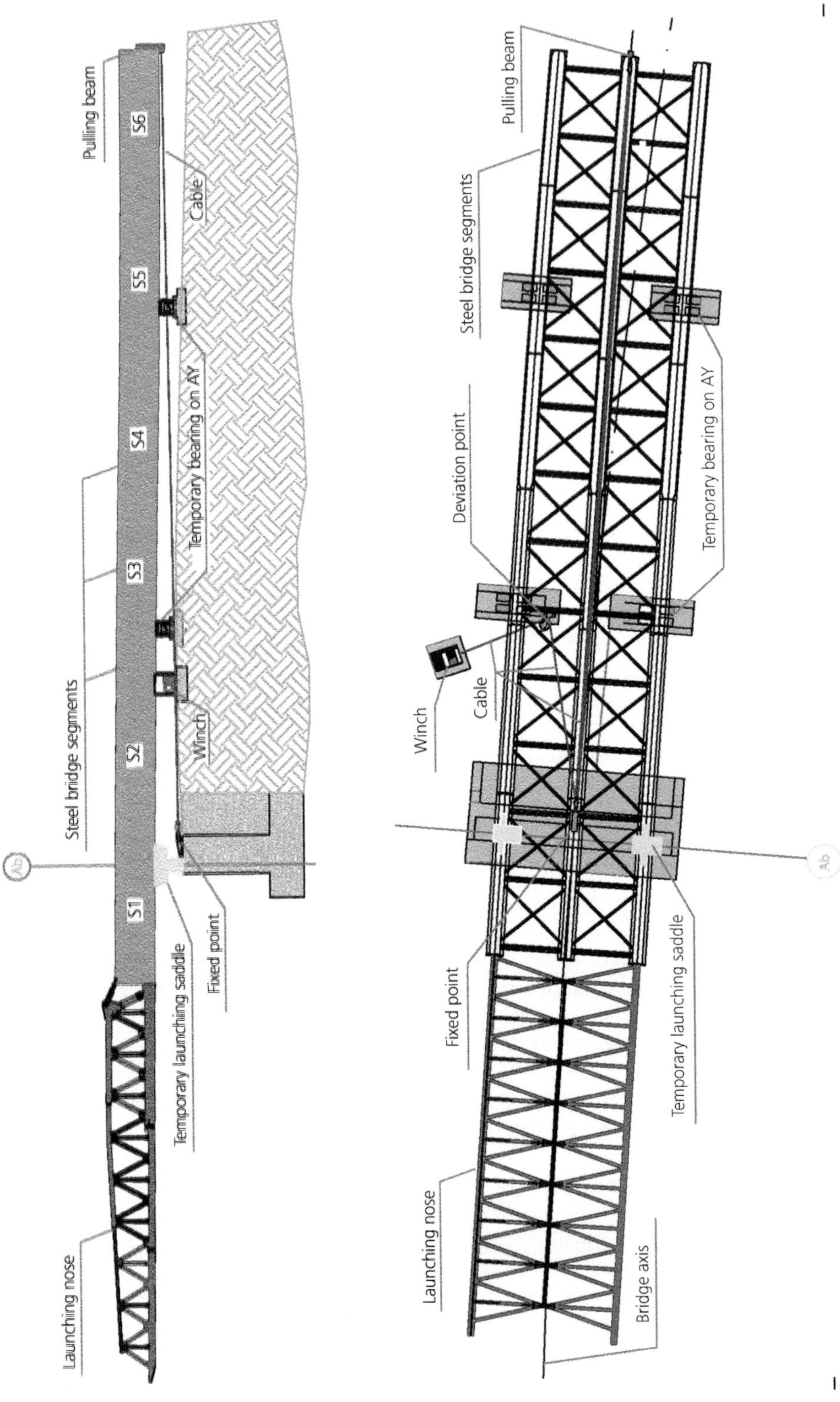

Figure 13.10 Launching of steel girders with winches (Freyssinet)

Figure 13.11 Local strengthening of the bottom slab near the cut out for the pulling stick (Freyssinet)

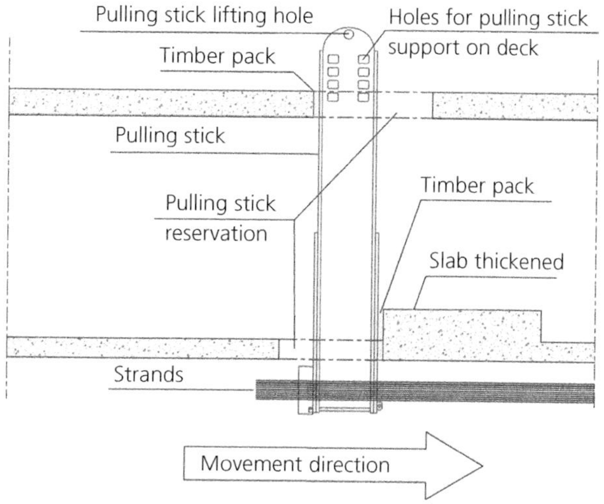

The load introduction of the vertical reactions during the launch into the steel girders and the concrete deck needs special attention. For more details refer to Section III, Chapter 15.

13.4.2.1 Sliding over final bearings

Final bearings are modified so they can be used for the moving of the bridge deck (refer to Figure 13.13). This is achieved by using a modified bearing steel top plate (tapered and, at its edges, chamfered). A stainless steel plate is fixed on top of the chamfered plate, on which the sliding pads bear and move during the sliding. Special devices are required to prevent abnormal rotation (shims beneath the bearing's top plate) and sliding between the bearing's top plate and its base in the case of free-sliding and guided bearings.

When the bridge is in its final position, it is jacked up to remove the stainless steel plate and the temporary fixity between the bearing's top plate and its base (in the case of free-sliding and guided bearings). The final connection between the bearing and the concrete deck is achieved by

- welding a compensating plate against a matching steel plate, embedded for this purpose in the soffit of the bridge deck, or
- building a robust and tight formwork on three sides around the bearing's steel top plate. Fill the space between the bearing and the bottom of the deck with dry mortar. Once the mortar starts hardening, the formwork and the temporary fixity between the bearing's top plate and its base in the case of free-sliding and guided bearings should be removed.

The first solution is difficult to build (matching plates, in-situ welding) and makes the final corrosion protection dependent on site-applied surface preparation and painting. The second solution is possible if, also for maximum horizontal forces, the concurrent vertical reactions are big enough to transfer the horizontal forces by friction.

Furthermore, use of final bearings is generally only possible for launching of concrete box girders if the bottom slab is laterally extended beyond the outer web face. For details, refer to Section III, Chapter 15.

Figure 13.12 Attachment details (Freyssinet)

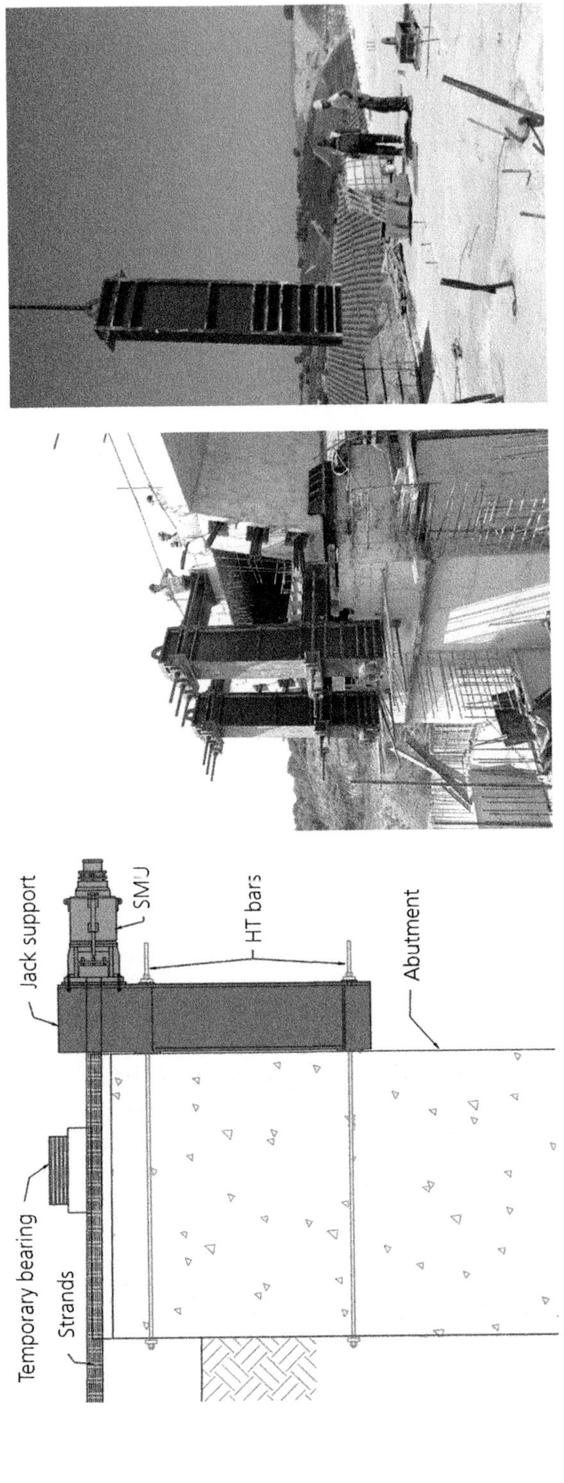

Vertical beams at abutment for fixity of SMUs

Vertical beams for fixity of strand bundles for the last launch

Pulling stick for fixity of strand bundles (normal launch)

Figure 13.13 Permanent bearing (modified) used during launching and during final stage (Freyssinet)

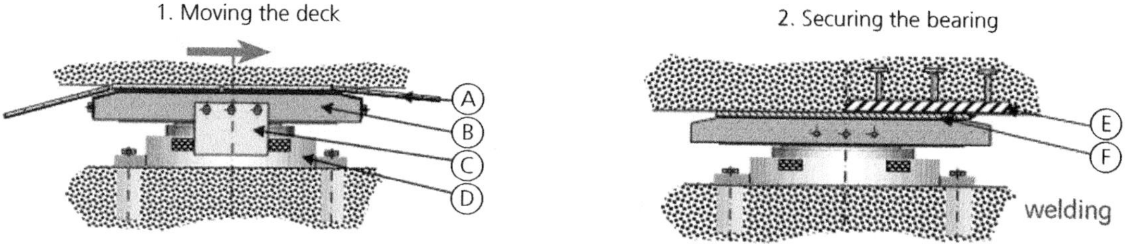

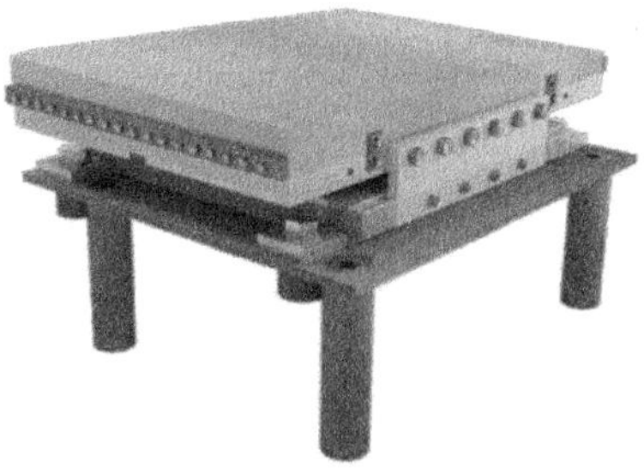

A. Sliding pads – neoprene/PTFE
B. Sliding chair – steel
C. Temporary stop – steel
D. Bearing
E. Fixing plate – steel
F. Compensating plate – steel

This solution is not often used because of the lead time usually required to design and procure the permanent bearings, and sometimes because of the matter of responsibility (the bearings are being used for a different function than in the final design).

13.4.2.2 Sliding over temporary sliding bearings

Sliding over temporary sliding bearings is the most popular solution for the launching of concrete decks. The temporary sliding bearing consists of an external steel frame filled with concrete. The top face is a properly finished horizontal surface chamfered at its ends and covered with a stainless steel plate (refer to Figure 13.14). These temporary bearings are installed on the supports like permanent bearings (that is with a layer of grout against the pier top face). However, unlike the permanent bearings, these sliding bearings are not installed horizontally but parallel to the deck's soffit.

Sliding pads are inserted between the deck and the temporary bearings during launching. When these pads come out from the bearing they are recovered and reinserted from the back. To ease the installation, these pads are chamfered at their ends.

When the launching is completed, the deck is jacked up and the temporary bearings are removed. They may be used for another bridge or stored for future use. The final bearings are then placed in position and fixed to the pier and the deck, for example as follows:

- inserting shear studs into box-outs in the deck and the pier and bolting them to the bearing's top and bottom plates
- grouting a plinth at the top and bottom between the bearing and the deck and the pier respectively.

The box-outs in the pier have to be temporarily covered for the launching duration.

Figure 13.14 Typical temporary sliding bearings for concrete decks (Freyssinet)

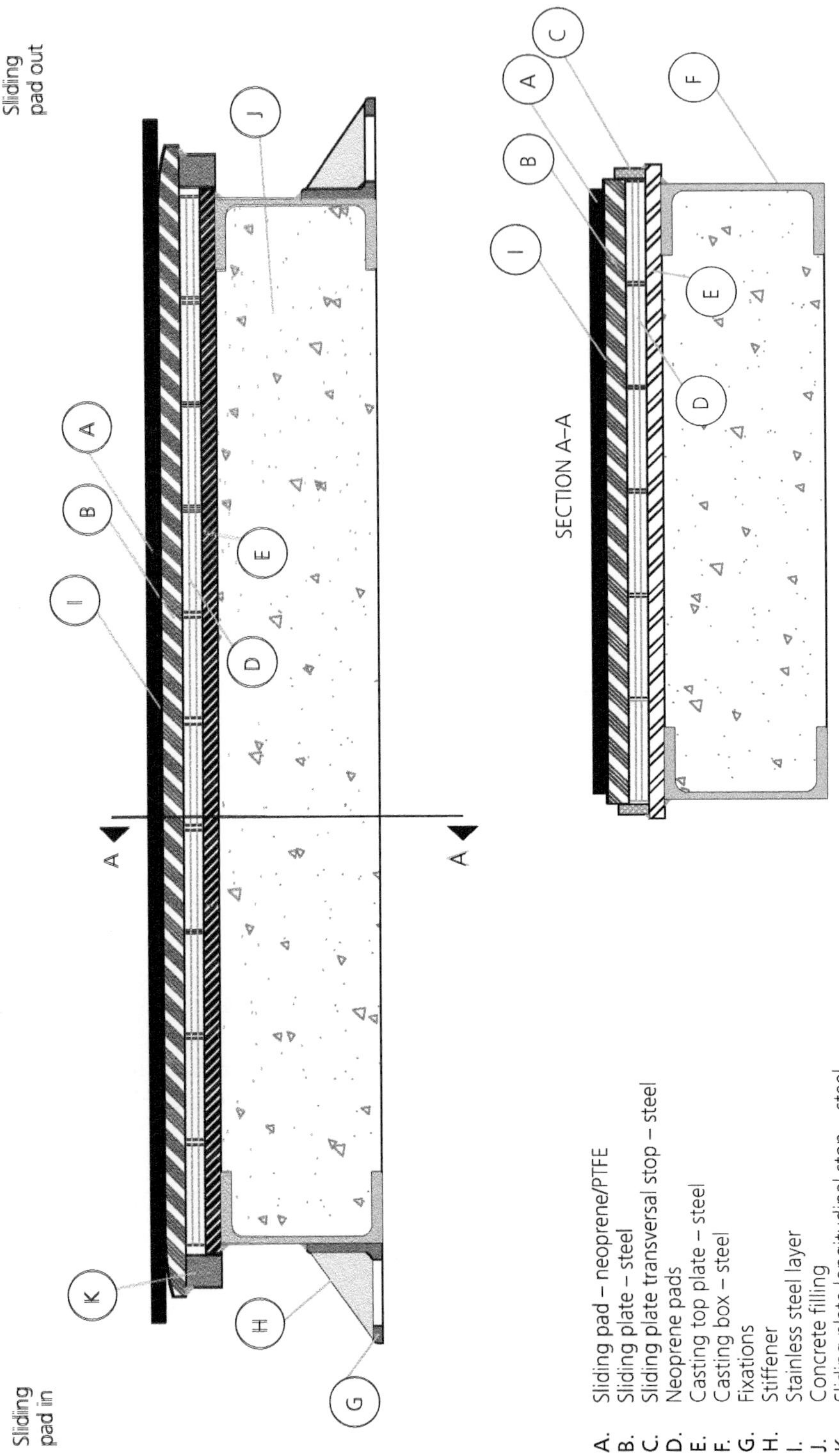

A. Sliding pad – neoprene/PTFE
B. Sliding plate – steel
C. Sliding plate transversal stop – steel
D. Neoprene pads
E. Casting top plate – steel
F. Casting box – steel
G. Fixations
H. Stiffener
I. Stainless steel layer
J. Concrete filling
K. Sliding plate longitudinal stop – steel

Figure 13.15 Temporary sliding bearing (sliding skid) for steel/composite bridges (Freyssinet)

The same principle is also used for temporary sliding bearings for launching of steel girders. But these bearings are normally made from steel and are articulated to cope with the much bigger angular variations of steel decks over supports during launching. They are called sliding skids. Refer to Figure 13.15.

13.4.2.3 Roller saddles

Roller saddles, as shown in Figure 13.16, are made up of a steel frame and rollers to ensure not only that the steel girder rolls over them with a very low friction, but also that the support reactions are

Figure 13.16 Hydraulic roller (left) and typical roller saddle (right) (left: LMV; right: SCL)

introduced into the steel girders at the right location and balanced. The most common type of roller saddle is the rocker one. The base of the device is articulated to ensure that the rollers, generally arranged in pairs, remain in contact with the bottom flange even if the deck experiences a significant rotation.

The number of rollers required depends on the maximum vertical reactions and on the capacity of the steel girder's bottom flange/lower part of web to take point loads. Generally rollers are used which can take 500 kN. Rollers with a bigger capacity of up to 1000 kN are available but are expensive.

Rollers are generally built from high-resistance steel, such as quenched and tempered steels. These steels are hardened and tempered (hardening followed by tempering at around 600°C) and are used to fabricate mechanical elements such as shafts, semi-axles, etc.

Sometimes hydraulic systems are used for particular structures, which makes it possible to measure the reaction on each bearing and to make possible some adjustment during launching (refer to Figure 13.16, left).

13.4.2.4 Sliding pads (for sliding bearings)

Sliding pads are made of reinforced neoprene with a PTFE layer on one side only. Their thickness is normally limited to about 20 mm. Refer to Figure 13.17.

Pads made of high-density polyethylene are approximately 15 mm thick, are cheaper and are an effective alternative to PTFE. However, their resistance and the creep under high compressive loads have to be investigated.

Special care has to be taken by the contractor during casting of concrete decks with regard to evenness of all the external surface areas, which are sliding against temporary bearings and lateral guides. Irregularities over a length of 2 m should not exceed 2 mm. Furthermore, the stainless steel surfaces need to be well taken care of to avoid them getting spoiled or scratched by dust, concrete or metal scraps. Special silicone/PTFE grease is used in order to lubricate the sliding surfaces.

13.4.3 Guiding system

An incrementally launched bridge has to be guided, whatever its alignment is (straight or curved). The guiding system may be quite small for a short straight bridge, whereas more sophisticated devices are necessary for curved bridges.

If the bottom deck soffit is inclined (for geometric reasons or for being sure that the deck is always touching the guides on the lower side), an additional force has to be taken into account when designing the guiding devices.

Figure 13.17 Typical sliding pad (Eberspächer)

Principle structure

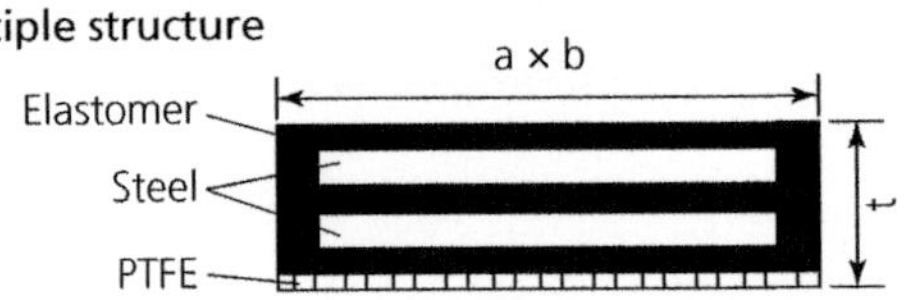

Technical data

Max. perm. pressure: 22.5 N/mm²
for launching bearings
up to 15 N/mm²
max. size: 1000 × 1000 mm
with lubrication cavities

13.4.3.1 Guiding devices for straight bridges

For straight bridges, the guiding devices are generally installed every two or three piers, or more frequently. The guiding force can be estimated as a small percentage of the vertical reaction on the launching bearing plus the reactions from wind acting on the deck.

At least one line of guiding devices has to be installed just after the casting or assembly yard in order to eventually adjust the transversal position of the rear part of the deck before the concreting or joining of the next segment. For illustration refer to Figure 13.18.

13.4.3.2 Guiding devices for curved bridges

For curved bridges the guiding force can be calculated as follows:

- Formulating the equilibrium of the deck (in a horizontal plane) under the guiding forces, the wind loading and the friction forces on the piers. This method can be used if there are only two guiding lines present (three equations for three variables: two guiding forces and the launching force).
- Modelling the deck as a curved beam supported on N bearings (N = number of guiding lines) under a system of tangential friction forces acting on the piers and the wind loading.

The commonly used guiding systems are made from steel brackets fixed on the sides of the piers by means of embedded HT bars or transverse prestressing cables. A stainless steel plate is used as contact surface to receive sliding pads passing from the rear to the front during the movement of the bridge deck. If external concrete buffers are present on the pier (which is quite common if the bridge is in a seismic area), the guiding devices can be just steel shims fixed against the buffers and stainless steel plates.

13.4.3.3 Guiding devices for steel/composite bridges

The most popular device to guide steel and composite bridge decks are rollers, fixed to the temporary roller saddle. Is it also possible to fix a rigid bracket to the temporary sliding skid (refer to Figure 13.19).

Figure 13.18 Typical guiding device (left) and sliding pad introduction phase (right) (Freyssinet)

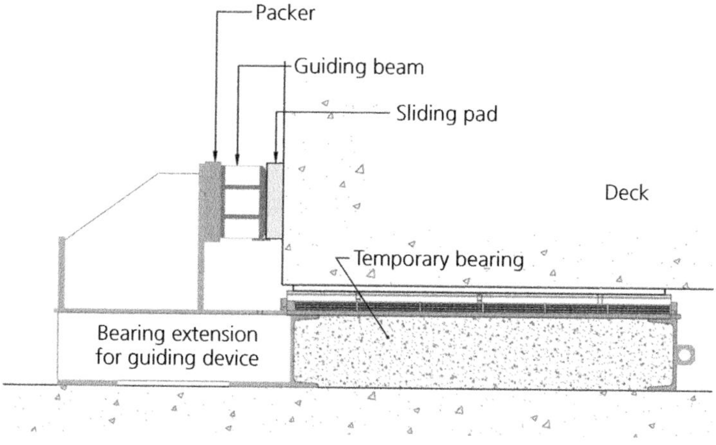

Figure 13.19 Typical roller guiding device (left) and guiding skid device with lateral guide (right) (Freyssinet)

13.4.4 Assembly and casting yard

The distance of the assembly area/casting mould from the abutment is selected to

- avoid excessive angular rotation at the rear end of the deck/steel girders after the initial launch, which makes it difficult to add the second segment
- enable installation of the launching nose in front of the first segment being assembled or cast
- minimise resetting of strand bundles, brackets or pulling sticks if concrete decks are launched by pulling with SMUs; minimize resetting of brackets or pulling sticks if deck/steel girders are launched by pushing with pushing jacks installed not behind the segment but between assembly area/casting bed and the abutment.

Sometimes the space available behind the abutment is limited or a lot of excavation would be needed to create enough space. In such cases launching systems may have to be used whereby segments are pushed or pulled from behind. To allow installation of launching noses and control the angular rotations mentioned above and in some cases even to ensure basic stability against overturning during the first launch, it may be necessary to build a temporary tower in the first span. Angular rotation can also be minimised and overturning prevented by placing counterweights onto the first segment during the initial launch.

The following explanations refer to steel/composite bridges only. For details about casting beds and casting yards for casting of segments for incrementally launched bridge decks, refer to Section III, Chapter 19.

The assembly yard comprising of the assembly area and the launching area is one of the most important factors when planning an ILM operation. A site visit at the beginning of the studies is strongly recommended.

If possible, the launching will be done upwards from one side only. The most important aspects which need to be investigated when planning the assembly area are as follows.

- Geometry of the assembly yard (length, width)
  - length: the length needs to be sufficient to assemble an element and to have enough distance between the assembly area and the abutment for reasons explained above

- width: the width needs to be big enough so that cherry pickers, trucks and cranes can circulate in the assembly yard beside the bridge deck being assembled (including, for example, welding cabins around main beams) and launched.

- Level for launching: at final level or elevated. Need for excavation or back fill. Normally it is better to launch at final level, which means that the upper part of the abutment wall can only be cast after the launching.
- Soil capacity: plate loading tests should be done to determine permissible soil pressure for the design of pad footings for temporary supports in the assembly area and between assembly area and the abutment. These foundations could also be provided with some lugs for easy dismantling and reuse. Normally a minimum permissible soil pressure of 0.2 MPa is required.
- Accesses to the assembly yard: site roads must have a gradient and a geometry that allows the passage of trucks loaded with the steel elements. Some reference values are: maximal gradient = 7%; minimum radius = 30 m.

Once the temporary foundations have been built, steel elements can be delivered to the yard for assembly of segments. For handling the steel elements, normally lifting pincers are used or lugs fixed (bolting is preferred) to the steel elements to be handled. During handling and assembly of the steel elements care must be taken not to damage the painting. Paint suppliers have products that can be used to touch up locally damaged areas.

The elements of the first segment are normally installed on temporary concrete blocks which are high enough to achieve the specified camber. It is recommended to make a topographic survey on segment length and on camber. Single elements are connected by means of bolted or welded joints. For the first solution, torque and control procedure must be defined. For the second solution, non-destructive tests on welding must be defined (type of control and extent).

At the same time the first segment is assembled, the launching nose is installed and connected.

For composite bridges with a cast-in-situ deck slab, 'lost' formwork panels are often used to avoid the need for form travellers. Such formwork panels are either 50 to 60 mm thick concrete planks or 5 to 6 mm thick steel plates. These panels are installed on the steel elements prior to the launching (except in the case of concrete planks, where on the first span no planks are installed to reduce deflections). Alternatively the precast planks can also be installed by a transport frame rolling on the upper flanges of the steel girders after the launching. An area for storage of these precast planks or steel plates must be provided in the assembly yard, together with an installation zone, so as not to interfere with other activities.

Sometimes painting activities are carried out before launching. After launching, touch-up will be done at areas where paint has been damaged.

For illustration of an assembly yard for a composite bridge deck, refer to Figure 13.20.

13.4.5 The launching nose

Technical data for typical launching noses are given in Table 13.4. For the main elements of typical launching noses, refer to Figure 13.21.

Fabricators of steel decks often own launching noses, which they recycle from project to project by making slight modifications.

Figure 13.20 Assembly yard for a composite deck (Steel Project Italia)

13.4.5.1 Concrete decks

The launching nose is generally made of a pair of built-up main beams, which are horizontally and transversely braced against each other with rolled sections (refer to Figure 13.21). Particular attention is to be given to

- the stiffness, which needs to be in line with the bridge designer's assumption
- the fixation to the deck
- the detailing of the tip of the nose with hydraulic jacks to allow the nose to land and slide over temporary bearings installed on piers; since the calculation of the deflection of the tip of the nose depends on many parameters, it is advisable to base the sizing of these jacks on increased theoretical deflection and theoretical force needed to lift up the nose.

For fixing the nose to concrete decks either HT bars or, if forces are very big, PT tendons can be used. Corrosion protection has to be applied, in particular for long bridges (with a long construction time). Sometimes the depth of the deck is at the front locally increased to allow use of a launching nose with a

Table 13.4 Technical data for typical launching noses

	Weight ratio g_{nose}/g_{girder}	Inertia ratio I_{nose}/I_{girder}	Length L_{nose}/L_{span}
Concrete girders	0.08–0.11	0.10	0.60–0.70
Steel girders	0.30	Not applicable	0.30–0.65

Figure 13.21 Typical launching nose for a concrete deck: main elements (Freyssinet)

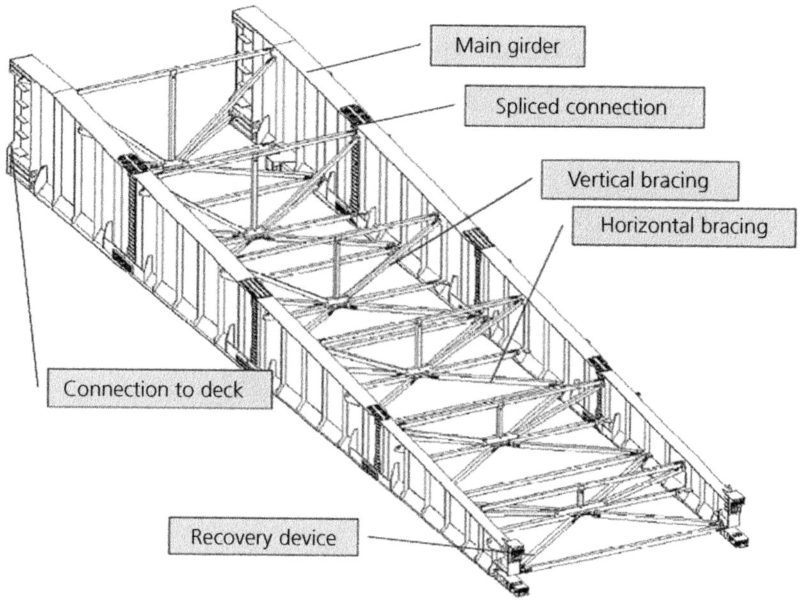

bigger depth than the deck. This temporary vertical extension can be fixed to the deck by vertical HT bars.

13.4.5.2 Steel decks

For steel/composite bridges the main beams of the launching noses may be trusses or beams, which are fixed to the bridge's steel girders by high-strength friction bolts (refer to Figure 13.22).

Figure 13.22 A 97-m-long truss launching nose (left) and a short swinging launching nose (right) (left: Seteco Ingegneria; right: SCL)

13.5. Kinematics

13.5.1 Concrete bridges

For concrete bridges the process begins with the construction of the first segment (S1), which is match-cast against the launching nose (LN; installed on temporary supports). The structure is built more or less at final level. When the concrete has reached the design strength, the centric PT in S1 and the HT bars connecting the LN to S1 are stressed. S1 + LN are launched forward so that the casting area is freed up for fabrication of the second segment. After casting and curing of the second segment (S2), the centric PT in S2 is stressed. The full structure is then launched forward by a segment length and the construction of S3 can start. These sequences are continued until the deck is completed. Subsequently the LN is dismantled, the deck is put on its final bearings (by jacking) and the continuity PT is installed and stressed. Refer to Figure 13.23.

Figure 13.23 Concrete bridge deck – typical sequence (Freyssinet)

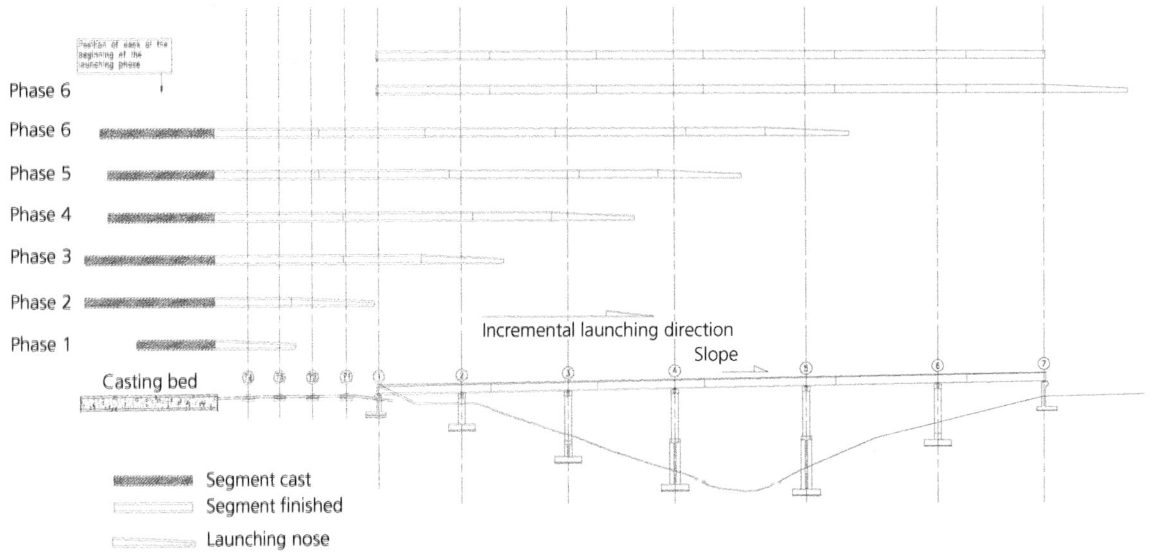

13.5.2 Steel and composite bridges

For steel and composite bridges, normally the maximum number of segments that fits into the assembly yard is installed on temporary supports (with crane or portal crane). Segments are connected to each other by bolting or welding. The launching nose (LN) is also installed and connected to the first segment. Connection of different segments must take into account the design camber. The steel girders are generally launched some distance above the final level on temporary supports installed on top of abutment and piers. Therefore, for the majority of cases a bigger part of the upper abutment can be built before the launching of the deck. Higher towers are, of course, more expensive and result in more time for lowering the deck to its final position after completion of the launching. Refer to Figure 13.24.

At the end of the launching, after the dismantling of the LN, the girders are lowered by means of hydraulic jacks to their final level onto the permanent bearings.

Figure 13.24 Composite bridge typical sequence (Freyssinet)

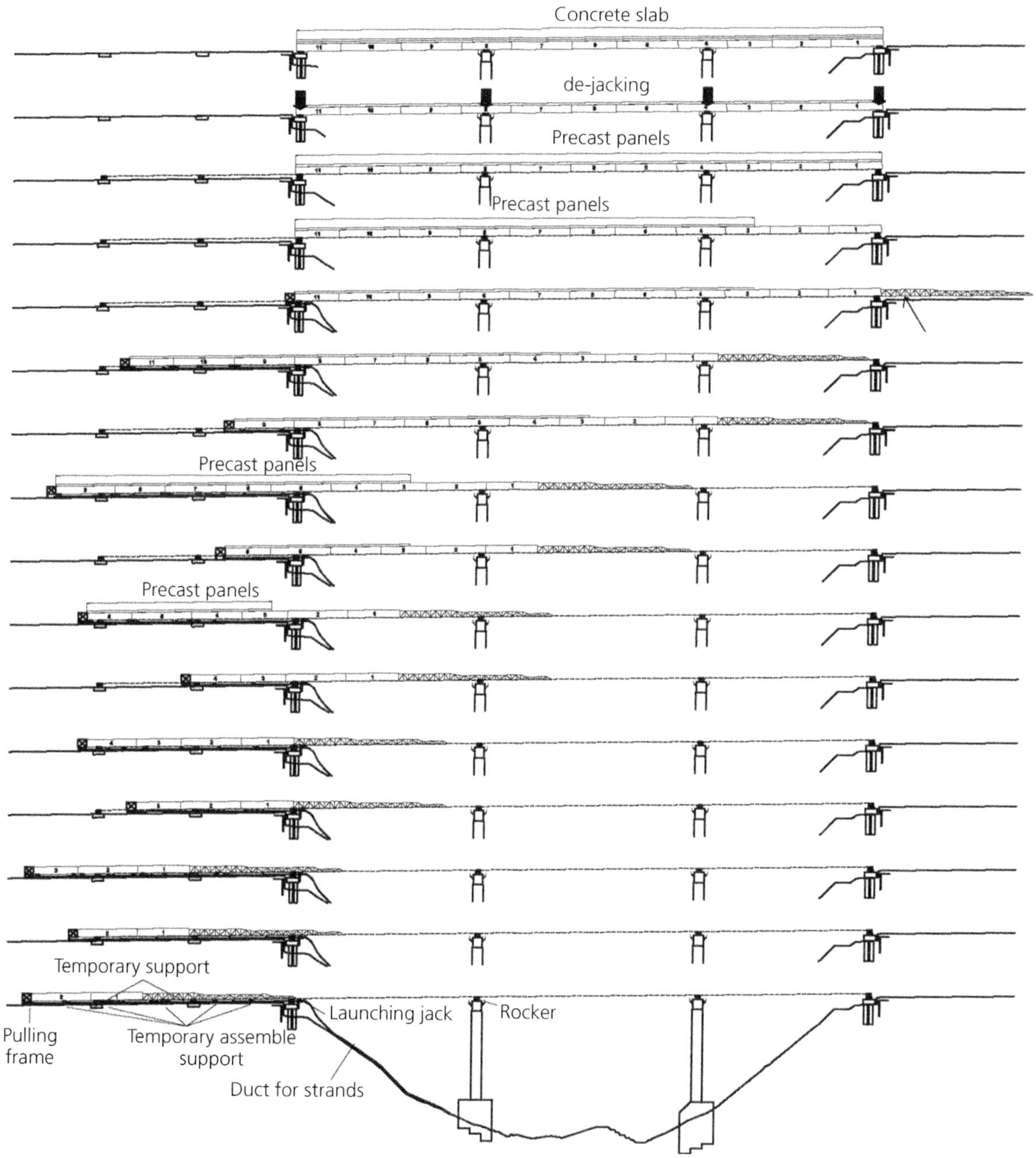

13.6. Permanent work interfaces

As explained in detail in Section III, Chapter 15, the following points need to be verified:

- Introduction of support reactions (from the temporary bearings) during launching into the soffit of the concrete deck or steel girders.
- Plan dimensions of pier caps and abutment support benches and vertical clearances against soffit of girders to be launched to ensure that there is enough space available for
  - temporary supports if any
  - temporary bearings
  - guiding devices
  - jacks to lift up deck during launching if sliding pads get jammed or damaged (location in front or behind the temporary bearings)
  - temporary storage of permanent bearings on top of the pier
  - jacks for removal of temporary supports
  - jacks for installation of the permanent bearings
  - attachment of platforms for workers.
- Capacity of abutment's foundations to take the launching force.
- Capacity of piers and their foundations to take the bending moments due to
  - the friction forces during launching
  - the lateral guiding forces
  - the eccentric vertical loading during jacking-up of the deck, when there is need to replace jammed or damaged sliding pads.

Additional points which need to be considered are as follows:

- When modelling the deck during launching, the launching nose needs to be modelled as well with its weight and stiffness.
- Loading during launching and lowering: there needs to be a discussion among the involved parties in order to define and agree about the loads to be considered during construction.
  - Climatic actions on deck (wind, snow, temperature): during the moving stages a reduced design wind speed can be considered. However for the stationary stages the full wind load is applied with the agreed reduction factor.
  - Earthquake action: in the case where seismic loading has to be considered, two different approaches can be adopted during stop phases
    I. transmit seismic actions to fixed point(s) at launching abutment or fixed piers
    II. isolate the bridge with neoprene pads positioned over jacks (see Figure 13.25).
  - Construction live loads (workers, precast/steel planks and reinforcement for composite bridge decks, temporary equipment on the deck).
  - Differential 'settlement' between piers, transverse differential 'settlement' between pair of temporary bearings on a support (refer to Figure 13.26).

13.7. Cycles

The cycle for bridge decks built by ILM is primarily controlled by the time needed to assemble the segments (in the case of steel girders) or to cast and cure a concrete segment. Typically a launching operation to move the deck by a segment length takes not more than one shift. For the cycle to cast a concrete segment, refer to Section III, Chapter 19.

Figure 13.25 Seismic isolator device for launching (Freyssinet)

Figure 13.26 Differential settlements (EN 1991 – Eurocode 1): limits in mm

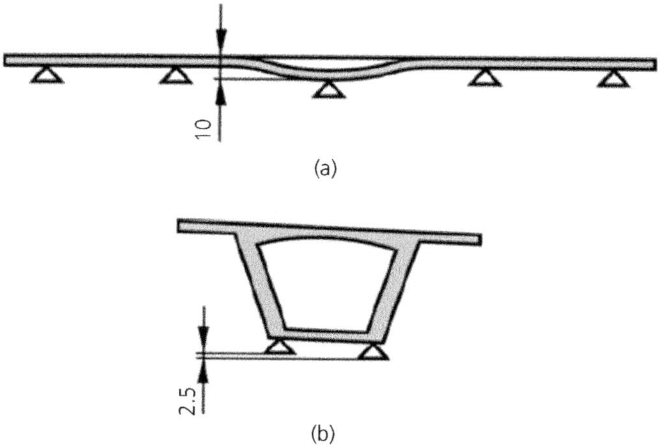

13.8. Risks

The list of risks given in Table 13.5 is by no means comprehensive, but addresses risks which are particular to ILM. For a specific project a main contractor needs to carry out a full risk assessment of the operations performed during construction phases, the erection sequence and the site constraints. For casting operation of concrete decks refer to Section III, Chapter 19.

Table 13.5 List of risks specific to the operation of the ILM

Activity	Risk	Mitigating measures	Concrete deck	Steel or composite deck
During Launching				
Insert of sliding pads on temporary bearing/ >guiding devices; jacking up of nose on piers; Bridge jacking	Vertical access and work at height: climbing up and down, work performed at height (at least 3 to 4 m above ground)	Access ladder fixed to stable base or access staircase; access platforms with handrails and kicker boards around pier caps; use of cherry picker	X	X
Wrong insert of sliding pads on temporary bearing/guiding devices	Jamming of launching operations	Emergency stop button at every temporary bearing location; Monitoring system of pier displacements. Provide space for jacks on top of the piers; provide spare pads	X	X
Working on top of and inside deck/LN	Falling down from deck/LN	Provision of temporary handrails around all exposed edges on top and inside of deck/LN	X	X
Insertion of sliding pads	If pads are not greased: high friction so higher forces on piers and on launching system	Use grease to reduce friction during launching	X	X
Splice connection design	Interference between bolted connections and launching bearings/rollers	Check before fabrication of steel components that splice connection geometry does not interfere with launching bearings/rollers	X (valid for LN)	X
Local load introduction	Local buckling of web under concentrated loads	Check during design phases that steel elements can resist local reactions from temporary bearings	X (valid for LN)	X
Local load introduction	Punching through of bottom slab	Centre temporary bearings at intersection of web's and bottom slab's centre line	X	NA
Rigid connection nose/deck	Opening of the connection between nose and deck	Make an SLS calculation to check that the section is always compressed during launching phases	X	X
Falling of splice bolt	Fall of untightened bolts	Check tightening of bolts; place bolts with the head in the upright position	X (valid for LN)	X

Table 13.5. Continued

Activity	Risk	Mitigating measures	Concrete deck	Steel or composite deck
Jack stroke/capacity on recovery device not sufficient	Stop of launching since the recovery device has not enough stroke or has not enough capacity to take the reaction load	Size jacks on increased theoretical values of deflection and reactions at the tip of the LN	X	X
Launching over 'in service' road/railway	Falling of debris, tools (possible)	Use a 10 cm kicker board plus a safety net (if necessary)	X	X
Downhill launching	Uncontrolled bridge deck movement	Design a braking system to block the bridge during launching	X	X
Assembly phases				
Temporary equipment handling near high-voltage cables	Contact between handling devices and high-voltage cables	Check carefully positioning of high-voltage cables; note this in the installation procedures	X	X
Pulling stick insertion	Breaking of the bottom slab in the load introduction zone of pulling stick	Verify whether a thickening of the bottom slab is necessary in the pulling stick insertion zone	N/A	X
Temporary bearing location and jack location	If temporary bearings or jacks are too close to pier cap side, local spalling of concrete in the non-reinforced zone can occur	Check that the position of temporary bearings and jacks is far enough from pier cap side	X	X
Temporary pier positioned on existing roads	Collapse of temporary pier and/or the bridge if a truck hit the structure	Protect temporary pier so that trucks cannot hit it	X	X
Placing of precast planks	Interference between steel studs and plank's starter bars	Preassembly of precast plank's reinforcement cage with a template	N/A	X
Installation of temporary bearings	Vertical access and work at height: climbing up and down, work performed at height (at least 3 to 4 m above ground)	Access ladder fixed to stable base or access staircase, access platform with handrails and kicker boards around pier cap; use of cherry picker	X	X

13.9. Performance criteria

Table 13.6 shows items that need to be addressed when drafting a performance specification for an ILM project.

Table 13.6 Items that need to be addressed when drafting a performance specification for an ILM project

	Concrete deck	Steel or composite deck
■ Deck		
– weight, dimensions, spans arrangement	X	X
■ Precast planks (if installed before launching)		
– weight and dimensions		X
■ Alignment		
– gradient and super-elevation, curvature in plan	X	X
■ Permanent work interfaces		
– available width below main girders or webs for temporary bearings	X	X
– area of deck's cross section available for lateral guidance of deck during launching		
– space constraints for openings for pulling stick or fixity of brackets	X	NA
– abutment and pier capacity against launching forces	X	X
– abutment geometry for installation of pulling devices	X	X
– stage construction of abutment to permit passage of deck during launching	X	X
– pier cap dimensions: pier support under webs; space for temporary bearings, jacks for removal of wrongly inserted sliding pads and jacks for installation of permanent bearings length, weight and stiffness of LN		
– fixity of LN to deck		
■ Assembly/casting yard		
– space available behind abutment to set up assembly/casting yard	X	X
– height of deck assembly above final level	NA	X
– delivery, temporary storage and handling of deck elements (steel) and planks	NA	X
■ Access		
– installation and stressing of PT tendons	X	
– rebar installation	X	X
– concreting	X	X
– accesses to piers/abutments during launching	X	X
– LN installation	X	X
– splicing of main girders including corrosion protection of steel work		X
– installation of planks		X
■ Clearances	X	X
■ Cycles		
– main contractor's work schedule	X	X
– single or double shift		
■ Reusability	X	X
■ Methods, time, crane requirements for assembly, relocation, dismantling of assembly/casting yard and launching equipment and accessories	X	X
■ Follow-up work: closing of holes, openings etc.	X	

13.10. Photo gallery

Figure 13.27 KM03, Turkey six highway bridges

Box girder, span 55 m, width 22 m, height 3.4 m, weight 380 kN/m, max length 650 m, min radius 1400 m. Launching system: pulling with SMU (Freyssinet)

Figure 13.28 BTZ, Algeria four railway bridges

Box girder, span 40 m, width 10.6 m, height 2.6 m, weight 210 kN/m, max length 600 m, min radius 1060 m, vertical gradient +2%/−3%. Launching system: pulling with SMUs and pushing (Freyssinet)

Figure 13.29 Approach viaducts of road bridge over Sungai Johor, Malaysia

Triple-cell box girder, span 48 m, width 21 m, height 3.6 m, weight 264 000 kN, total length 485 m per approach viaduct, vertical gradient 3%, straight. Launching system: lifting/pushing on every support (VSL International)

Figure 13.30 Road bridge over Tampines Expressway at Senkang East Road Interchange, Singapore

Box girder, max span 51 m with temporary pier, width 17.5 m, height 2.2 m, weight 28 000 kN per deck (twin bridges), total length 100 m, vertical gradient ~0%, radius straight but with skew supports at 8.5 degree. Launching system: pulling with SMUs (VSL International)

Figure 13.31 Iron Cove, Australia highway bridge

Box girder, span 53 m, width 16.8 m, height 3.3 m, weight 300 kN/m, max length 475 m, radius 1030 m. Launching system: lifting-pushing (Freyssinet)

Figure 13.32 Gilloy-Johannesburg highway interchange H, South Africa

Box girder, max span 40 m, width 12 m, height 3.1 m, weight 230 kN/m, total length 247 m, radius 137 m. Launching system: lifting-pushing (Steel Project Italia)

Figure 13.33 High Speed Line Oued Tlelat/Tlemcen, Algeria – viaduct PK96 – railway bridge

Steel truss, span 130 m, width 13.2 m, height 13.75 m, steel deck weight 130 kN/m, total length 1780 m, straight alignment. Launching system: pulling with strand jacks/pushing on rollers (Società Italiana per Condotte d'Acqua)

Figure 13.34 Millau Viaduct, France – highway bridge

Steel multibox girder, max span 342 m, width 28 m, height 4.2 m, steel deck weight 150 kN/m, total length 2460 m, radius 20 000 m. Launching system: lifting-pushing on every support (Freyssinet)

Figure 13.35 Serra viaduct, Italy – highway bridge

Twin-girder composite bridge, span 115.5 + 119.5 m, width 13.2 m, height 4.57.5 m, steel weight 80 kN/m, length 235 m. Launching system: pulling by winch; existing bridge used as launching nose for the new bridge (Seteco Ingegneria)

Figure 13.36 Tevere viaduct, Italy

Twin-girder composite bridge, max span 166 m, width 19.7 m, height 7 m, weight 100 kN/m, total length 344 m. Launching system: pulling by bar jacks, launching nose about 97 m long (Seteco Ingegneria)

REFERENCES

Figures 13.1, 13.5, 13.6, 13.7, 13.9, 13.10–13.15, 13.18, 13.19, 13.21, 13.23, 13.24, 13.25, 13.27, 13.28, 13.31, 13.34: Freyssinet
Figures 13.2, 13.3, 13.4: Placidi, M 'La construction au sol et la mise en place par déplacement', course script from ENPC (École Nationale des Ponts et Chaussées), 1998
Figures 13.8a, 13.8b, 13.29, 13.30: VSL International
Figures 13.16 (left): LMV; http://www.lmvworld.com/projects/machines-and-equipment/42-rollers-for-bridge-launching
Figures 13.16 (right), 13.22 (right): SCL; http://sclitaly.weebly.com/
Figure 13.17: Eberspächer; http://www.eberspaecher-hydraulik.de/en/produkte/verschubanlagen
Figures 13.20, 13.32: Steel Project Italia; http://www.steelprojectitalia.it/index.php
Figures 13.22 (left), 13.35, 13.36: Seteco Ingegneria; http://www.setecoge.it/
Figure 13.26: EN 1991 Eurocode 1: Actions on structures. Part 1–6 General actions. Actions during execution
Figure 13.33: Società Italiana per Condotte d'Acqua; http://www.condotte.com/en/index.aspx

Bridge Deck Erection Equipment
ISBN 978-0-7277-6193-4

https://doi.org/10.1680/bdee.61934.283

Chapter 14
Prefinished bridge installation methods

14.1. Introduction

The Prefinished Bridge Installation method (PBI) is a traffic-friendly method using jacking and skidding techniques or transporters. With the PBI, the bridge deck, the superstructure of a bridge or even the entire bridge with its foundations is first built away from its final location and is subsequently moved into its final position, under or over an existing infrastructure corridor that needs to be crossed.

These methods are not very new but their application is an effective and fast way to minimise disruption of mobility of people or goods, with their associated costs to society and risks for the safety of road-users and operators.

This chapter presents four different jacking & skidding and transportation techniques and a number of installation solutions for a safe and rapid installation of bridge decks or bridges which have been completely prefabricated and pre-finished before installation.

The full-span construction method used to build decks of long viaducts is basically also a PBI method. This method is dealt with in Section II, Chapter 8.

As shown in Figure 14.1, there are three options where pre-finished bridge decks can be installed.

1. Building above final position: jacking down in the Z-axis
2. Building beside final position: skidding along the Y-axis
launching along the X-axis
rotation around the Z-axis
3. Building at a certain distance from the deck's final location: Transportation with modular transporters

The PBI method is appropriate for building bridges and underpasses in, under and above existing infrastructure corridors when disruption of mobility during construction has a significant impact. The method can be used for small- to medium-sized structures with little restrictions to structural configuration and design. The equipment required for some of the PBI methods is not very heavy and can therefore be easily mobilised, even in congested areas. If the structure permits, the bridge deck can be completely finished prior to the installation operation, since the cost of the equipment is not very sensitive to the weight of the object.

Most of the bridge decks dealt with in this chapter are ready for traffic use directly after their installation. The installation takes place during a very short closure time of the existing traffic lanes or rail tracks. It requires reliable techniques and the management of a multidisciplinary process.

Figure 14.1 Different moving directions of bridge decks (DeBoerDC/deBoer International)

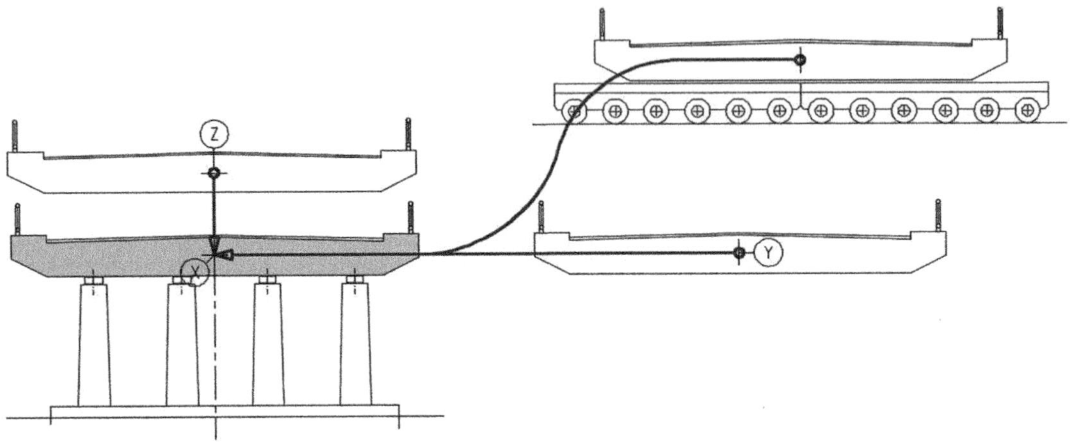

In developing the system, it is important to consider the interaction between permanent works design and the installation method. The range of available applications for the PBI method is extensive. Seven different design and installation concepts will be presented.

Moreover, it is worth mentioning that the PBI method has a very good reputation with regards to safety for operators and road users. Historically there have been very few incidents. But experience and craftsmanship of the parties involved is crucial.

Last but not least, all these construction methods can, of course, be used in a reverse way for dismantling bridges.

14.2. Moving techniques

PBI decks are always heavy structures, because it is not a question of handling just components, but complete, finished structures with weights which can exceed 100 000 kN. In most cases, craning is not an option. Although there are a lot of developments in this field, the capacity of a crane is always limited. The extra space needed to deploy the crane also causes disruption. This limits its use, which is why use of cranes is not dealt with in this chapter.

For general comments about available equipment to move bridge elements, refer to Section III, Chapter 17.

For the moving of complete viaducts, bridges and underpasses, the following four techniques are used:

- Hydraulic jacking systems for vertical movements
- Strand-jacking systems for vertical movements
- Skidding or sliding systems for horizontal movements
- Transportation with SPMTs

In practice, these different techniques will often be combined on a project.

14.2.1 Hydraulic jacking systems (for vertical movements)

For the raising and lowering of heavy loads, hydraulic jacking systems are available, utilising solid ram jacks with almost unlimited capacity. For examples refer to Figures 14.2 to 14.4 and for details about solid ram jacks refer to Section III, Chapter 17.

Figure 14.2 Sketch of a jacking configuration with jacking and support towers (DeBoerDC/deBoer International)

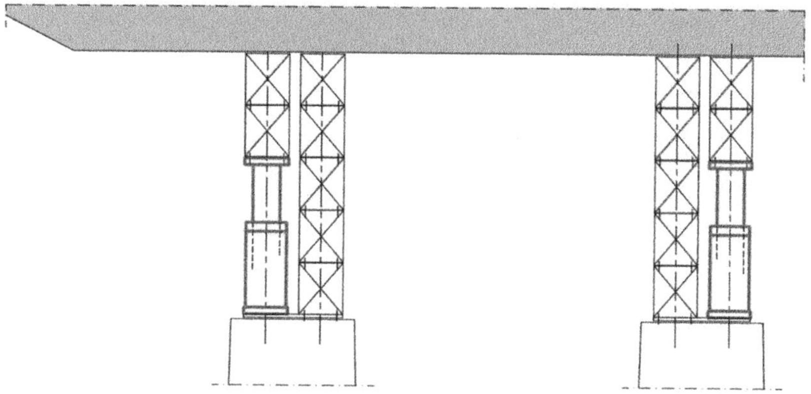

Figure 14.3 Group of three 1100 kN jacks with locking nut, stroke = 400 mm, with packers h = 340 mm (DeBoerDC/deBoer International)

Figure 14.4 Group of four 3000 kN jacks with locking nut, stroke = 200 mm, stacking elements h = 150 mm (Freyssinet)

To guarantee the safety and precision of the installation of bridges, hydraulic jacks have lock nuts fitted to the hydraulic ram. A jacking system is much more than a number of jacks, it consists of

- hydraulic jacks with a stroke and capacity which suits the specific operation
- shims and packers to achieve the overall lifting height
- foundations and temporary supports on which the jacks can be installed
- load-spreading plates/temporary bearing under the bridge deck
- power pack units and control systems for the movement of the bridge within the specified tolerances
- easy access facilities for the operators to every working station.

There are two choices for the working pressure of the hydraulic systems: a medium pressure system of 450 bar or the high pressure system of 700 bar. High pressure jacks have less diameter and weight for the same capacity. But if the stroke is long, the ratio between height and outside diameter becomes unfavourable and stability may become controlling for the dimensioning of the jack. Another aspect is the reduction of the contact pressure to permissible values: high pressure jacks need more or thicker load-spreading plates on the top of the piston.

A new development in jacking technique is the use of jacks with lifting heights of more than 1.50 m, enabling a lot of jacking operations by one single stroke (refer to Figure 14.5). When utilising jacks with such long pistons, it becomes more challenging to ensure stability of the system. Another important aspect is the use of electronic measuring and other IT devices for the piloting of jacking systems.

Figure 14.5 Group of six 4000 kN jacks with locking nut, stroke = 1.6 m (DeBoerDC/deBoer International)

14.2.2 Lifting and lowering with strand-jacking systems

For the lifting (or lowering) of heavy loads, strand-jacking systems are available with almost unlimited capacity. For examples, refer to Figures 14.6 to 14.8 and for details of Strand Moving Units (SMU) refer to Section III, Chapter 17.

A strand-jacking system allows quicker lifting or lowering than a jacking system with hydraulic solid ram jacks and takes the load by tension and not by compression, eliminating stability and buckling

Figure 14.6 Sketch of a strand-jacking configuration with SMUs and support legs (DeBoerDC/deBoer International)

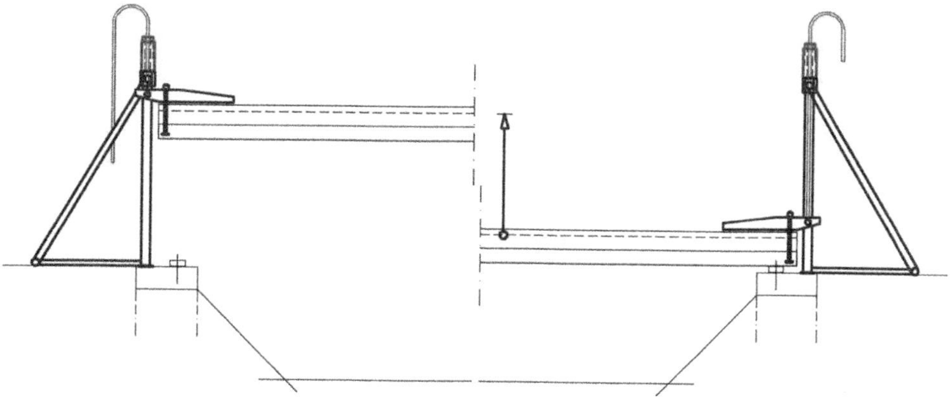

Figure 14.7 Lifting of a 42 000 kN bridge span with 16 strand jacks, length 221 m and a lifting height of 50 m (Freyssinet)

Figure 14.8 Lifting of a 6500 kN steel bridge with a 4 × 2000 kN jacking system with HT bars; lifting height 8 m (DeBoerDC/deBoer International)

issues. It is not normally necessary to equip a strand moving unit with a locking nut on the plunger to hold the load if it has to be suspended for an undetermined time, as is usually the case with a solid ram jack. The load in a strand moving unit can at any time be transferred from the anchorage sitting on top of the plunger to the lower anchorage located below the jack within the jack's chair.

A good system consists of

- strand moving units with a capacity which suits the specific operation
- suspension cable (most of the time made of strands)
- foundations and temporary supports on which the jacks can be installed
- connection pieces to the load
- power pack units
- control systems for the movement of the bridge within the specified tolerances
- easy access facilities for the operators to every working station.

An advantage of this moving technique is that there is no limitation of the height over which to lift or lower, the speed is high and once installed no heavy stacking material (packers) has to be handled. It is important to stabilise horizontally the top of the support towers or legs.

14.2.3 Skidding or sliding

For horizontal movements, skidding beams with stainless steel and sliding bearings with PTFE (e.g. Teflon) are generally used. Other combinations of materials with a higher friction, for example PE on steel or even steel on steel, are also possible.

The friction coefficient is a variable parameter and depends on the bearing pressure and on the lubrication of the contact surfaces. With PTFE/stainless steel it will usually vary from 2 to 5%. A pushing system is generally designed for at least 8% in order to overcome the static friction. It is important to note that the friction can also be nearly zero.

For sliding jobs on air cushion, the friction falls to under 1%. For an example, refer to Figure 14.11.

For information about friction coefficients and associated maximum bearing pressure, refer to Table 17.2 in Section III, Chapter 17.

A skidding system includes

- skid beams with double webs to cope with eccentric loading (for an example, refer to Figure 14.9)
- sliding bearings or rollers, which can be equipped with vertical hydraulic jacks for compensation of deformations of the skid beams
- hydraulic horizontal push/pull units with enough capacity to overcome the static friction and allow for the steering of the bridge or use of strand units for pulling (for an example of a gripper push unit, refer to Figure 14.10)
- lateral guidance: the friction of the sliding material is very low in all directions, and therefore lateral movement must be restrained
- power pack units
- control systems for the movement of the bridge within the specified tolerances

Figure 14.9 Cross section of a skid beam with eccentric loading (DeBoerDC/deBoer International)

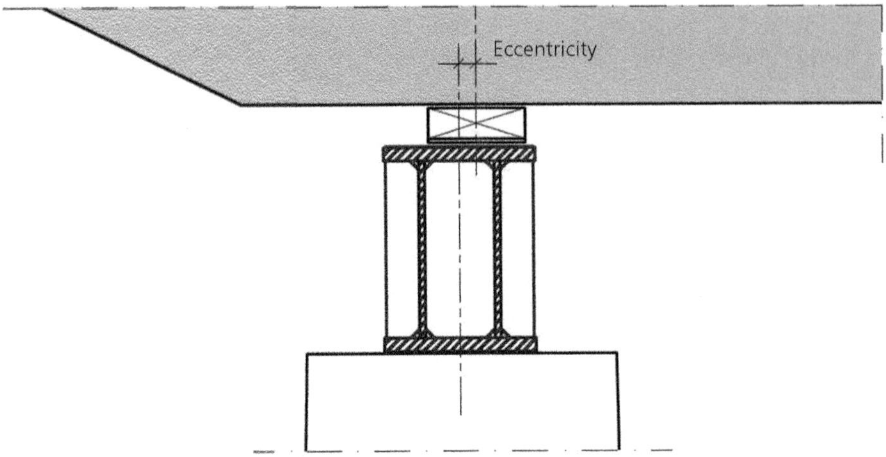

- easy access facilities along the entire length of the skid beams for inspection, greasing and maintenance.

Skidding in up- or downhill inclination

It is preferable to install the skid tracks horizontally. When this is not possible, a brake system has to be provided to cope with a scenario where the friction is nearly zero.

Figure 14.10 Skid system with a gripper push unit (DeBoerDC/deBoer International)

Figure 14.11 Air cushion system with push and pull unit for sliding forwards and backwards, integrated level compensation (Freyssinet)

With the modern skidding equipment, piloted by computerised control systems, heavier and more complex bridge decks can be moved. This also creates the possibility of moving bridge decks with their substructure (for an example, refer to Figure 14.12).

14.2.4 Self-propelled modular transporters

A Self-Propelled Modular Transporter (SPMT), as shown in Figure 14.13, is a wheeled transport system with nearly unlimited capacities in dimensions, movements and loads.

This moving technique, provided by specialist transport companies, is a useful option for complex bridge-moving operations over longer distances.

The basic component of a SPMT is a multi-axle platform vehicle with eight or twelve pendulum axles and the ability to steer each pendulum axle independently. The units can be coupled (mechanically and electronically) to create one big moving platform. The necessary components are as follows:

- A number of standard SPMT units.
- Power pack units with digital steering and control systems.
- Load-spreading structures: the SPMT units require a uniformed-spread loading, but the bridge decks often have specific pick-up points near the bearings.
- SPMTs can be equipped with jacking towers; with such a system the transport and jacking operations are integrated.

Figure 14.12 Skidding of a five-span bridge deck, including piers and pad footings; sliding 2% downwards, so there is need for a brake system (Strukton/Aron Nijs)

The transport road needs to be levelled and to have enough bearing capacity. A temporary road, compacted to a specified minimum bearing pressure and covered with steel plates, can provide an appropriate solution. The cylinders incorporated in the SPMT are able to compensate for minor local unevenness.

For an example of use of an SPMT, refer to Figure 14.14.

Figure 14.13 Heavy-duty SPMT module with six axles: payload 250 kN/axle, Hmin is 1.25 m, lifting height >300 mm (DeBoerDC/deBoer International)

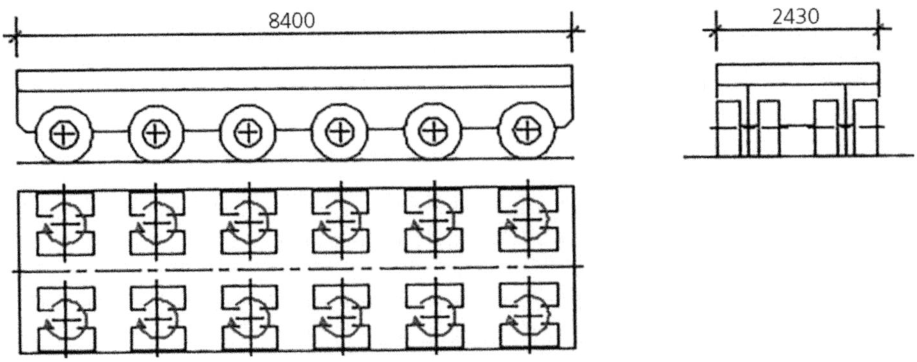

Figure 14.14 Working with SPMTs: installation of bridge elements with substructure – steel structure for load spreading with the pick-up points near the piers; steel plates on temporary road (Mammoet)

14.3. Impact on the structure of the permanent works

The big advantage of the above described moving techniques is their very low impact on the permanent works. There are no limitations of weight or dimensions. Architects and engineers can look at a multitude of structural schemes.

The basic requirement that bridge decks or bridge superstructures have to meet to be installed by PBI methods is that the element which shall be fully or partially pre-finished off site has to be a self-supporting structure during the installation operation. This is easily achievable for bridge decks if their final support is on bearings. Therefore it is no surprise that practically all bridge decks installed by PBI are bearing-supported decks. There is no need for bearings if the bridge deck is installed including its piers and substructure.

Optimum solutions can only be achieved if the installation method has already been considered during the basic design stage.

Bridge bearing replacement

If a designer considers the replacement of bearings when designing a new bridge, very small modifications to the structure will be required to make it buildable by a PBI method. Hence some explanations are presented in this subsection on what provisions have to be made in the design and detailing of bridges to permit easy inspection and replacement of bearings.

When bearings have to be replaced, a jacking system, as described in subsection 14.2, is required.

Figure 14.15 Stiffened areas for hydraulic jacks (IABSE)

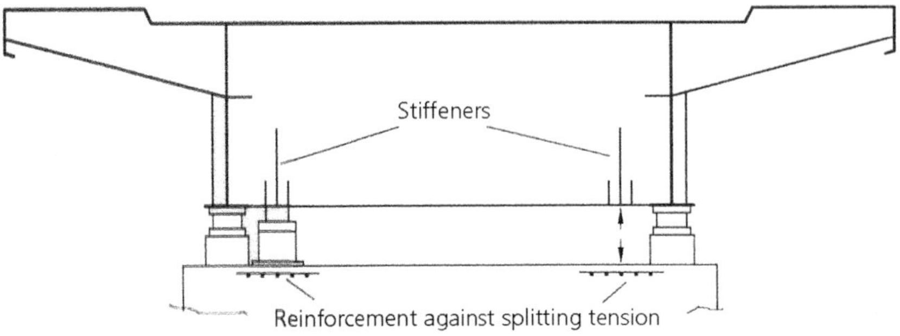

Bridge bearings are the interface between the bridge deck and the piers and abutments. They are, apart from expansion joints, the weakest structural element of a bridge, with a significantly shorter life span than the bridge itself. There is a need for inspection at regular intervals.

All bridge designers need therefore to give provision for bearing inspection and replacement during the conceptual design stage of their projects. The replacement of bearings will almost always take place while traffic flows are present. Therefore well-designed jacking points are defined on the general arrangement drawings at each bearing support point with regard to location but also with regard to maximum jacking force and permissible lifting height.

It is important that the designers provide enough space for installation of jacks which are able to take the dead and live load, making it possible to replace bearings without disruption of the traffic above. Enough space also has to be given for proper access to inspect the bearings and to replace them.

IABSE has published guidelines on required conditions and clearances. These are shown in Figure 14.15 and Table 14.1.

When the vertical clearances are smaller than the minimum dimensions indicated in Table 14.1, for normal hydrdaulic jacks it may be possible to use hydraulic flat jacks for the jacking operation. They have a very low construction height but are not equipped with locking nuts to mechanically take the load during the actual replacement activities (refer to Table 14.1 and Figure 14.16). But this will of course not solve the problem of access to the bearings.

Table 14.1 Required clearance for hydraulic jacks

Force: kN	Required clearance	
	Normal hydraulic jack: mm	Flat jack: mm
500	300	150
1000	360	180
2000	450	200
5000	600	250

Figure 14.16 Jacking system with flat-jacks, poor access conditions (DeBoerDC/deBoer International)

The last option left is to install the jacks beside the pier or abutment, either sitting on brackets which are attached to the pier or abutment or on steel props sitting on the pile cap. This facilitates the use of jacks with locking nut on the plunger/piston. For round piers, heavy prestressed steel clamp brackets will facilitate jacking operations. For examples, refer to Figures 14.17 and 14.18.

Figure 14.17 A series of 500 kN jacks on brackets attached to the wall of a pier (DeBoerDC/deBoer International)

Figure 14.18 Clamp bracket designed for a 10 000 kN load on a pier of 2 m diameter (DeBoerDC/deBoer International)

14.4. PBI installation methods

In this subsection seven installation methods based on different bridge-moving techniques are presented.

- Jacking-down method
- Rotation method
- Launching along bridge axis
- Launching over temporary girders along bridge axis
- Installing bridge decks with SPMT
- Skidding method
- Sliding of underpass bridges

A 'method' always defines the structure itself and the way it is built (and removed).

14.4.1 Jacking-down method

For explanation of the jacking-down method, refer to Figures 14.19 to 14.21 and Table 14.2 on pp. 297–298.

Figure 14.19 Cross section of a bridge deck – kinematics of jacking-down method (DeBoerDC/deBoer International)

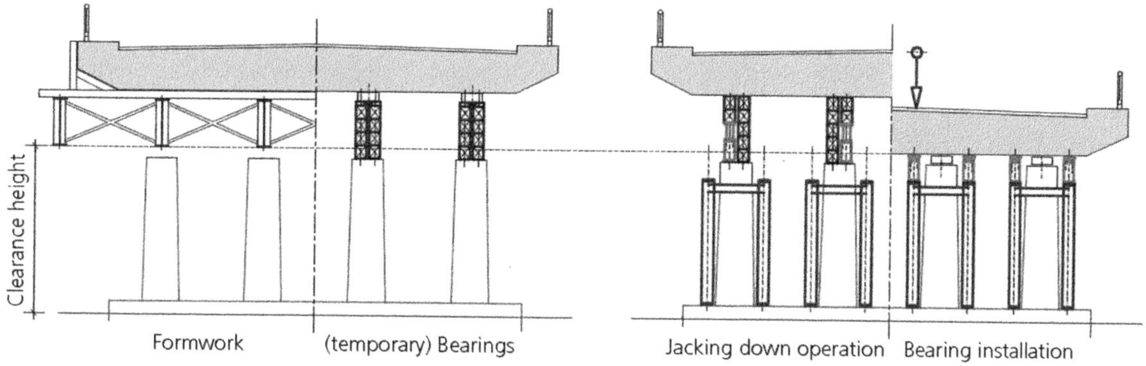

Figure 14.20 Jacking-down of a U-shaped prestressed concrete flyover with a length of 120 m and a mass of 38 000 kN. Jacking-down height 2.6 m. During construction the trains continued to pass underneath. (DeBoerDC/deBoer International)

Figure 14.21 Detail of a jacking installation (VSL International)

Table 14.2 The jacking-down method

Application	Lack of vertical clearance below a cast-in-situ concrete deck for installation of falsework and soffit formwork
Type of bridge decks	Suitable for every cast-in-situ bridge deck with one or more spans, especially for decks with a complicate design such as a: skew crossing, curved alignment, random distribution of the supports, architectural features
Kinematics	1. Building substructure: piers and abutments 2. Construction of temporary supports at the bearing locations, whose height corresponds to the jacking-down height; installation of the formwork system in elevated position with clearance for the undergoing traffic 3. Installation of the bridge bearings on the temporary supports 4. Construction of the bridge deck, prestressing 5. Removal of the formwork etc. 6. Installation of temporary props and jacks, jacking-down of the bridge deck 7. Positioning of the deck, grouting bridge bearings 8. Load transfer from the jacks to the bearings
Equipment	Support towers under every bearing Jacking system on temporary props Hydraulic power pack units and computerised control system
Permanent work interfaces	The big advantage of this method is the absence of interfaces with the design, as long as the deck is designed with proper provision of jacking points for installation and replacement of the bearings The bridge deck is just constructed at a higher level on its original bearings
Additional remarks	Before prestressing of the deck, the bearings need to be activated to allow the movements of the deck The shortening of the deck due to the behaviour of the concrete, prestressing and temperature changes has to be taken into account

14.4.2 Rotation method

For explanation of the rotation method, refer to Figures 14.22 and 14.23 and Table 14.3 on pp. 298–299.

Figure 14.22 Rotation method: top view of a bridge, with rotation bearing on the main pier and a temporary stability support on a round sliding track at the back (DeBoerDC/deBoer International)

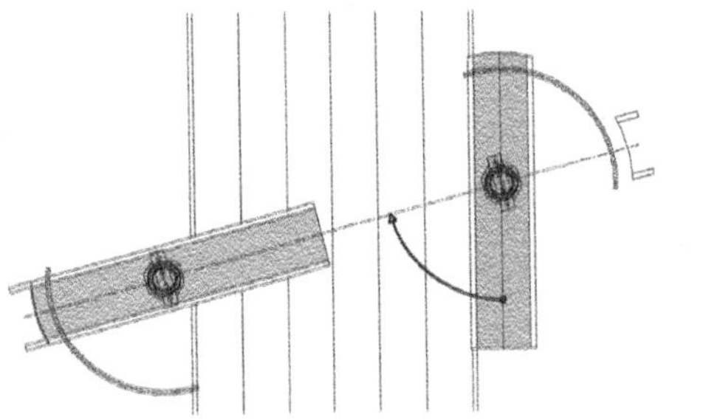

Figure 14.23 Rotation of two balanced cantilevers which were constructed parallel to a shipping canal; final main span 120 m, weight 2 × 40 000 kN (DeBoerDC/deBoer International)

Table 14.3 The rotation method

Application	Construction of the main span of a multispan bridge deck crossing an infrastructure corridor (road/rail/water) at minimum vertical clearance and without access from underneath
Type of bridge decks	Steel, concrete or composite decks with one longer main span, segmented longitudinally so that the main span plus part of the adjacent span can be built as a more-or-less balanced cantilever around one or two piers of the main span
Kinematics	1. Building substructure, piers and abutments 2. Installation of the rotation bearing system 3. Construction of the bridge decks on falsework beside and parallel to the existing infrastructure to be crossed 4. Rotation and positioning of the bridge decks 5. Construction of the main-span stitch, for concrete decks' wet joint and continuity prestressing 6. Installation of the final bearing system
Equipment	Rotation-bearing system around the main piers Option for the application of a temporary stability support on sliding tracks at the back Jacking system for the rotation and positioning and the installation of the bearings
Permanent work interfaces	The basic design of the bridge needs to facilitate this method The dimensions of the pile cap of the main piers need to allow for the installation of the rotation-bearing system
Additional remarks	Rotation can be done about one or both piers of the main span Stability during rotation has to be ensured

14.4.3 Launching method along bridge axis

For explanation of the launching method along bridge axis, refer to Figures 14.24 to 14.26 and Table 14.4 on pp. 299–300.

Figure 14.24 Kinematics of the launching method; launching nose attached to the bridge deck (DeBoerDC/deBoer International)

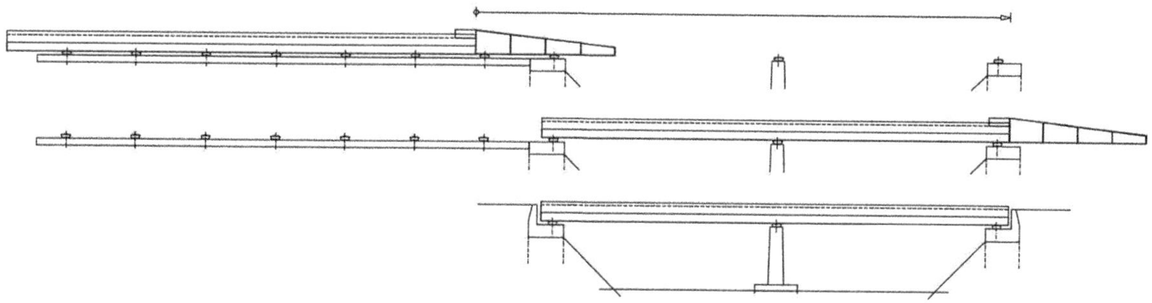

Figure 14.25 Launching of a prestressed concrete overbridge with a three-span system; total length = 70 m and weight = 10 000 kN; during launching, trains continued to run (DeBoerDC/deBoer International)

Figure 14.26 Launching of a Warren truss bridge span 2 × 108 m, with the nodes as the only supporting points (DeBoerDC/deBoer International)

Table 14.4 Launching method along bridge axis

Application	Bridge decks with several spans, which cannot be built in final position but with sufficient space behind one abutment to pre-finish the whole deck prior to launching
Type of bridge decks	All types of steel or concrete bridge decks with a straight or constant-curved alignment; the difference with ILM (Incremental Launching System, see Section II, Chapter 13) is that in this case the deck is completely finished before starting the launching operation
Kinematics	1. Building of the substructure and the supports for the launching bearings in the fabrication or assembly yard just behind the abutment; these temporary supports can be combined to a continuous sliding track/formwork foundation 2. Construction of the bridge deck, prestressing if the deck is made of concrete 3. Installation of the launching nose in the yard behind the abutment 4 Jacking-up of the bridge, installation of the launching bearings, push/pull system, lateral guides, testing of the system 5. Launching the bridge along the bridge axis 6. Transfer of the bridge to hydraulic jacks on the permanent piers and abutments; positioning of the bridge and installation of the bearing system 7. Load transfer of the bridge deck to the permanent bearings
Equipment	Launching nose with vertical jacking device at its tip Sliding bearings with lateral guides Pull or push traction system Jacking system for bearing installation
Permanent work interfaces	This method needs to be considered during basic design, since during the launch the deck is subjected to positive and negative bending moments of similar magnitude along its entire length: this requires a modified prestressing layout for concrete decks. For details refer to Section III, Chapter 15. For the alignment, the same rules as for ILM are applicable (refer to Section II, Chapter 13 and Section III, Chapter 15)
Additional remarks	This method is similar to the ILM described in Section II, Chapter 13; however, unlike a typical ILM the deck is completely finished before starting the launching operation. When launching along a slope (upwards or downwards), a brake system is required

14.4.4 Launching over temporary girders along bridge axis

For an explanation of launching over temporary girders along bridge axis, refer to Figures 14.27 to 14.29 and Table 14.5 on pp. 301–302.

Figure 14.27 Kinematics for launching over temporary girders along bridge axis (DeBoerDC/deBoer International)

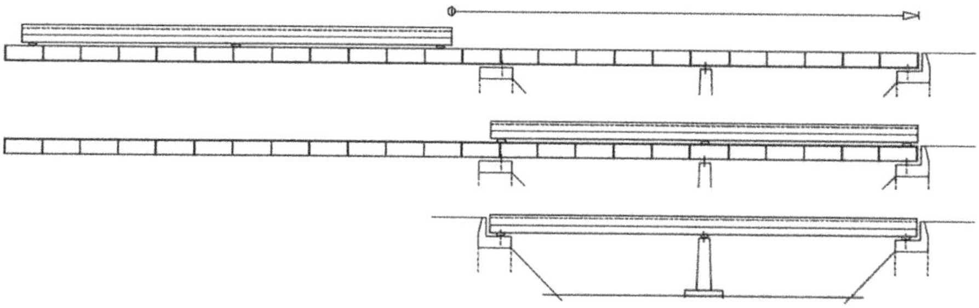

Figure 14.28 Launching of a one-span bow arch on temporary girders, with an air pad system, over ongoing traffic underneath; weight of the bridge 23 000 kN, launching distance 70 m, jacking-down over 3.5 m; the temporary girders are made from concrete (Freyssinet)

Figure 14.29 This is an example of launching over temporary girders: demolishing a continuous multispan prestressed concrete bridge deck where this technique has been used in the reverse way; weight 24 000 kN, jacking-up height 2.80 m, de-launching distance 65 m with a radius of 250 m and slope of 1.6%; the temporary girders are launched underneath the jacked-up bridge deck (DeBoerDC/deBoer International)

Table 14.5 Launching over temporary girders along bridge axis

Application	Bridge decks with one or more spans, which cannot be built in their final position but with sufficient space behind one abutment to pre-finish the whole deck prior to launching. The difference with the launching method 14.4.3 is that with this method the bridge deck is during launching supported at the same support points as designed and detailed for the final stage
Type of bridge decks	Any type of steel or concrete deck structure No specific requirements concerning shape of the deck's soffit
Kinematics	1. Building of the substructure and the supports for the temporary girders in the fabrication yard just behind the abutment 2. Installation of the temporary girders between the yard and the abutment and over the spans to be bridged 3. Construction of the bridge deck, prestressing if deck is in concrete 4. Jacking-up of the bridge deck, installation of the sliding bearings, push/pull system, lateral guide systems, testing of all systems 5. Launching the bridge deck over the temporary girders into final position 6. Transfer of the bridge deck on hydraulic jacks onto piers and abutments; removal of the temporary bearings and girders by de-launching, jacking-down to the final level 7. Positioning of the bridge deck and installation of the bearing system 8. Load transfer to permanent bearings
Equipment	Temporary girders spanning from pier to pier over entire bridge length and into assembly/casting yard (designed to support final bridge deck with final articulation moving along its entire length), with integrated scaffolding; alignment of the temporary girders has to be straight or needs to have a constant curve Temporary piers/bearings for these girders Sliding bearings with lifting jacks and lateral guides Pull or push traction system Jacking system for jacking-down and bearing installation
Permanent work interfaces	The big advantage of this method is that the bridge deck is always supported at or near its final bearing points: no impact on the structural behaviour of the deck No further interfaces, as long as the deck is designed with proper provisions for the bearings and jacking points for bearing replacement
Additional remarks	Working above ongoing traffic The sliding bearings are equipped with hydraulic jacks to compensate the deflection of the temporary girders

14.4.5 Installing bridge decks with SPMT

For explanation of installing of bridge decks with SPMT, refer to Figures 14.30 and 14.31 and Table 14.6 on pp. 303–304.

Figure 14.30 Installing a bridge deck with SPMTs; the jacking system in the casting yard, which allows the deck to be loaded onto the SPMTs, is not shown (DeBoerDC/deBoer International)

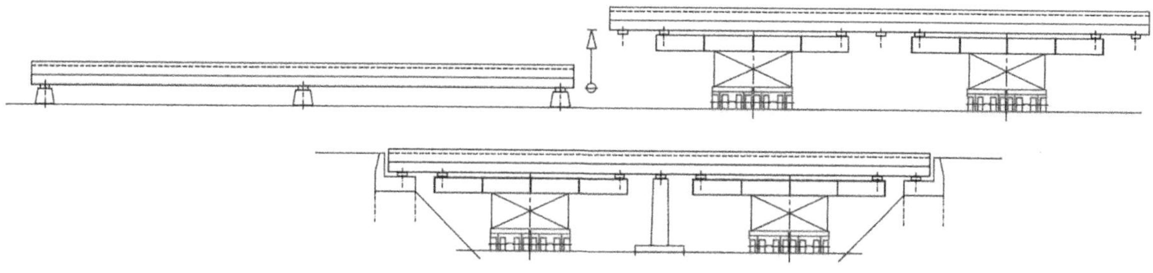

Figure 14.31 Installation with SPMTs of a 50 000 kN prestressed concrete flyover during a 60-hour closing of the undergoing highway (Mammoet)

Table 14.6 Installing bridge decks with SPMT

Application	Overpasses for rail or road corridors with one or more spans
Type of bridge decks	Any type of steel or concrete bridge deck going over a relatively solid and accessible area
Kinematics	1. Construction of the substructure and the temporary supports in the casting/assembly yard 2. Construction of the bridge deck in the yard and construction of the substructure, piers and abutments 3. Jacking-up of the bridge to the planned height above ground 4. Installation of the SPMTs with temporary supports and bearing points; transfer of the load by lifting the bridge deck with the hydraulic system of the SPMTs 5. Construction of the road between the casting/assembly yard and the bridge location 6. Transport of the bridge deck to its final position 7. Transfer of the bridge deck on to hydraulic jacks on the piers and abutments; positioning of the bridge deck and installation of the bearing system 8. Load transfer to permanent bearings
Equipment	Jacking system in casting yard SPMTs with power packs and control system Temporary steelwork between the top side of the SPMTs and the pick-up points for the bridge deck Steel plates (or similar) for the construction of the temporary road Jacking system for positioning and installation of the bearings
Permanent work interfaces	The temporary bearing points are at a certain distance from the permanent bearing points; the deck has to be designed for this load-case No further interfaces, as long as the deck is designed with proper provisions for the bearings and jacking points for the bearing replacement
Additional remarks	Precondition: enough space is needed near the site to cast or assemble the bridge deck and to create a corridor for the transport of the bridge deck to its final position Damage to the bridge deck due to exceeding the permissible level differences between various support points in the case of statically indeterminate articulation

14.4.6 Skidding method

For explanation of the skidding method, see Figures 14.32 to 14.35 and Table 14.7 on pp. 305–306.

Figure 14.32 Replacement of an existing bridge deck – skidding method (DeBoerDC/deBoer International)

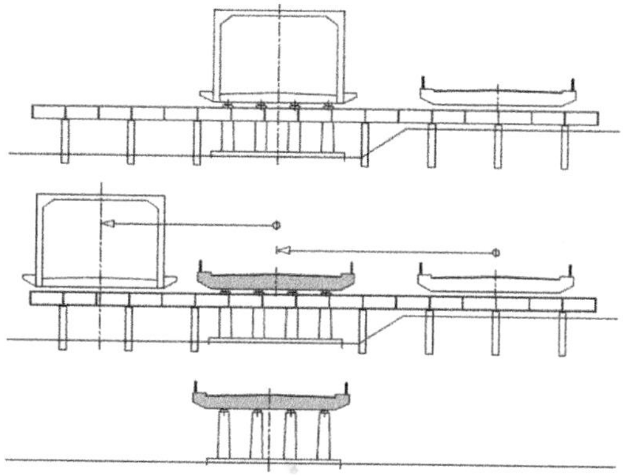

Figure 14.33 Replacement of a pair of single-track steel-truss railway bridge decks by three prestressed concrete bridge decks; operation performed during a 52-hour possession of the railways

Figure 14.34 Detail of the pulling system for a railway bridge deck (Freyssinet)

Figure 14.35 Installation of a railway deck during a weekend closure of the train operations (DeBoerDC/deBoer International)

Table 14.7 Skidding method

Application	Building of new bridges or replacement of existing bridge decks, with parallel support lines
Type of bridge decks	All types of steel and concrete bridge decks, preferably with a well-designed bearing system
Kinematics	1. Construction of a new substructure or adaptation of the existing piers and abutments 2. Installation of the skidding beams on temporary piers and foundations 3. Construction of the new decks parallel to the highway/railway corridor 4. Skidding operation during a very short interruption of the traffic 5. Transfer of the bridge deck on to hydraulic jacks on the piers/abutments or substructure; positioning of the bridge deck and installation of the bearing system 6. Load transfer to the permanent bearings
Permanent work interfaces	This method needs to be considered during the basic design stage. Connection of the new bridge deck to the piers/abutments or substructure can be made with bearings or wet joints
Equipment	Skidding beams with temporary supports Sliding bearings with or without lifting jacks Push/pull traction system Computerised distance control and monitoring
Additional remarks	This method is generally used for building new bridge decks or for the replacement of existing ones. A more recent development is the skidding of bridge decks with their complete substructure

14.4.7 Sliding of underpass bridges

For explanation of sliding of underpass bridges, see Figures 14.36 to 14.39 and Table 14.8 on pp. 306–307.

Figure 14.36 Cross section of an underpass bridge – sliding method with internal skid system (DeBoerDC/deBoer International)

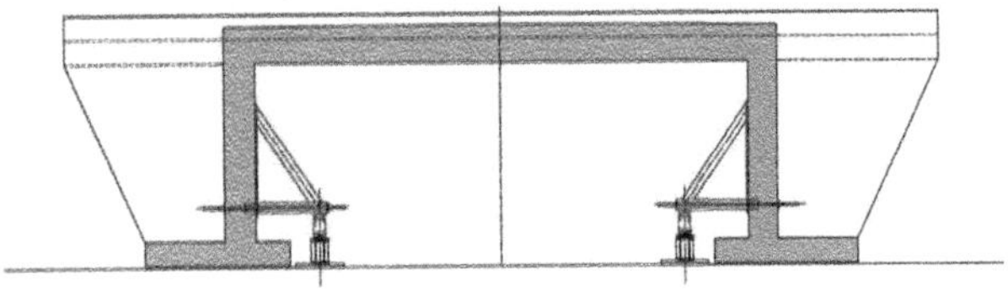

Figure 14.37 Sliding of an underbridge with internal sliding system; mass 18 000 kN, sliding distance 45 m (DeBoerDC/deBoer International)

Figure 14.38 Sliding tunnel segment with external sliding system (VSL International)

Figure 14.39 Underpass ready to be skidded on an air-cushion system (Freyssinet)

Table 14.8 Sliding of underpass bridges

Application	Construction of underpass bridges through embankments of existing road or rail corridors; soil conditions permitting pad footings
Type of bridge decks	Concrete underpass bridges with abutments/retaining walls of any dimension and design
Kinematics	1. Construction of the complete structure next to highway/railway's embankment 2. Installation of the sliding system; jacking-up and test sliding; 3. *During closure of the highway/railroad*: 3.1 Excavation works, foundations or subsoil improvement 3.2 Extension of the skid beams 3.3 Sliding and positioning of the structure 3.4 Jacking of the structure down to its prepared subsoil 3.5 Back-fill and reconstruction of the road or rail
Equipment	Brackets (permanent, in concrete, if outside; temporary, in steel, if inside) as pick up points plus temporary props for stability if needed Sliding tracks with load-spreading track sleepers Sliding bearings with vertical jacks Push/pull cylinders with hydraulic control system
Permanent work interfaces	Parallel walls of any shape and inclination are possible; this method needs to be considered during basic design; engineering of the pick-up points
Additional remarks	The sliding equipment can be placed inside or outside the structure When placed outside it is possible to make a floor slab between the walls; but in this case the dismantling of the sliding equipment is on the critical path of the operation, see Figure 14.38 Settlement of the skidding beams: within certain limits the vertical jacks compensate deformations This kind of operation can also be performed with SPMTs, presenting the advantage that the prefabrication yard can be at a greater distance from the site. See Figure 14.14

14.5. Engineering of PBI methods with jacking and skidding techniques

Every operation needs a coherent engineering report and a method statement in which the engineers give their instructions for the sequence of the operation and the safe use of the equipment. These documents must be based on the local and/or agreed standards. Usually the standards for steel construction are used. The movements during PBI are so slow that an installation can be considered as a static structure.

Considering the specific nature of this kind of construction, the following guide lines are recommended as minimum requirements.

14.5.1 Hydraulic working pressure

The number of hydraulic jacks should be chosen in such a way that the actual working pressure does not exceed 85% of the safe value given by the manufacturer. This also applies for the effective stroke. The reason for this reduction lies in the uncertainties in the planning phase, eccentric loadings and the need for additional capacity for steering options. The effective stroke should also be reduced to compensate for the settlement in stacks of shims and elastic deformation in props and neoprene bearings.

14.5.2 Fundamental aspects of hydraulics

A number of similar jacks placed around a bridge pier or a bearing have to form hydraulically a group. Their combined action is located in their centre of gravity, on condition that they are coupled hydraulically to guarantee the same oil pressure. The free oil flow within such a group of jacks is fundamental. Therefore installation of check or brake valves on the individual jacks is, in cases where the jacks are hydraulically linked, not recommended.

14.5.3 Support reactions

During installation, decks are supported at multiple points. For a single-span deck, a three-point support system is generally possible, as long as the load is taken by hydraulic jacks, which can be hydraulically linked during an operation to form three independent groups. For movement of multi-span decks, level differences between support points need to be kept within limits the structure is able to cope with.

14.5.4 Factor of safety

The steel codes give factors of safety for the dead-weight and the live-loads. The temporary structures as part of a jacking and skidding system can follow the same rules. During an operation the structural system changes. All the phases need to be considered. Although the speed of movement is very low, there are some dynamic effects, especially with statically indeterminate systems. With the hydraulic system the loads can be measured, but this does not work with jacks with engaged safety rings. Furthermore, every operation has its own specific constraints that must be considered on a case-by-case basis.

To compensate these uncertainties, and to ensure the robustness of the system, an increase of the factor of safety given within the relevant code should be considered, based on the level of uncertainty for any given operation.

14.5.5 Stability

The stability of a jacking and skidding system is mainly governed by the impact of an accidental horizontal load on the bridge deck. This load is transmitted to the different jacking units in accordance with their stiffness. Refer to Figure 14.40 for illustration.

Figure 14.40 Stability: It is recommended to take as the minimum horizontal load at least the highest value of: $F_h = 2\%$ of the mass of the bridge deck or $F_h = 1\%$ of the mass + F_h wind + F_h impact (>100 kN) (DeBoerDC/deBoer International)

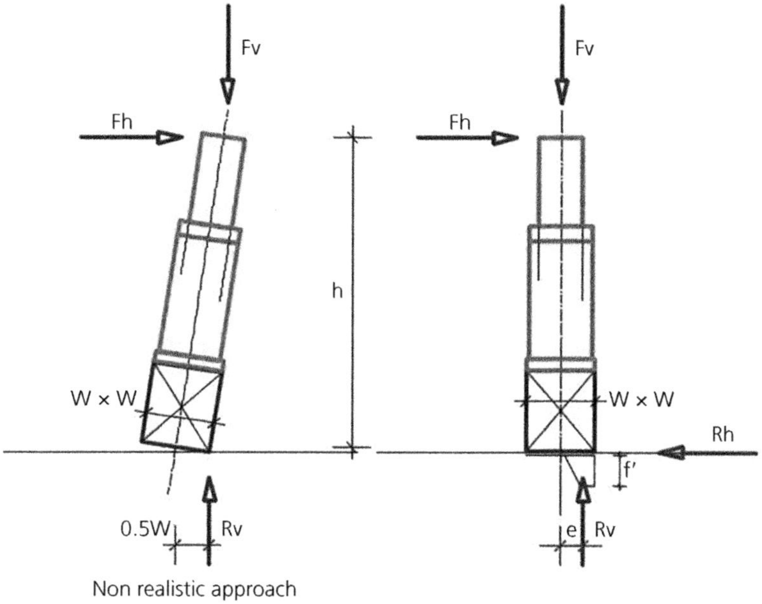

Horizontal forces need to be evaluated carefully. They can range from 2% to 10% of the moved load. Jacking systems must be designed and detailed to be able to transfer these horizontal forces.

Stability is not just to check the equation $F_h \times h < F_v \times e$ with $e = \frac{1}{2}w$. The eccentricity has to be reduced to the centre of the pressure area of the contact surface.

With the normal safety factors of, for example, 1.5 the width of the pressure area ($=X_u$) is defined by f'_b (= the effective compressive strength of the contact materials):

$$X_u = 1.5\ Fv / \tfrac{3}{4}.\ W.\ f'_b \text{ check: } 1.5 \times F_h \times h < 1.5\ F_v \times e_d \text{ with } e_d = \tfrac{1}{2}\ W - [7/18 \times X_u].$$

The ultimate stability check is usually done with $2.0 \times F_h$ and $0.9\ F_v$, and with the width of the pressure area calculated using f'_{ck} (= the ultimate compressive strength).

Each situation must be evaluated properly since the horizontal force could reach up to 10% of the vertical loads.

These recommendations do not apply to verification of the stability of SPMTs.

14.5.6 Eccentricities

(For illustration, refer to Figure 14.41.)

A jacking tower is a stacking of packers, shims and a jack. Theoretically all these elements are centred on one vertical axis, but this can not be achieved in the field. See also Figure 14.9.

Figure 14.41 Eccentricities: Assumptions for eccentricities: bolted connections of standard elements e = 0 mm; not bolted connections of standard elements e = 10 mm; stacking of different elements e = >20 mm (DeBoerDC/deBoer International)

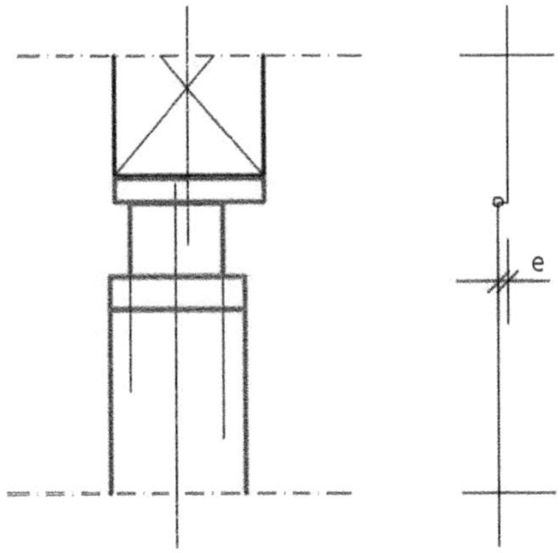

When checking the stresses in a jacking tower, it is necessary to assume a nominal eccentricity whose magnitude is determined by looking at achievable fabrication and installation tolerances.

14.5.7 Static model of jacking towers

The head of the jack's piston shall be considered as a flexural hinge, even when no spherical disc is incorporated. Hence the stability of a jacking tower against overturning is provided by the fixity of the tower or the jack to the bridge deck and/or the supporting structure, which can take the tower's vertical reaction at a certain limited eccentricity (refer also to 14.5.6).

Figure 14.42 Static models of jacking towers (DeBoerDC/deBoer International)

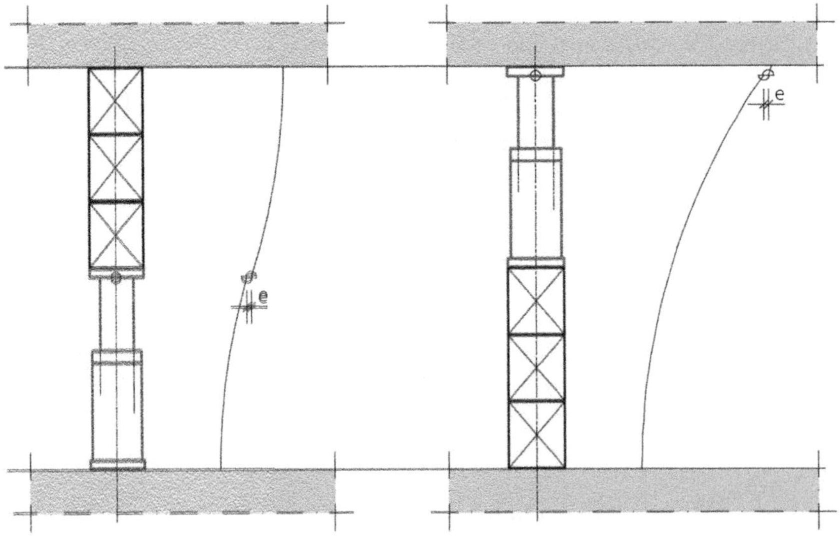

The schematics shown in Figure 14.42 are only accurate when the bridge decks and the piers/abutments and substructure can be assumed to be indefinitely stiff during the operation. This is usually the case with concrete decks but not always with steel structures. If not, the rotational spring stiffness of these contact areas needs to be checked and considered.

The modelling of the jacking towers is based on the assumption that all the jacks are extending simultaneously within the given tolerances.

14.6. Risks

The list of risks given in Table 14.9 is by no means comprehensive, but addresses risks which are particular to PBI. For a specific project, a main contractor needs to carry out a full risk assessment of the operation performed using a PBI method, considering the actual features of the bridge, the selected PBI method and techniques, the kinematics and the site constraints.

14.7. Conclusions

The PBI installation methods and moving techniques presented in this chapter should be approached as a challenge and require specialist know-how concerning:

- the checking of the permanent works during the different stages of construction and installation
- engineering of the auxiliary equipment and hydraulic systems
- planning of the moving operation of a bridge deck during a short possession time.

Experience and know-how are a prerequisite for an effective operation with minimal disturbance for road/rail users and the neighbourhood of the construction site.

Table 14.9 Summary of the risks addressed in this chapter

Activity	Risk	Mitigating measures
Working above ongoing traffic during PBI	Fall of objects	Working with appropriate and cleaned equipment, protected working conditions for operators, monitoring of the movements
Moving of bridge decks or bridges (general)	Excessive load redistribution between support points resulting in damage to the bridge or the temporary support system or even in collapse	■ Minimise the number of individually operated jacking points by sensible hydraulic linking of jacks, but maintaining stability (e.g. three-point support system for statically determined bridge decks) ■ Definition of clear criteria with regard to permissible differential level differences among support points ■ Automated recording of levels and pressures and setting of limits with automatic stop of movements whenever limits are reached
	Misalignment during moving	Guide systems
	Excessive deformations of sliding supports	Sliding sledges equipped with vertical hydraulic jacks for compensation of deformations
	Loss of horizontal stability	■ Use of vertical jacks with adequate horizontal load-bearing capacity ■ Use of engineered shim stacks (flatness, self-centring devices) ■ Check that the temporary support system is stable based on actual support conditions at the top and bottom of supports (flexurally fixed or hinged, etc.) ■ Check of local stability within support towers (avoidance of more than one intermediate 'hinges' or none if both ends are hinged)
Jacking up or down (bridge deck 'rigidly' propped up during movement; generally over small heights only at slow speed)	Instability of a jacking or support tower	Considering the appropriate horizontal load for checking contact pressures
	Insufficient capacity of a jacking tower to transfer horizontal loading	Installation of an *external* guiding device
	Excessive load redistribution due to level differences between the jacking points	Computerised height measurement and control system
	Instability due to eccentricities in the different components of a jacking tower	■ Bolted connections between packers and shims ■ Use of shims with centring pins
	Failure in the hydraulic system, loss of oil pressure resulting in increasing level differences between jacking points with associated excessive load redistribution (for statically in determined structures)	■ Working with jacks equipped with locking nuts on the pistons which can be kept always only a few millimetres above the jack's cylinder ■ Use of brake valve if a single jack is used per independent support point

Table 14.9. Continued

Activity	Risk	Mitigating measures
Lifting with SMU (bridge deck suspended by 'flexible' tension members during movement; generally over medium to large heights at 'high' speed)	Failure of a strand	See Section III, Chapter 17.
	Excessive load redistribution due to level differences between the jacking points	Computerised height measurement and control system
	Failure in the hydraulic system, loss of oil pressure resulting in increasing level differences between jacking points with associated excessive load redistribution (for statically indeterminate structures)	■ Limiting stroke to permissible difference in levels between jacking points ■ Working with jacks equipped with locking nuts on the pistons which can be adjusted during jack movements ■ Use of brake valves in combination with automated control system (automatic stop of movement once pressure and/or level threshold are exceeded)
Skidding	Misalignment of the bridge	Lateral guide system
	Movement is out of control due to gradient	Installation of brake systems
	Twisting of the skid beam due to transverse eccentricities	Use double-web skid beams
	Problems with sliding sledges	Create facilities to jack up the bridge for replacement of the sliding sledges at any position
Working with SPMTs	Unacceptable settlement of the temporary road	■ Geotechnical assessment of ground conditions ■ Construction of a temporary road compacted to specified bearing pressure and covered with steel plates or similar
Underpass bridge sliding	Settlement of the sliding beams	■ Geotechnical assessment of ground conditions ■ Compaction of ground to specified bearing pressure ■ Design and detailing of sliding beams stiff enough and wide enough at base to transfer loading to ground at permissible bearing stress level; use of load-spreading sleepers ■ Equipping the sliding sledges with lifting jacks
	Insufficient soil-bearing capacity, resulting in excessive settlement and subsequent tilting of bridge being moved, requiring abandonment of the operation	Thorough soil investigation as part of planning the operation, which may result in any or a combination of the following measures: Soil improvement, larger load-spreading sleepers, lowering of the groundwater level
Operations performed during a short closure time of the rail or road	Not able to complete work within given closure of traffic corridor	Well thought-out method statement and planning, 'overkill' of equipment, experienced and fully trained operators and staff

REFERENCES

IABSE Structural Engineering Document No. 6, 'Structural bearings and expansion joints for bridges'; https://www.iabse.org

Figures and tables

Figures 14.1–14.3, 14.5,14.6, 14.8–14.10, 14.13, 14.16–14.20, 14.22–14.27, 14.29, 14.30, 14.32, 14.33, 14.35–14.37, 14.40–14.42: (DeBoerDC/deBoer International)
Figures 14.21, 14.38: VSL International
Figures 14.4, 14.7, 14.11, 14.28, 14.34, 14.39: Freyssinet
Figure 14.12: Strukton/Aron Nijs
Figures 14.14, 14.31: Mammoet
Figure 14.15: IABSE
Table 14.1: IABSE

Section III

Design Aspects and Precasting

Bridge Deck Erection Equipment
ISBN 978-0-7277-6193-4

https://doi.org/10.1680/bdee.61934.317

Chapter 15
Detailing aspects of bridge decks built by BDE

15.1. Introduction

This chapter covers detailing aspects for bridge decks built with BDE addressed in this book. It is divided into the following subsections.

- Prestressing
- Precast girders (for concrete bridges made up of longitudinally cut decks)
- Segmental concrete decks (for concrete bridges made up of transversely cut decks)
- Incrementally launched concrete, steel and composite bridge decks
- Stay-supported bridge decks made from cast-in-situ or precast concrete, steel or composite segments.

15.2. Prestressing

Prestressing is a technical necessity for concrete bridges exceeding 20 m span because it:

- minimises span/depth ratio
- increases stiffness by minimising cracking of concrete
- makes it possible to use high-strength steel, which takes up much less space in the deck's cross section than ordinary reinforcement (its tensile capacity exceeds the capacity of passive steel by about 4).

In addition, the use of prestressing allows the transverse cutting of a concrete bridge deck into precast segments without the need for wet joints and reinforcement, with overlaps between adjacent segments.

Nowadays strands with breaking loads of up to 279 kN per strand are almost exclusively used for the prestressing of bridge decks. There are two main classes of prestressing (refer to Figure 15.1): in the case of **pre-tensioning**, the strands are stressed before concrete is cast, which requires a precasting bed provided with 'dead men' at both ends, against which the strands can be stressed. After the concrete has reached sufficient strength, the prestressing forces are transferred from the 'dead men' to the girder to be prestressed through a bond between the strands and the surrounding concrete at the two ends of the girder. The only corrosion protection for the pre-tensioned strands is provided by the alkaline environment of the girder's concrete. In the case of **Post-Tensioning (PT)**, the strands are only stressed after the concrete has been cast and has achieved adequate strength. For this purpose the strands need to be movable against the concrete during stressing and are therefore either placed in cast-in ducts (internal post-tensioning) or in ducts located outside of the cross section but anchored and deviated in concrete elements of the girder to be post-tensioned (external post-tensioning). The corrosion

Figure 15.1 Comparison between pre-tensioned and post-tensioned girders (VSL International)

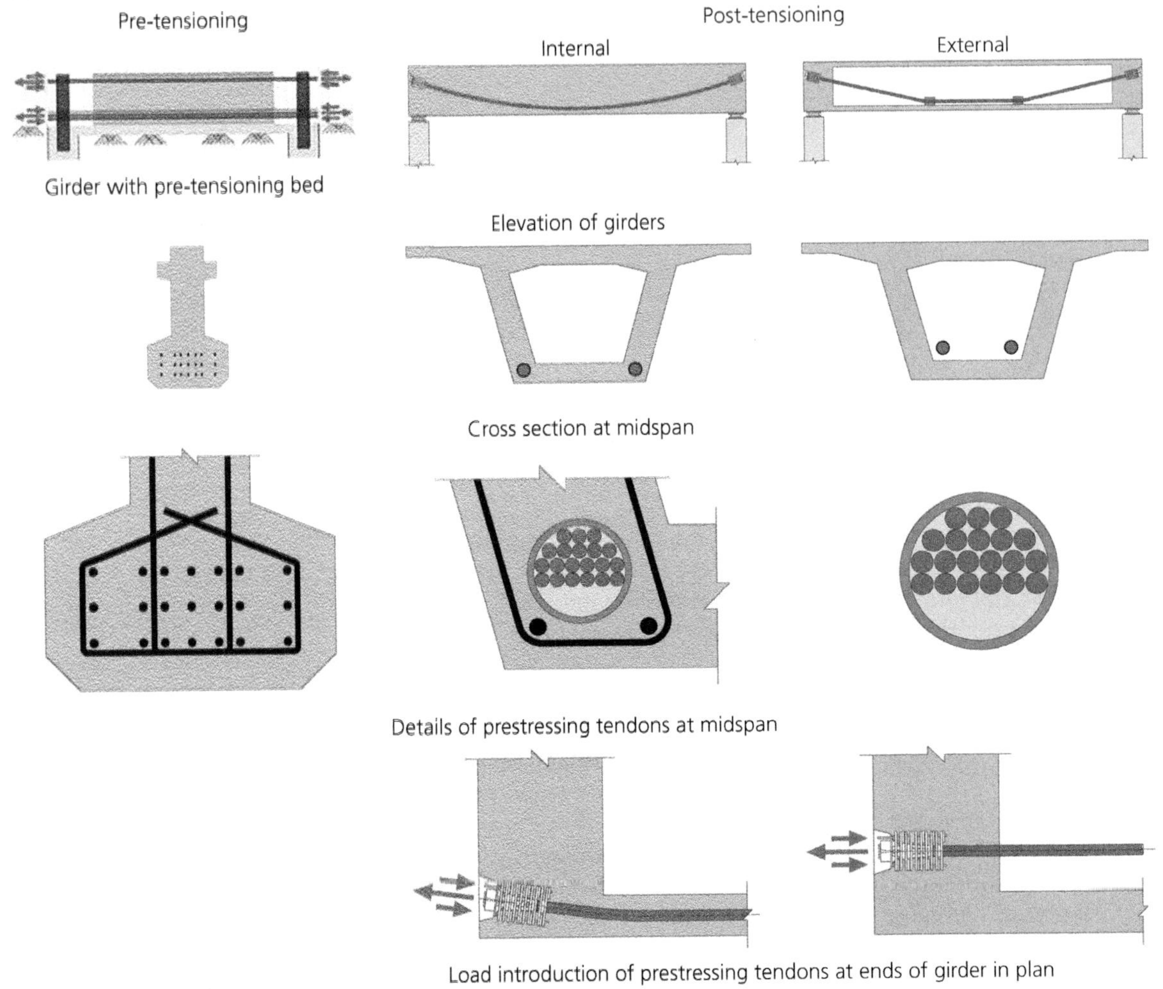

protection for the post-tensioning strands is provided by the alkaline environment of the cementitious grout injected after the stressing, and the encapsulation of the duct if made from plastic ducts. In some countries, a flexible filler (grease or wax) instead of cementitious grout is used for corrosion protection of external tendons. Such protection is, however, not recommended for internal prestressing, since structural redundancy would be unnecessarily compromised. The use of cementitious grout is a reliable corrosion protection if injected by a qualified PT specialist company.

15.3. Precast girders

As explained in Section I, Chapter 2, decks can be broken down into elements longitudinally or transversely.

The bridge deck shown in Figure 15.2 is made up of longitudinal precast elements, called precast girders, which are supported on cast-in-situ piers and T-type cross heads. The longitudinal elements

Figure 15.2 Bridge deck broken down into longitudinal precast elements (Meyer 2009)

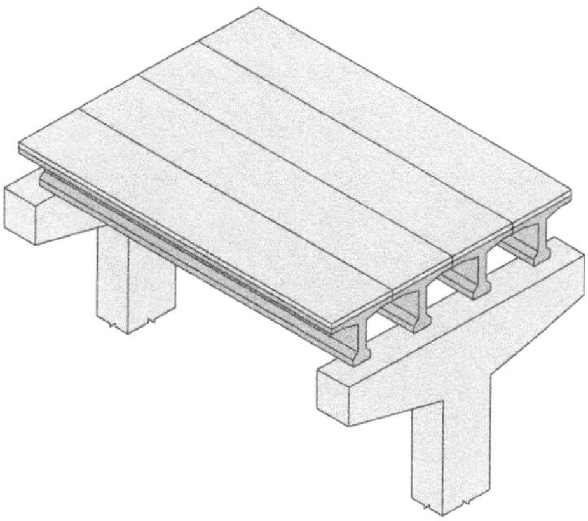

in Figure 15.2 are wide flange T-girders, which remove the need for deck slab formwork. They are connected transversely through a cast-in-situ top slab. However, the part of the deck that cannot be built off site is still significant. The cross heads or portals needed to support the girders making up a span extend over the full width of the deck and generally require, therefore, ground-supported falsework and formwork systems.

The longitudinal elements of precast concrete decks are I, T (bulb Tee), wide flange T (deck bulb Tee) or U girders (refer to Figure 15.3). Girders are either pre-tensioned, post-tensioned or prestressed by a combination of pre- and post-tensioning. Precast T-girders with deck slab as part of the T-girder's precast top flange may give savings in concrete, but the in-situ stitches between the girders are difficult to build. Alternatively, for decks with constant cross fall, box girders can be used whose top flanges are part of the deck slab. For such girders, transverse deck continuity is achieved by casting longitudinal in-situ stitches between the top flanges of adjacent girders and by stressing of transverse PT.

Figure 15.3 Typical cross sections of decks built with precast girders: precast girders in yellow, lost formwork in brown and cast-in-situ concrete in grey (VSL International)

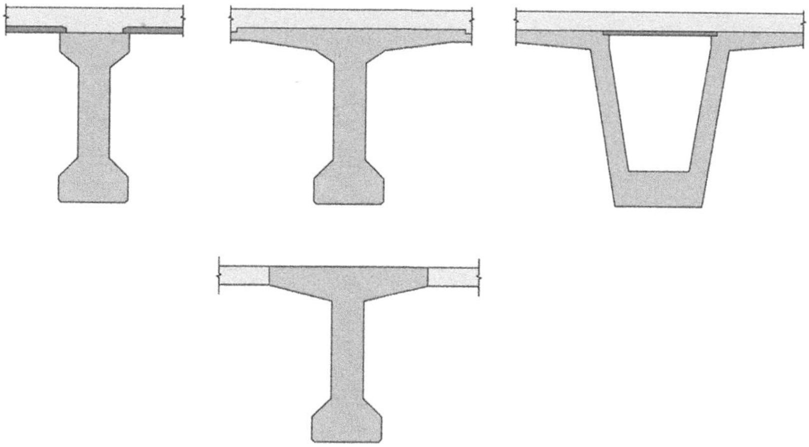

Figure 15.4 Span range as a function of precast girder type according to AASHTO, with the grey bar for pre-tensioned girders and red box for post-tensioned girders (AASHTO)

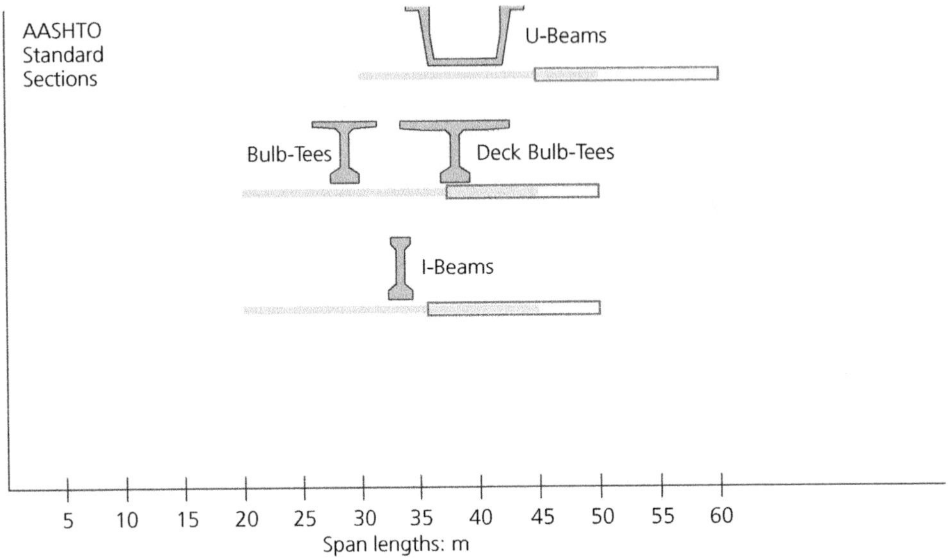

Decks with precast girders are used for both road and railway bridges. Their construction tends to be relatively simple: since the girders are cast in one pour and bridge the entire span, the girders' formwork is not very complicated and the girders can generally be erected by a pair of cranes. The girders are either cast on site in purpose-built precast yards or off site in permanent precast yards. They act as support elements when constructing the cast-in-situ top slab, therefore eliminating the need for ground-supported formwork. With girders of similar section, variation in deck width can easily be accommodated by changing the number of girders per span. Different live-load requirements can be coped with by changing the transverse spacing between the girders. Since girders are straight in plan, they are not really suited to be used for alignment with considerable curvature in plan both from a structural as well as from an aesthetic point of view.

Design and detailing of precast girders have been standardised, for example in the USA by AASHTO and several owners (Florida DOT, Texas DOT, Nebraska DOT) and in Australia (Super-Tees). AASHTO gives a span range as a function of precast girder types (refer to Figure 15.4).

Use of precast girders exceeding 40 m in length poses specific challenges for transport and erection. A U-girder with a span of 60 m weighs, for example, more than 2000 kN.

Particular attention needs to be given to the support detailing for the precast girders. In Figure 15.5 the following three configurations are shown.

- The precast girder and the cross head both have vertical end faces.
- Either the precast girder or the cross head have a vertical end face and the other one has a half-joint end face.
- The precast girder and the cross head both have half-joint end faces.

Figure 15.5 Support configurations of the precast girders and the cross heads/portals (VSL International)

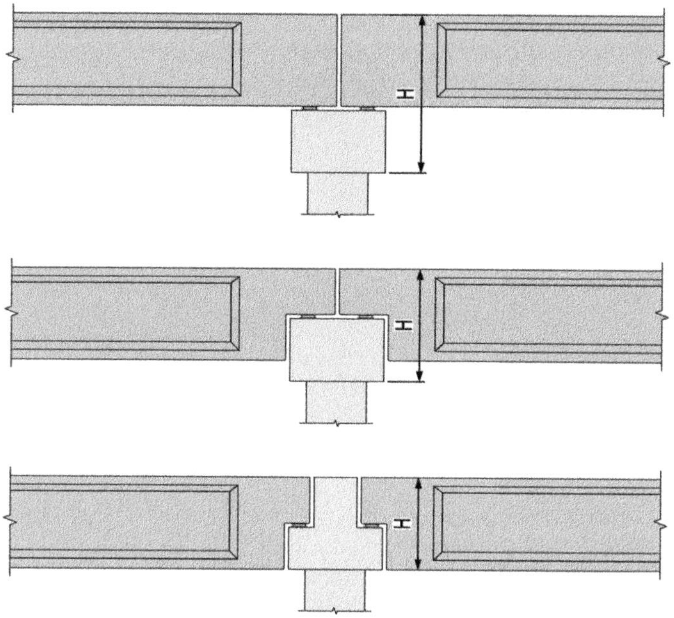

The combined structural height of the deck and the cross head/portal at the pier location is biggest for the first, smallest for the last and somewhere in between for the second option. Vertical end faces are more efficient from a structural point of view. A girder's maximum shear force does not have to be hung up by reinforcement links, as is required when having a half-joint. Deck continuity, if specified, can more easily be achieved. On the other hand, half-joint end faces on both cross heads/portals and the precast girders give better support condition when precast girders are erected with an overhead gantry. The bearings shown in Figure 15.5 can be temporary if the deck is monolithically connected to the cross heads/portals by in-situ stitches, or they can be permanent. When the bearings are permanent, the configurations with half-joints provide very restrictive access for inspection and replacement of the bearings and are therefore prohibited in some countries.

The main reason why construction with precast girders is perceived to be simple is that the girders are always erected as simply supported. Whereas such an articulation may be ideal as a permanent system for railway bridges, it is not desirable for highway bridges due to the need for expansion joints at all pier locations. Hence decks should be made continuous for road viaducts for final stage in the longest lengths that global temperature variation will permit. Such continuity may be for bending between spans or to achieve frame action between the deck and the substructure. Often such continuity is only achieved by provision of passive reinforcement. Since such reinforcement tends to be dense, it is important that the designer gives proper attention to its detailing. The steps required to create a bridge with continuity between spans or even frame action with the substructure when using simply supported precast girders complicate the construction of such precast girder decks significantly.

15.4. Segmental concrete decks

A segmental concrete deck, whereby the concrete deck is, for production purposes, broken down transversely into elements, which are called segments, is shown in Figure 15.6.

Figure 15.6 Segmental concrete deck: a bridge deck broken down into transverse elements (Meyer 2009)

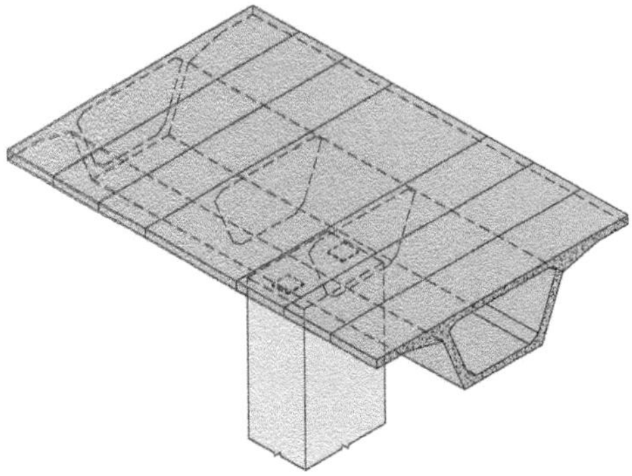

Segmental concrete bridge decks have either cast-in-situ or precast segments. These segments extend over the full width of a road's single or dual carriageway or over the full width of a single- or double-track railway line.

15.4.1 Cross sections

Typical cross sections for segmental bridge decks for road bridges with single-cell box girder segments are shown in Figure 15.7.

Use of single-cell box girder segments with a pair of main webs is particularly suitable for precast segmental construction with regard to segment storage and segment handling during erection. Transversely segments can always be supported or suspended at two points only (statically determinate support conditions in transverse direction). In addition, a closed box girder gives access during erection for:

- the installation and the removal of temporary bottom stress bars
- the application of the epoxy glue
- the installation of the permanent post-tensioning if it is external.

This would not be the case with an open double T-section.

Single-cell box girders are also ideal cross sections for bridge decks built with Movable Scaffolding System (MSS) or by Incremental Launching Method (ILM), as long as they do not have internal struts. Multi-cell box girders with more than two main webs are options for very wide bridge decks built with MSS or by ILM.

If a dual carriageway is supported by a single-deck viaduct which has variable curvature in plan, the variation in cross fall results in variation of the cross section's outer geometry, unless it is possible to provide a common cross fall over the full width of the dual carriageway. The use of twin decks instead of single decks for dual carriageways is therefore a better choice for precast segmental road viaducts and road decks built with MSS.

Twin decks can be supported on single piers (twin bridges) and on top of T-type cross heads. For precast segmental viaducts a third option is to connect the spans monolithically with a T-type cross head (refer to Figure 15.8).

Figure 15.7 Typical cross sections for segmental bridge decks for road bridges with single-cell box girder segments (VSL International)

			Ideal for	
		Precast segmental	MSS	ILM
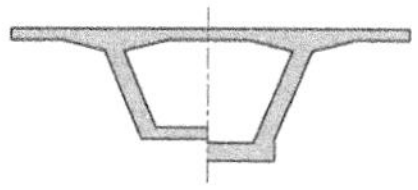	Support of two traffic lanes plus shoulders	Yes	Yes, constant depth	Yes, constant depth
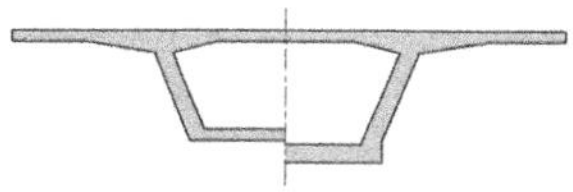	Support of three traffic lanes plus shoulders	Yes	Yes, constant depth	Yes, constant depth
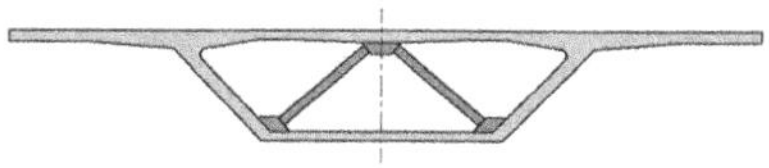	Support of four traffic lanes plus centre divider and shoulders	Yes	No	No
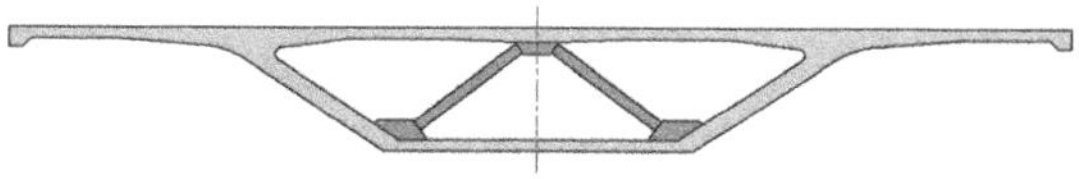	Support of six traffic lanes plus centre divider and shoulders	Yes	No	No
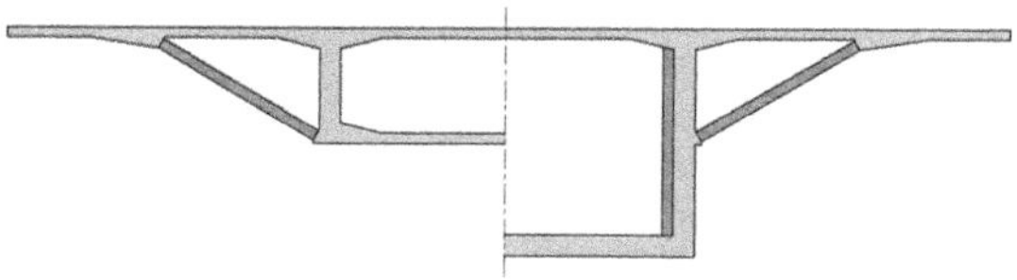	Support of six traffic lanes plus centre divider and shoulders	Yes	Yes, constant depth	Yes, constant depth

For the latter two solutions, the transverse out-of-balance moments introduced into the pier and foundation during the deck erection need to be checked, taking into account the actual weights of the deck erection equipment.

For road bridges built by MSS, the open double Tee cross section per carriageway offers a lot of advantages, since the entire formwork is directly supported by the MSS's main girders (refer to Figure 15.9). Inclined webs at a minimum of 1 to 2 degrees are recommended to ease stripping and to avoid damage to the formwork on the MSS. If required, pier diaphragms can be built after the forward launching of the MSS with an independent set of formwork.

For segmental rail bridge decks, box or U-sections are used (refer to Figure 15.10).

Unlike road viaducts, decks with U cross sections are quite common for viaducts of mass rapid and light transit rail systems, since the deck width of rail viaducts is in general limited to accommodate a maximum of two tracks. In addition, the structural up stands of the U-section can also double up as noise barriers, support for cabling and emergency walkways. Single-track U-girders are structurally particularly efficient, since the bottom slab has only a short transverse span. The girders can be fully precast due to their limited weight and width. In addition, the production and erection of precast

Figure 15.8 Twin decks – different support options (VSL International)

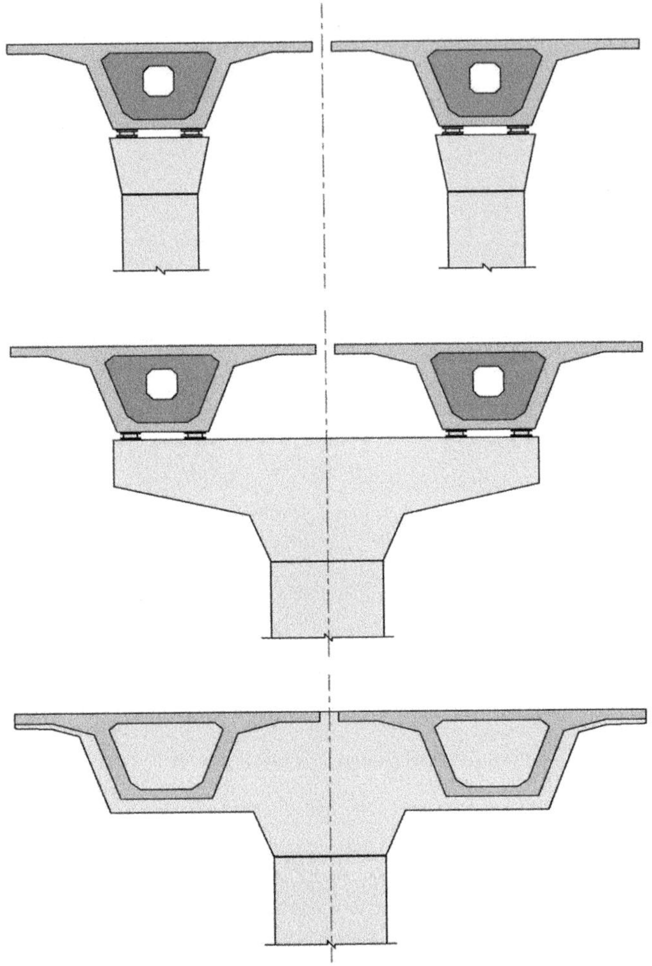

segments with U-sections is simpler than with precast box girder segments. On the other hand, box girders have a number of advantages over U-girders. They can:

- cope much better with tight plan curvature
- simply be deepened/haunched to bridge over spans exceeding the standard span lengths, which occur at big road traffic junctions and skewed road or river crossings
- handle bifurcations without need for special sections
- offer the possibility to make the deck continuous over three to four spans in order to improve structural efficiency and cut down on maintenance costs, of particular importance to the end user.

15.4.2 Deck construction methods

Cast-in-situ decks or precast segmental decks are made self-supporting for their self-weight by any of the four following methods.

Figure 15.9 Twin road bridges with open double Tee cross sections (VSL International)

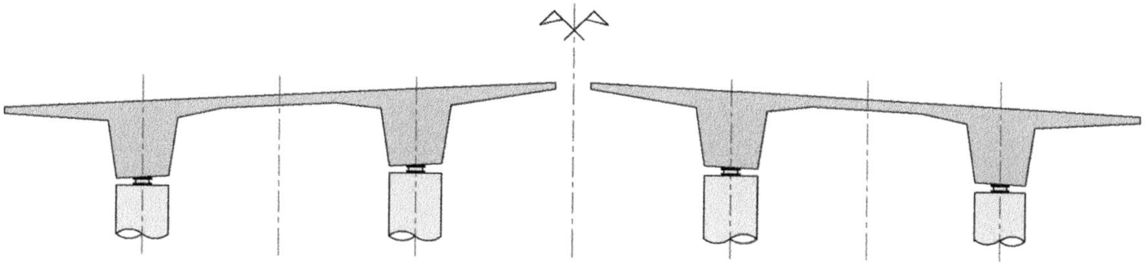

Figure 15.10 Cross sections for segmental rail bridge decks (Meyer 2009)

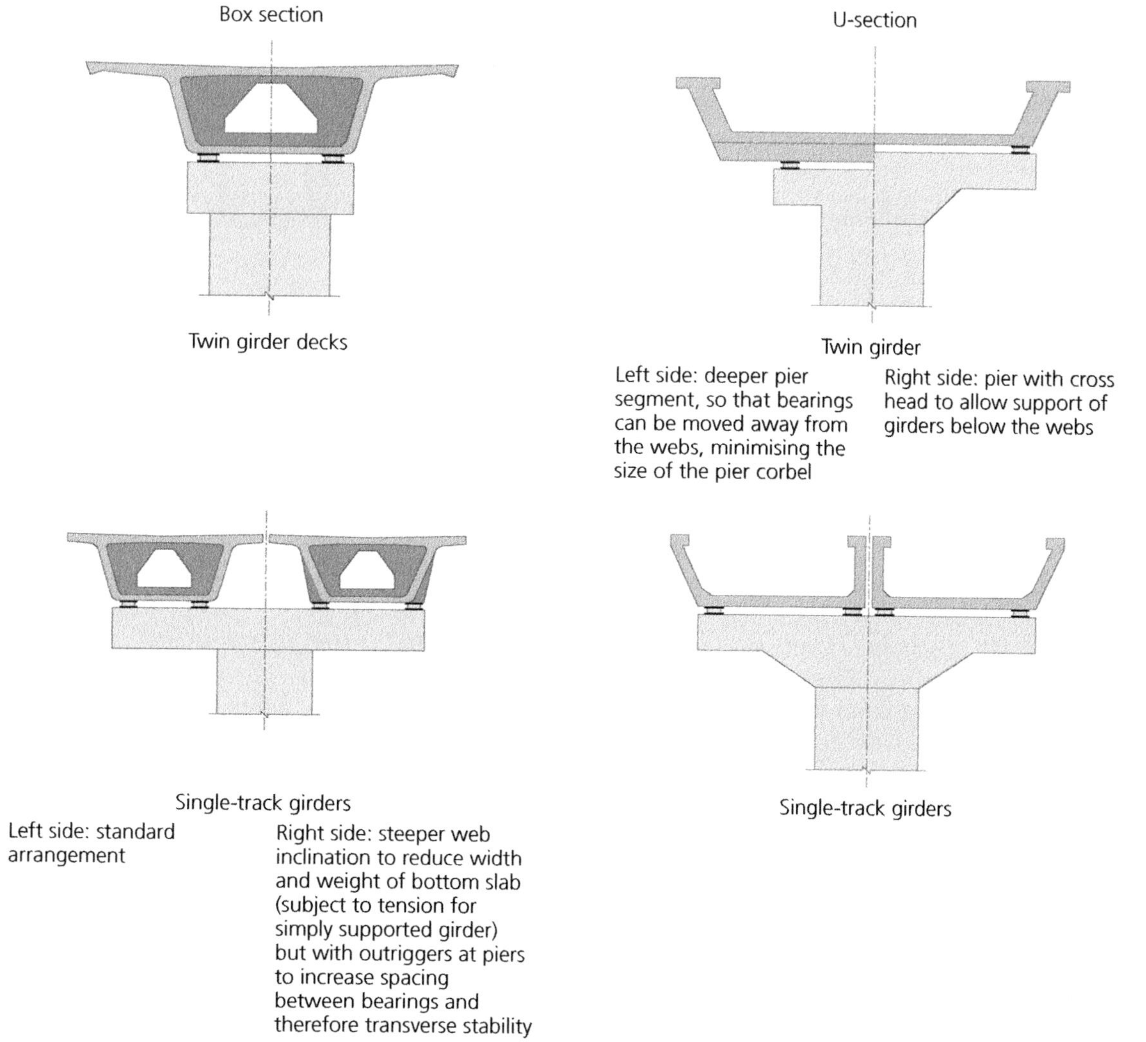

- The Balanced Cantilever Method (BCM) (refer also to Section II, Chapters 5 and 9): The deck is progressively activated in cantilever mode segment by segment, starting from a pier table and extended until the centre of the spans on either side of the pier table has been reached (or in the case of an end span until the out-of-balance part of the deck in the end span has been reached). In general, the deck is cantilevered symmetrically about the pier to minimise the out-of-balance moment at the pier. Stitches are cast between adjacent cantilevers to achieve deck continuity. A special case of this method is the Progressive Cantilever Method (PCM; refer also to Section II, Chapter 10) where, with the help of a temporary mast and temporary stays, an entire span is erected by the cantilever method using the previous spans as anchorage.
- The Span-by-Span method (SBS) (refer also to Section II, Chapters 6 and 12): The deck is either cast span by span in situ on falsework or on an MSS. The length of cast-in-situ segments extends over the full span length with the construction joint either over the piers for simply supported decks or at about 1/5 of the span for continuous decks. Alternatively the spans can be made up of precast segments, which vary in length between 2 to 4 m and are supported during erection either by falsework or a BDE.
- The Full Span Method (FSM; refer also to Section II, Chapter 8) is a method where the entire span is one precast segment.
- The Incremental Launching Method (ILM; refer also to Section II, Chapter 13): The deck, which is progressively built segment by segment behind one of the bridge deck's abutments, is pushed or pulled into its final position.

Deck construction methods define the deck segmentation and longitudinal PT layout (refer to Figure 15.11).

Incrementally launched decks in concrete, together with incrementally launched steel and composite decks, are covered in subsection 15.5, 'Incrementally launched concrete, steel and composite bridge decks'.

15.4.3 Decks built by the Balanced Cantilever Method (BCM)

The box girder is the ideal cross section for decks which are activated by BCM. The box section can be detailed to always have enough bottom flange area to fully take the compression forces from the negative bending moment generated by the deck's self-weight during the construction of the deck. This makes it possible to achieve ductile flexural failure mode for any span lengths without the need for excessive girder depth. With a box section it is possible to keep the webs free of continuity tendons and these tendons can be easily curtailed along the span and anchored inside the box. For spans up to 50 to 60 m, the depth is generally kept constant and the thickness of the bottom slab is stepped up towards the supports, whereas for larger spans not only the thickness of the bottom slab is increased from the centre towards the supports but also the depth of the girder. The webs are either vertical or inclined. Inclined webs offer advantages for precast segments with regard to lowering of rebar cages, but add considerable complexity to the formwork, if they occur in combination with variable girder depth. For large spans and wide decks, inclined webs give more economy, since they allow a reduction in the width of the bottom slab at the centre of spans, where bending moments are positive.

Alternatively, for box girders of medium span length, where the girder depth is kept constant at the centre of the span over about 50% of the span length before being increased linearly against the pier table, it is possible to have fully inclined webs at the centre of the span only. The girder's height is increased towards the pier table not by extending the web's outer face downwards at the web's inclination but by a vertical 'extension'. Hence the extension of the web's outer face is located in a vertical plane and the width of the bottom slab remains constant. Such a configuration looks aesthetically

Figure 15.11 Deck construction methods define segmentation (in black) and PT layout (in red) (VSL International)

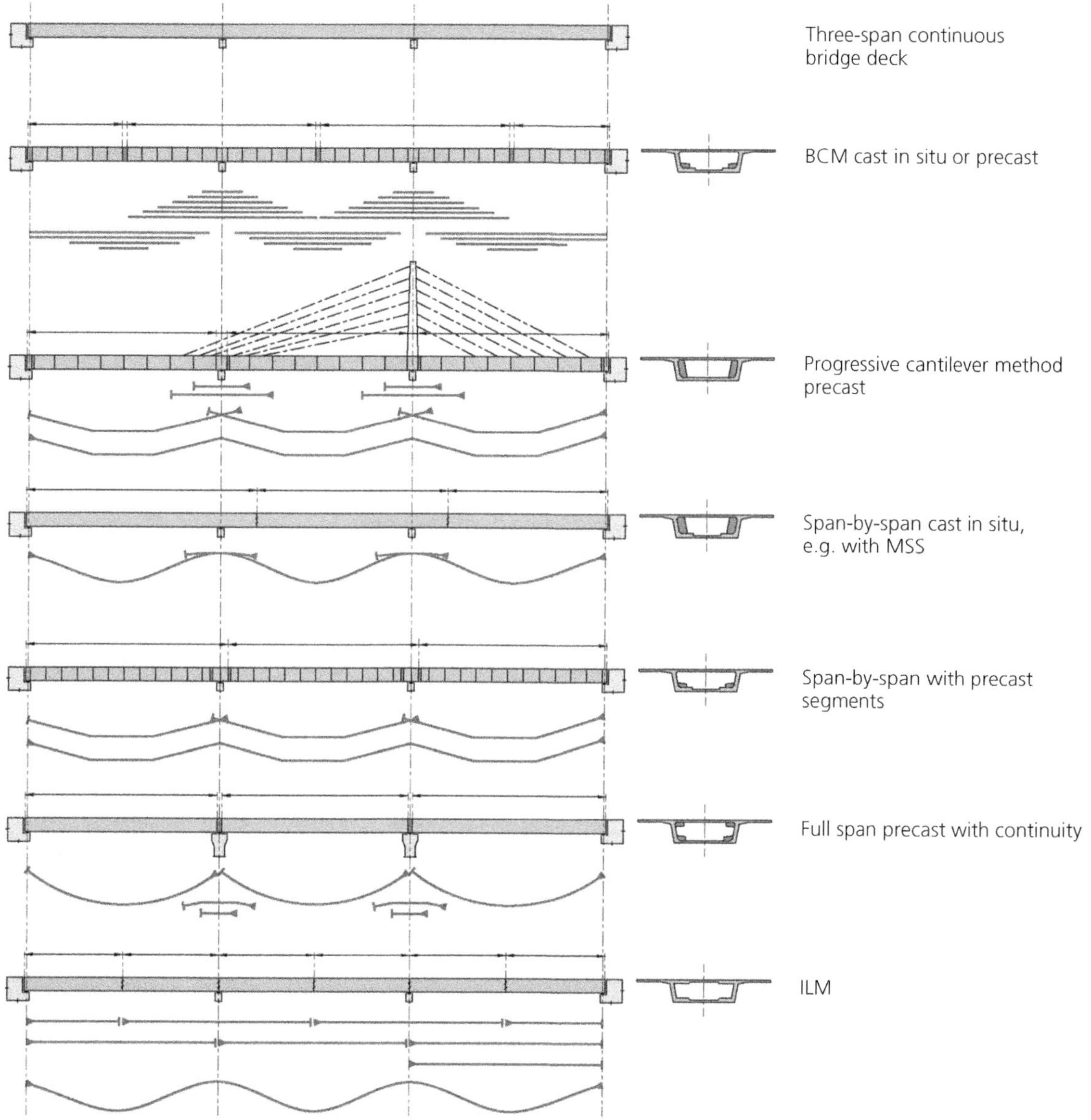

attractive and simplifies production of segments in case of precast bridge decks. It is however structurally efficient only as long as the increase in depth is limited to less than about 1.5 m, since the web's inner face should not be kinked.

In Figure 15.12, typical cross sections of a box girder are shown at support and at midspan.

In the support section of Figure 15.12, cantilever prestressing tendons, which are located in the deck slab, are shown not anchored in the web but in deck slab haunches on either side of, and close to, the

Figure 15.12 Cross sections of a box girder at support and midspan, with permanent and temporary PT elements (VSL International)

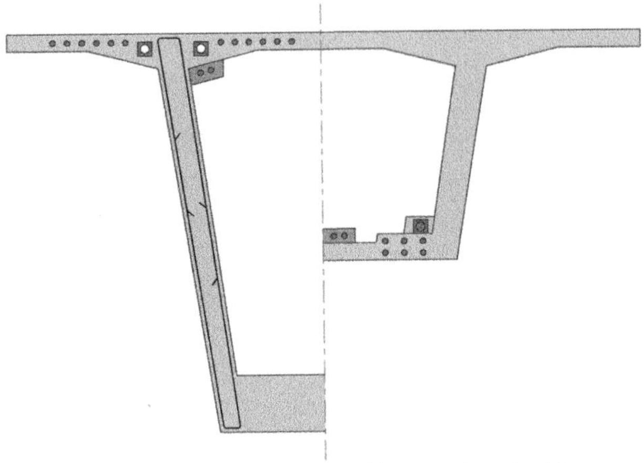

web. With such an arrangement the cantilever tendons do not have to cut through the web's shear reinforcement (structurally preferred and simpler to reinforce in the field), but a direct load transfer of longitudinal shear forces from the webs into the deck slab to build up the tension in the cantilever tendons is still achieved. It is important to check that the haunches are deep enough to accommodate the PT anchorages, their Local Zone Reinforcement (LZR) and the deck slab's longitudinal reinforcement, observing the specified concrete cover. Internal top blisters should be avoided for decks cast in situ by form travellers, since they complicate the traveller's inner form, but may be used for precast segmental decks to anchor temporary top stress bars.

Continuity tendons, shown in Figure 5.12's midspan section as internal prestressing tendons located in the bottom slab (bottom tendons), are anchored in internal blisters detailed next to the webs. This simplifies the load transfer of longitudinal shear forces from the webs into the bottom slab to build up the tension in the bottom tendons. Furthermore, the bottom slab is thickened locally near the web, so that the bottom tendons can be placed over a limited bottom slab width in two layers with effectively anchored vertical confinement links. For anchoring temporary bottom stress bars, required for precast segmental decks, a concrete blister at the centre of the bottom slab is a common solution.

Use of external PT is theoretically a suitable option for continuity PT, since the resulting PT effect at midspan is better when compared with internal bottom slab PT due to the smaller secondary moments. However, structural advantages are often offset by the need for heavier pier diaphragms and rather bulky midspan deviators, which take up a considerable part of the cross section's core. Hence external PT in combination with internal bottom slab PT is often more of interest for medium to long spans, haunched decks and decks with a minimum girder depth at midspan of at least 2.2 m.

For girders with haunched depth, particular attention has to be given to the detailing of the bottom prestressing tendons at segment joints when these tendons are located in the bottom slab (refer to Figure 15.13). The deck is cast in segments only 2 to 5 m long and, for practical reasons, the detailer tries to set out the bottom tendons at fixed vertical distance from the soffit. The detailing has, however,

Figure 15.13 Detailing of bottom PT located in bottom slab of haunched girders built by BCM (VSL International)

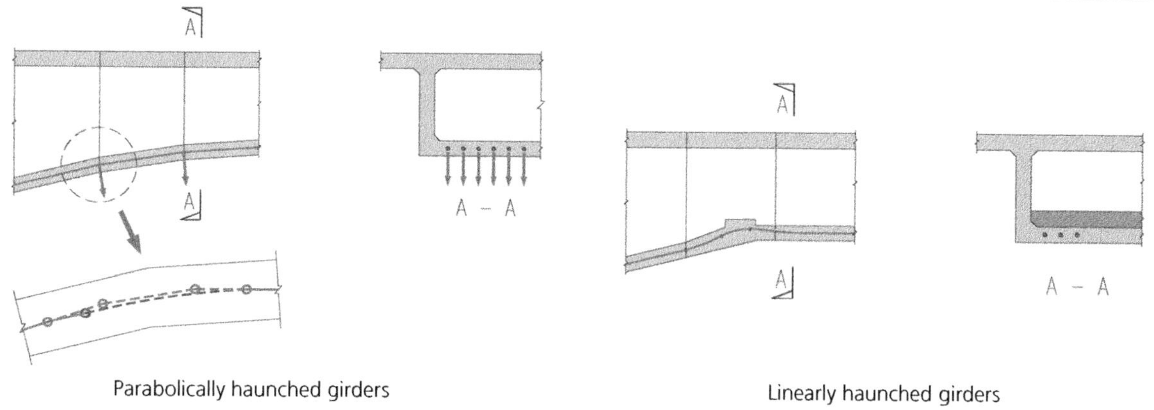

to ensure that tendons are deviated at the permissible radius of curvature and that radial deviation forces can be taken by the bottom slab and transferred back to the webs.

In a 'parabolically' haunched girder, the soffit is not smoothly curved but made up of straight lines between segment joints. Hence at every segment joint the bottom slab is subjected to radial deviation forces from the bottom tendons. The local thickening of the bottom slab next to the webs, as shown in Figure 15.1, allows the engineer to cluster the bottom tendons together and confine them with properly anchored vertical links, which can effectively resist the deviation forces.

In a linearly haunched girder, the bottom tendons are deviated at one location at a considerable angle. Generally the bottom slab is too thin to resist the forces arising from tendons' curvature at this point. This can be resolved by provision of an internal transverse rib in the bottom slab and an adequate vertical tendon profile at this location.

In elevation the cantilever tendons can be arranged either symmetrically or asymmetrically about the pier axis (refer to Figure 15.14). For a symmetrical arrangement, a pair of field segments must be ready for installation and stressing of a new set of cantilever tendons, whereas for an asymmetrical arrangement the next set of cantilever tendons can be stressed, whenever a field segment is ready.

For a symmetrical arrangement at least one pair of tendons per pair of a cantilever's field segments has to be provided. The minimum number of tendons for a symmetrical cantilever girder is therefore the number of pairs of field segments times two. For an asymmetrical arrangement at least one pair of tendons per a cantilever's field segment is required. The minimum number of tendons for an asymmetrical cantilever is therefore the number of field segments times two.

Symmetrical tendon arrangements are by far the most frequently used. Asymmetrical arrangements are sometimes used for cantilevers cast in situ by a traveller to reduce the cycle time and/or to better spread the workforce.

With a symmetrical segmentation, stressing can only commence after concrete in both pairs of the leading segments has achieved adequate strength, whereas with an asymmetrical segmentation tendons

Figure 15.14 Arrangement of cantilever tendons about pier axis (VSL International)

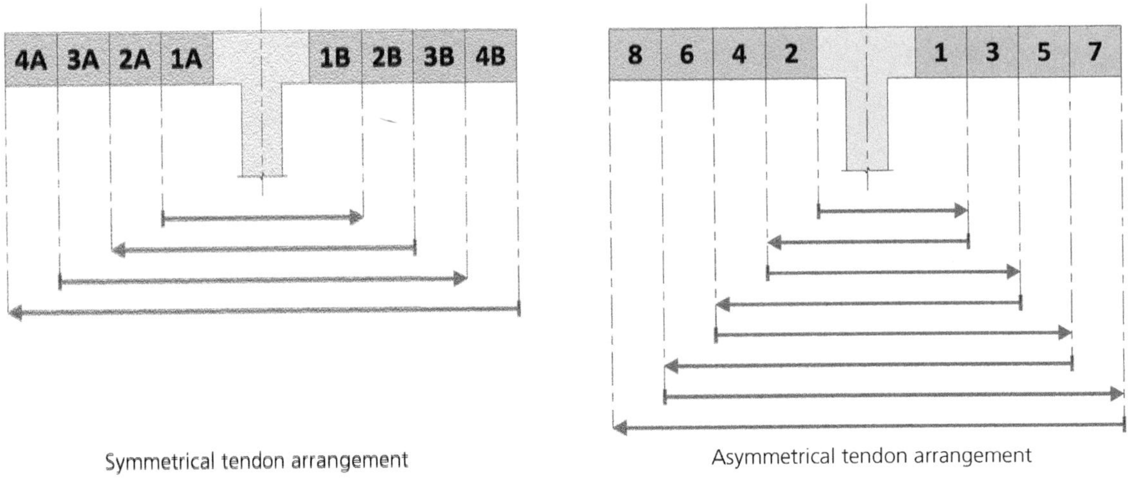

can be stressed as soon as the concrete in one segment has achieved adequate strength. This helps to cut down the cycle by one day, if only one segment can be cast per day. Asymmetrical segmentation also permits the workforce to be spread more efficiently, since there is always some need for activities at one of the two work fronts, even during the time the concrete cures. If the pier table is additionally cast unsymmetrically by half a segment length, the out-of-balance moment during cantilevering can be reduced. On projects where more than one pair of cantilevers is being built concurrently, the spreading of the workforce does not depend on an asymmetrical segmentation any more, since it is achieved between piers and not between up and downstream cantilevers.

A structural drawback of asymmetrical segmentations is the doubling of the ducts for the cantilever ducts. For bridges with medium to large spans, this may require the tendons to be arranged in two or even three vertical layers in the deck slab. Such tendon arrangements require, of course, the provision of properly anchored vertical links to avoid the risk of horizontal splitting and often result in the need to thicken the deck slab beyond what is structurally required.

The starting point for any deck construction by the cantilever method is the pier table. Pier tables need to be stabilised to cope with the out-of-balance moments occurring during cantilever construction. No special temporary measures are needed if the pier table is flexurally connected to the pier for the final articulation of the bridge. The geometry of a pier table depends on whether it is part of a cast-in-situ cantilever or, in the case of precast field segments, is cast in situ or made up from precast segments (refer to Figure 15.15). If the whole deck is cast in situ, the diaphragms and the end faces of the pier table are cast vertical regardless of the deck's gradient. If only the pier table is cast in situ but the field segments are precast, the diaphragms are cast vertical but the end faces are preferably cast perpendicular to the deck's gradient. If the entire deck is made up of precast segments, the diaphragms and the end faces are perpendicular to the deck's gradient.

The pier tables for cantilevers with precast field segments can either be cast in situ (option 1), made up of fully precast pier segments (option 2) or made up of precast pier shell segments (option 3). The three options are shown in Figure 15.16.

Figure 15.15 Geometry of pier table (VSL International)

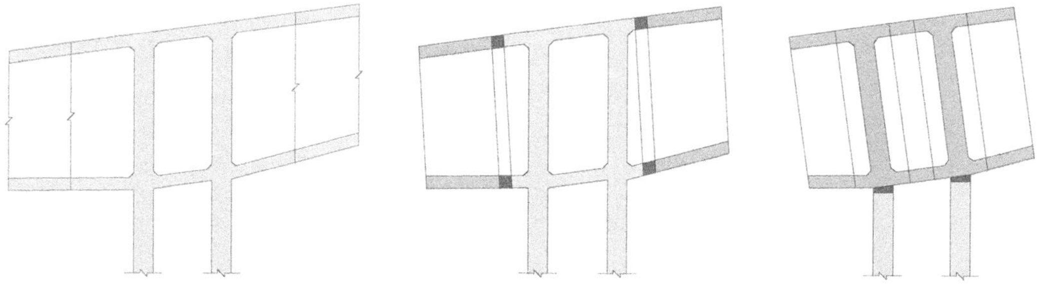

Cast in situ cantilever   Cast in situ pier table with precast field segments   Entire deck made up of precast segments

The cast-in-situ pier table of option 1 requires wet joints between the pier table and the first pair of field segments. These wet joints are typically 200 to 300 mm in length and are precompressed like match-cast joints, since they are too short to permit effective overlapping of reinforcement starter bars.

In the example for option 2 in Figure 15.16, the pier table is made up of a pair of pier segments fully precast with the diaphragms. The first pair of field segments are match-cast against the pier segments. The monolithic connection to the pier is achieved by a layer of in-situ grout and by pairs of vertical U-type prestressing tendons.

Option 3 in Figure 15.16 shows a pier table that consists of a pair of precast pier shell segments against which the first pair of field segments are cast. All elements of the two pier segments are precast except for the inner part of the pier table's diaphragm, which is cast in situ over the full height of the deck. The monolithic connection to the pier is achieved by vertical starter bars protruding from the top of the pier into the centre part of the pier table's diaphragm and the stitch concrete inside the precast part of the diaphragm and below the segments. The precast part of the diaphragm contains all the transverse reinforcement of the pier table and has shear keys at inner faces to improve shear transfer between the precast and the cast-in-situ concrete.

Option 1 is the slowest and option 2 the fastest with regard to deck erection and the amount of work and concreting to be carried out on site. Option 3 is a good compromise between options 1 and 2, when the pier table gets too heavy to be fully precast.

Figure 15.16 Pier table options for cantilevers made up of precast field segments (shown with monolithic connection to the pier) (VSL International)

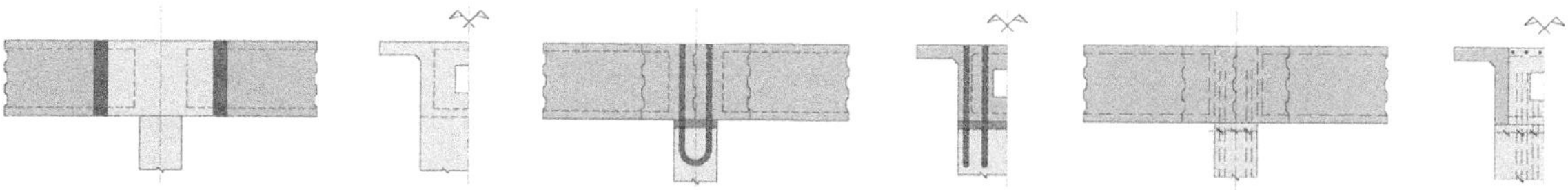

Option 1: cast in situ pier table   Option 2: fully precast pier table   Option 3: pier table made up of precast shell segments

Figure 15.17 Temporary stabilisation of a pier table on bearings against overturning during cantilever construction (VSL International)

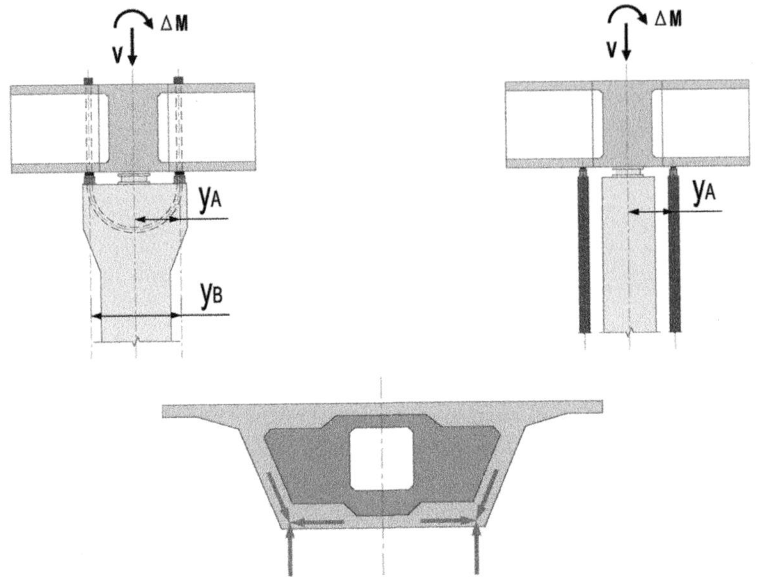

Provision of wet joints between the pier table and the first pair of field segments is also possible for options 2 and 3. Such wet joints give more options to cope with construction tolerances affecting the deck's final vertical and horizontal geometry.

Options 1 and 2 can also be used for non-monolithic connections between deck and pier, where the deck sits on bearings.

Decks with pier tables which are not monolithically tied into piers have to be stabilised against overturning and longitudinal and transverse horizontal movement during construction. Figure 15.17 shows two options of how such stability against overturning can be achieved.

In case 1, the pier head is wide enough to provide space for two lines of temporary bearings at spacing y_B and distance y_A from the centre of the pier. If the permanent bearings are activated only after completion of cantilever construction, the reactions in the two lines of temporary bearings are $V/2 \pm (\Delta M/y_B)$. Temporary bearings can be hydraulic jacks with safety rings, sand jacks (for cast-in-situ pier tables only) or fixed shims. For this form of stabilisation, the bending strength of the pier must also be checked.

In case 2, the pier head is not wide enough to provide space for lines of temporary bearings on either side of the permanent bearing axis. Hence temporary supports are provided by props installed on either side of the pier standing on the pier's pile cap. If permanent bearings are already activated during deck construction, the out-of-balance moment ΔM introduces downwards reaction of $\Delta M/y_A$ into the props either on the upstream or downstream side of the pier, depending on the orientation of

Figure 15.18 Examples of different methods for the construction of the out-of-balance part of an end span (VSL International)

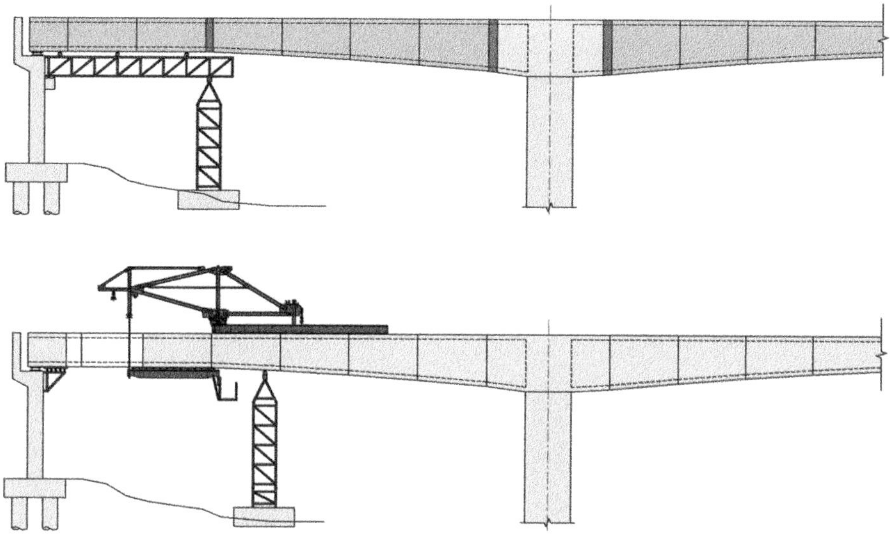

ΔM. Special measures, for example the provision of hydraulic jacks between props and deck, need to be put in place to prevent the props from taking part of V, which would overload them.

Since for both cases generally no diaphragm is available at the section where the temporary bearings are activated, the temporary bearings have to be centred under the webs to avoid local bending and punching in the bottom slab.

End spans of decks built by the cantilever method are often longer than half the internal span length. Hence they can only be partially built by the symmetrical cantilever method about the first internal pier. The out-of-balance part of the end spans can either be built on falsework or by cantilever method with a temporary tower and additional temporary cantilever PT anchored behind the first internal pier table's diaphragm.

Examples of different methods for the construction of the out-of-balance part of an end span are shown in Figure 15.18. In the first case, the respective precast field segments are erected on falsework, whereas in the second case the out-of-balance part of the end span is cast by the form traveller and with a temporary prop, which is activated after the symmetrical deck construction about the first internal pier has been completed.

Long viaducts may require intermediate expansion joints. To avoid complex half joint arrangements, which are hard to construct, difficult to block during BCM construction and offer restricted access to the bearings, expansion joints are best located at piers and not away from piers at about 1/5 to 1/6 of the span. In order to avoid the need for falsework or temporary props, precast segmental decks over expansion joint piers can be treated temporarily like continuous decks, whereby the pair of expansion joint segments are temporarily stitched against each other to enable the erection of the two end spans by symmetrical cantilever construction (false cantilever construction). Alternatively, the out-of-balance part of the end spans can be suspended from a launching gantry, if designed for the appropriate loading. For illustration of these two methods refer to Figure 15.19.

Figure 15.19 Erection of precast segmental decks over internal expansion joints (VSL International)

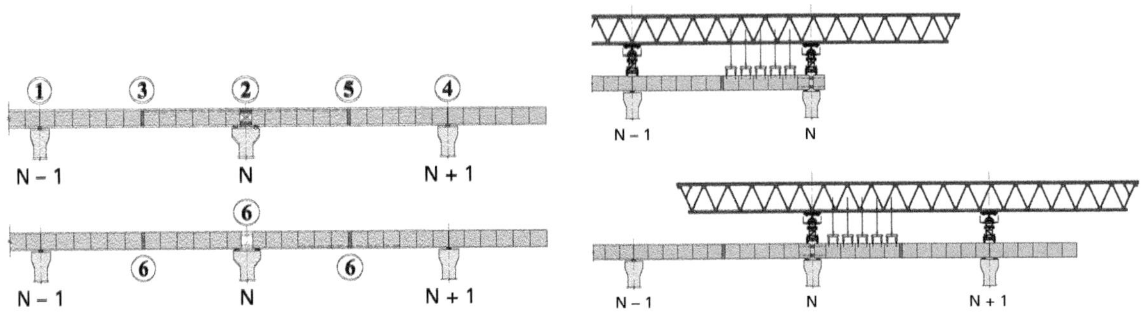

False cantilever construction

1. Construction of cantilever over pier N − 1
2. Construction of false cantilever over pier N
3. Casting of closure pour and stressing some of the bottom PT in span N − 1/N
4. Construction of cantilever over pier N + 1
5. Casting of closure pour and stressing some of the bottom PT in span N/N + 1
6. Deactivation of temporary cantilever prestressing over pier N and simultaneous activation of remaining bottom/continuity prestressing in the two end spans, followed by removal of the stitch between the two expansion joint segments

Support of precast field segments by a gantry

- Stage 1: After completion of cantilever construction over pier N − 1, suspension of upstream field segments at pier N in gantry, followed by casting of upstream closure pour and activation of continuity prestressing in end span N − 1/N
- Stage 2: After completion of cantilever construction over pier N + 1, suspension of downstream field segments at pier N in gantry, followed by casting of downstream closure pour and activation of continuity prestressing in end span N/N + 1

Tips of adjacent cantilevers are joined by cast-in-situ closure pours, which are 200 to 300 mm long for precast segmental decks and about 2 m long for cast-in-situ decks. It is important that the two cantilevers are fixed against each other to prevent any vertical or rotational displacement during pouring and curing of the closure pour concrete due to the weight of the poured concrete, mobile equipment, wind and particularly temperature variation (global and gradient). Rather than provide a temporary fixity which can take the static forces from the worst combination of these load cases, it is better to let restraint forces from the daily temperature variation between day and night introduce favourable stress states. Due to the rising sun, which heats up the top of the deck more than the soffit of the deck, a free cantilever deflects downwards during the first half of the day and deflects upwards once the sun's intensity starts to drop. If a closure pour for a bridge in Singapore, for example, is cast early in the night and finished by about 10 pm when cantilever tips have already moved upwards considerably, the closure pour concrete can be designed to have enough initial strength after nine hours to take the compression in the deck slab when the sun rises again the next morning at 7 am. The tension which develops in the bottom slab can easily be compensated by stressing some of the continuity PT tendons located in the bottom slab, but anchored far back in much older segments.

15.4.4 Decks activated by the Span-by-Span method (SBS)

Decks built by the span-by-span method can be cast in situ or made of precast segments. The full span method is also a span-by-span method, whereby the span is not broken down by transverse cuts into several precast segments but is precast as one self-supporting unit.

15.4.4.1 Cast-in-situ decks

If decks are cast in situ, they are either built conventionally on falsework or on Movable Scaffolding Systems (MSS). For decks with bending continuity over the internal piers, construction joints are preferably located at about 1/5 of the span length in front of the leading pier. End spans should not be longer than 80% of an internal span, so that the end span and the first internal support do not govern the design and its construction does not control the design and detailing of a MSS.

For box section viaducts built by MSS, basic external cross-sectional geometry is best kept constant, since geometrical variation of a cross section's width, depth and internal core dimensions requires special configurations. If the deck is an open double T section, it is not too hard to design an MSS which can cope with variation in depth of deck. Variation in the span length of box girders should also be minimised, since for each change in span length the inner form has to be adjusted and the form has to be pre-cambered differently. Particular attention needs to be given to design, detailing and timing for casting of pier diaphragms since they obstruct the longitudinal movement of the inner form panels. If an overhead MSS is used, its front support is centred over the leading pier and is sitting on steel props at the location where the pier diaphragm has to be cast.

The first-stage longitudinal prestressing needed to make the deck self-supporting for self-weight is normally internal, with prestressing tendons either overlapped over internal piers or coupled at the construction joints (refer to Figure 15.20). In the case of coupled tendons, at least 1/3 of all the tendons running through a construction joint should not be coupled.

External prestressing is generally used only for the second-stage prestressing, since at the time of the first-stage stressing the inner formwork, which would obstruct the installation of external tendons, still has to remain in place.

When defining the layout of the longitudinal and transverse PT tendons in plan it is important to avoid locations where temporary openings are needed during construction. Such openings are required for support of the MSS and its formwork, such as for the rear suspension of an underslung MSS (through openings in wing slabs) or for the front support of an overhead MSS (through holes in the pier diaphragm).

The substructure does not only have to be designed and detailed for the final stage but also for the load cases 'MSS construction'. Pier and foundations have to take the longitudinal loading during launching of the MSS, the longitudinal restraint forces due to temperature variation when the MSS is stationary (with longitudinal fixity/restraints at both front and rear support) and the vertical reactions during concreting (weight of the span and the MSS). It is important that the pier head's longitudinal dimensions are big enough to compensate for the loss in cross section due to recesses for brackets or openings for needle beams and to enable safe load introduction of the MSS's support reactions.

Careful stage-by-stage analysis of the deck is required to avoid overstressing the deck but also to be able to predict deformations. This analysis has to consider all the reactions from the MSS during launching and during concreting but also the change of reactions during the stressing of the longitudinal PT tendons. If the cross section is cast in two pours with partial stressing of the longitudinal PT after the first pour, the stage-by-stage analysis becomes rather complex. Concrete loading from the second pour will be taken both by the MSS and the first cast part of the section according to stiffness ratio.

15.4.4.2 Precast segmental decks

If decks are made of precast segments, they can be built by the span-by-span method in three different ways (structural systems 1 to 3 in Figure 15.21).

Figure 15.20 Layout of longitudinal PT in continuous bridge decks cast span by span in situ on falsework-supported formwork or on MSS (Menn, 1981)

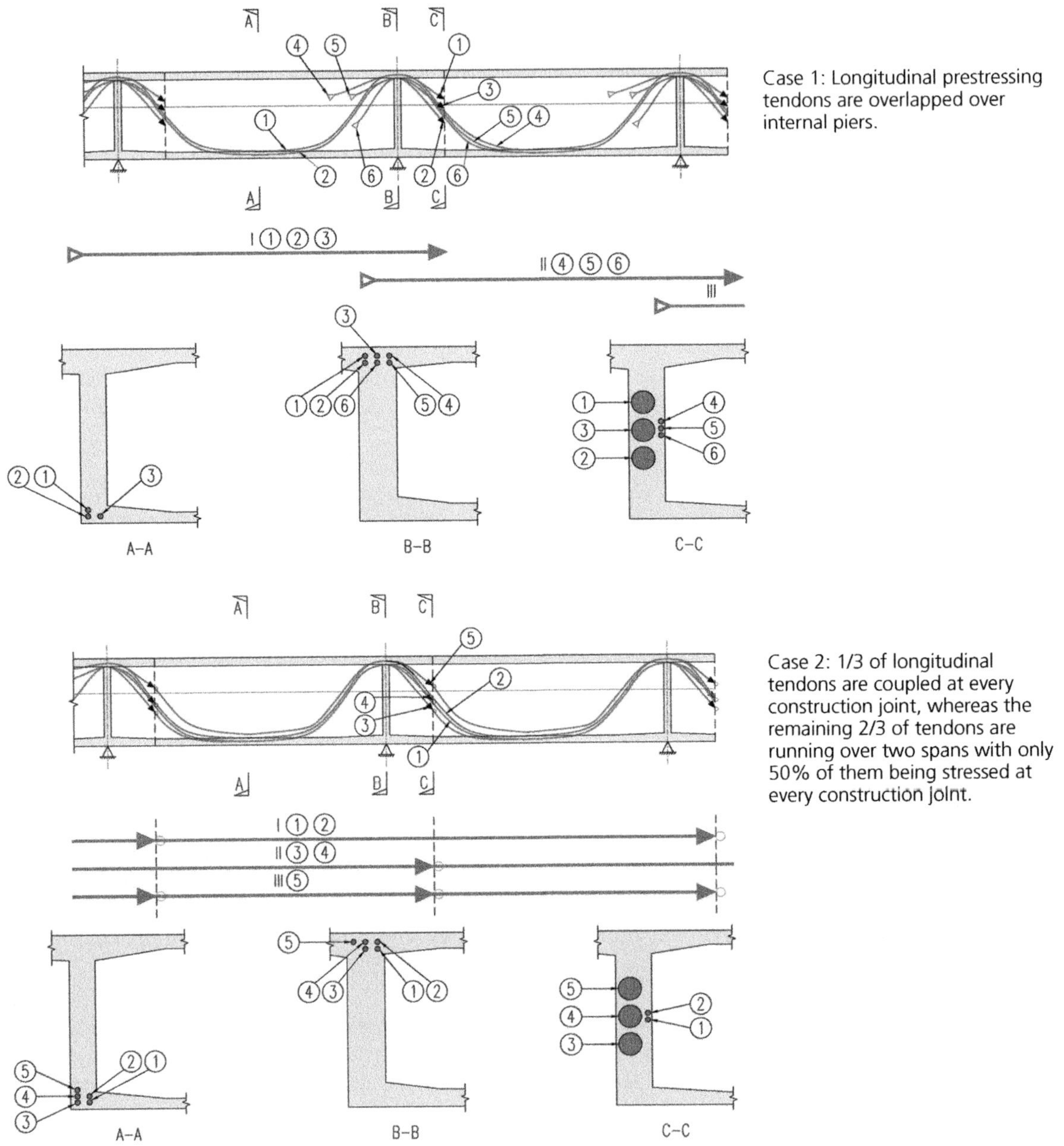

In system 1 the deck carries its self-weight as an end span of the part-built deck, typically with two wet joints per span (on either side of the pier segment). The construction joint is at the front face of the downstream pier segment of the span being built. The longitudinal tendons are overlapped over the pier segment's diaphragm.

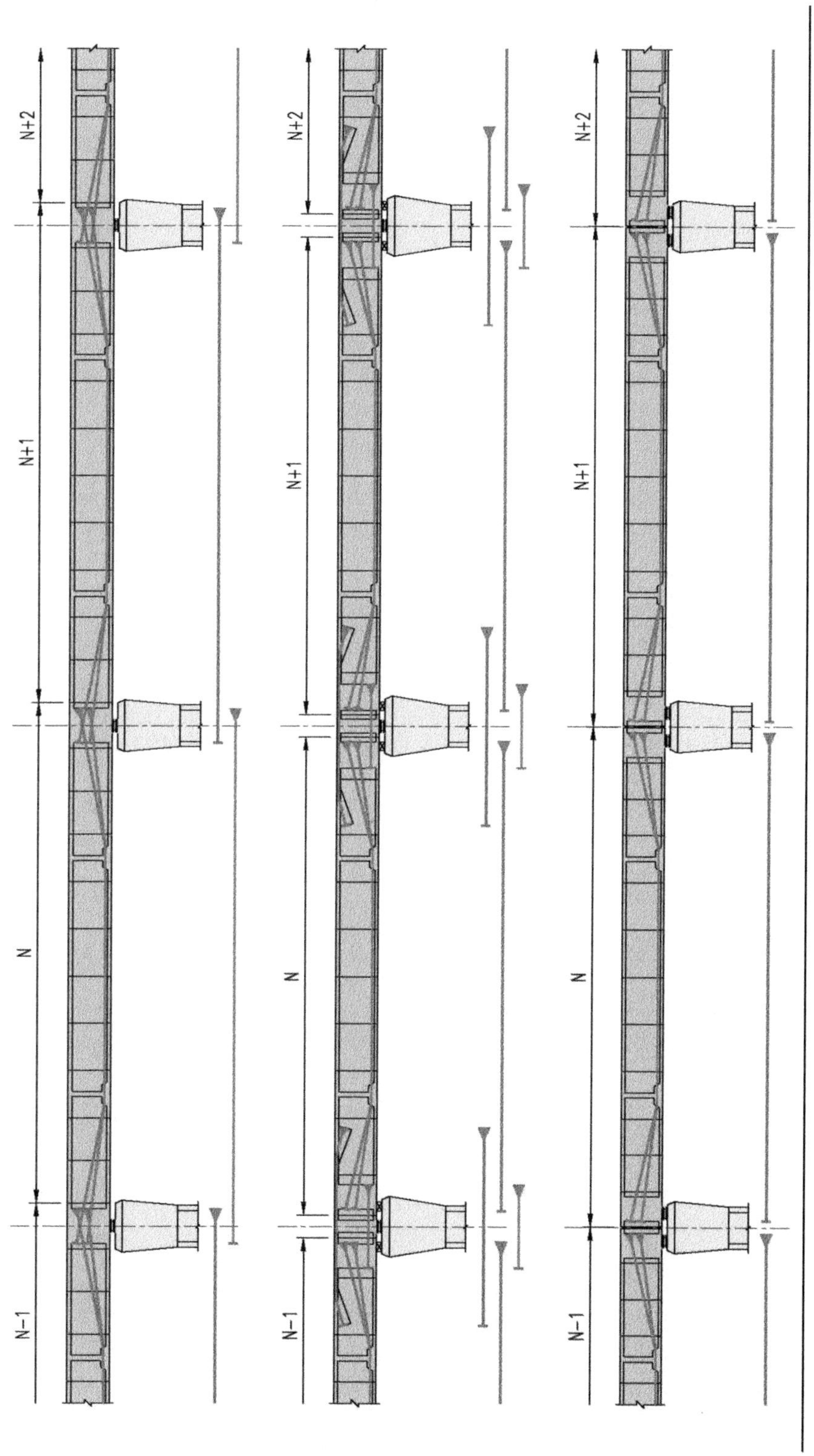

Figure 15.21 Three structural systems (1 to 3) to build precast segmental bridge decks by the span-by-span method (Ganz and Meyer 1997)

In system 2 the deck carries its self-weight as a simple beam on temporary bearings. The flexural deck continuity is established after the BDE has been moved into subsequent spans. This is achieved by closing the gap between adjacent spans with a precast pier segment, by casting wet joints between this segment and the first field segments, and by installation of continuity PT tendons over the pier. The pier head has to be long enough to provide space for three support lines (permanent bearings at the centre with temporary bearings on either side).

In system 3 the deck carries its self-weight and imposed load during service as a simple beam. There is no flexural deck continuity for the final stage.

System 1 requires casting of the wet joints to make the deck self-supporting. Hence the cycle to erect a span will be one shift longer than the deck-erection cycle with systems 2 and 3. However such benefits do not come without drawbacks. System 2 gives deck continuity, but is structurally not optimum due to the need for three diaphragms and bottom prestressing at internal piers. System 3 is by far the fastest to erect, but is the most unfavourable with regard to the long-term maintenance.

The design and detailing of the pier segments and the pier head require special attention to provide enough space for the temporary support of the decks, the stressing of the tendons and, in the case of external prestressing, the potential replacement of tendons in the future. How much space is required depends on which of the three systems has been selected for construction. Four options for pier head and segment arrangements, A to D, are shown in Figure 15.22.

In option A, full deck continuity over the internal spans is achieved during the deck erection. In the example shown in Figure 15.22, the first-stage longitudinal prestressing is external and overlapped over the internal piers. The capping tendons over the internal piers are internal.

With option B, full deck continuity over the internal piers is achieved only after the initial deck erection. In the example shown in Figure 15.22, the first-stage prestressing is external, anchored in diaphragms provided in the first/last field segment. The second-stage prestressing consists of internal capping and bottom tendons over the internal piers.

In option C, there is no girder continuity over the internal piers but only continuity of the deck slab through a link slab cast in situ after the deck girder erection.

Figure 15.22 Four options for pier head and pier segment arrangements and PT layout (VSL International)

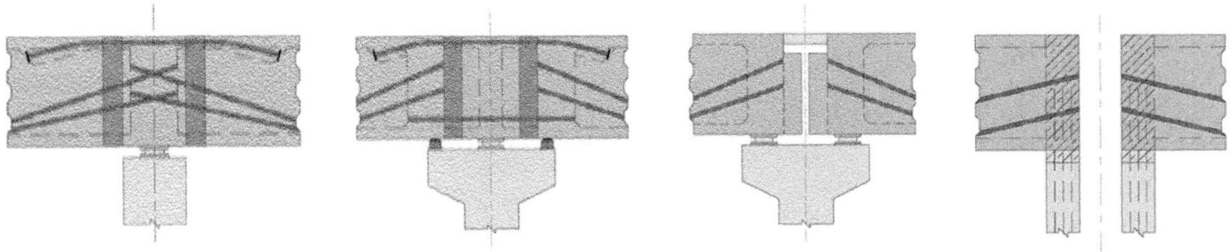

Option A for system 1 | Option B for system 2 | Option C for system 3 | Option D for a system similar to 2

Figure 15.23 Space requirements on pier heads for system 3 segmental bridge decks (VSL International)

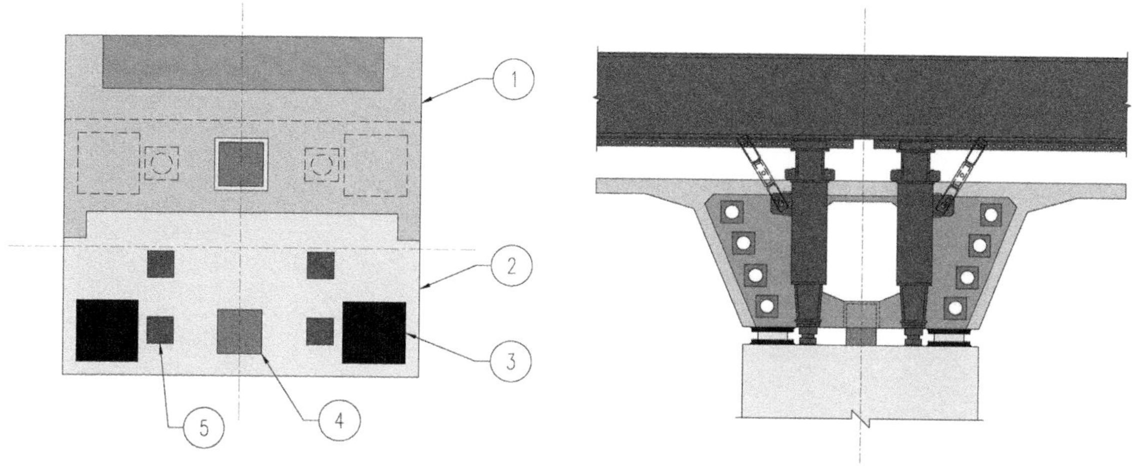

Plan at leading pier during erection with an overhead gantry

1. Downstream pier segment
2. Top of pier head
3. Permanent/temporary bearing location
4. Shear key
5. Support frame for launching gantry's front LCB

Cross section at leading pier during erection with an overhead launching gantry. Note that the legs of the gantry's LCB support frame have to be moved inwards to create enough space for stressing of the first-stage prestressing tendons

In option D, the decks are simply supported during erection and are single-span frames for the final stage. In the example shown in Figure 15.22, the monolithic connection between the deck and the piers is provided by vertical reinforcement starter bars protruding from the pier into the cast-in-situ part of the pier segment's diaphragm. The prestressing is internal, running parabolically in the webs of the deck.

For system 3, the pier head dimensions are not only governed by the need for two bearing axis, but also by the need to have adequate space on the front half of the pier head for support of the BDE, if an overhead launching gantry is being used (for details of this type of launching gantry and the definition and description of its elements, refer to Section II, Chapter 6). Furthermore, the legs of the support frame for the front Lower Cross Beam (LCB) have to be moved inwards to create enough space for the stressing of the first-stage prestressing tendons. For illustration, refer to Figure 15.23.

Use of external prestressing is very suitable for precast segmental decks activated by the span-by-span method. The weight of the deck can be minimised, since the thickness of the webs once free of tendon ducts is governed purely by global shear and transverse bending and the ability to place and compact the concrete. Casting of field segments is simplified, since no ducts have to be placed at an exact location in the rebar cage and secured against the conjugate segment and the bulkhead during concreting. The pier segments, where the external tendons are anchored, and the deviator field segments, where the external tendons are deviated, are best designed and detailed using strut tie models as indicated in Figure 15.24. Since the pier segment's diaphragm and field segment's deviators crack during stressing of the PT tendons, sizing and detailing of the reinforcement on the basis of stresses, which have been derived from a linear elastic analysis, do not guarantee satisfactory crack width and crack distribution at Serviceability Limit State (SLS). Strut and tie models, on the other hand, give good

Figure 15.24 Load introduction of external PT forces: typical arrangement of pier and deviator segment with indication of struts and ties (simplified) (VSL International)

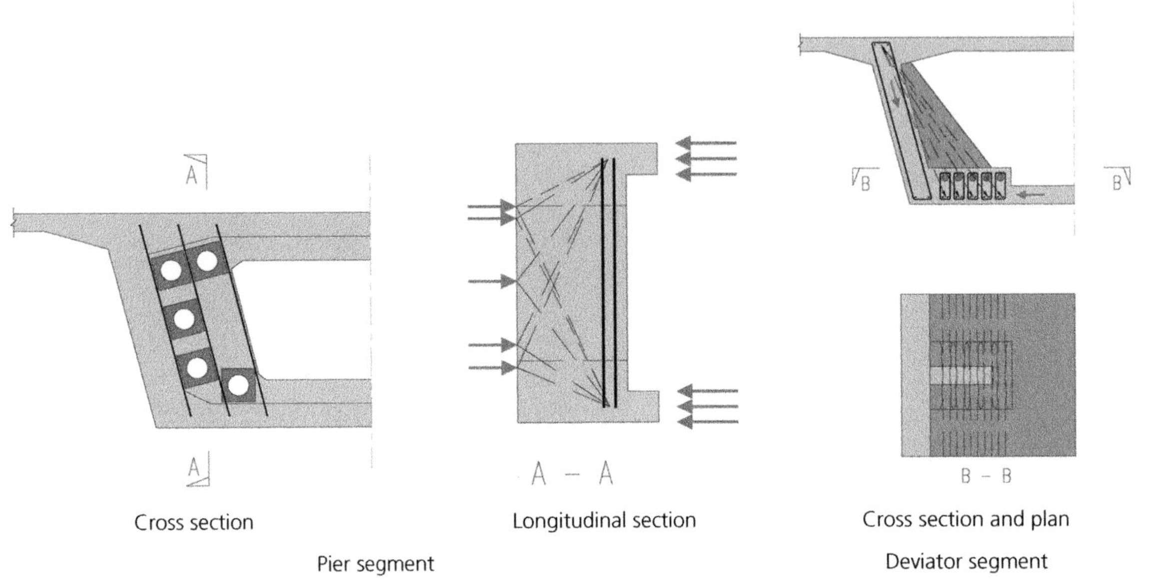

technical and economical solutions, since they not only give reinforcement content but define precisely the diameter, position, shape and length of the reinforcement needed to control cracking of the diaphragms and deviators within permissible limits at SLS.

The length of the pier segment's diaphragm, where the external tendons are anchored, is typically between 1.2 and 1.5 m. Anchorages are arranged as close as possible along the inner faces of the web, the top and the bottom slab. They are aligned to create 'vertical' corridors for positioning of the vertical tie reinforcement of the diaphragm, unobstructed over the full depth of the segment.

Vertical components of external tendons at deviators are best introduced into the deck's cross section by strut-tie action and not by a cross beam at the bottom slab level, which spans between the webs and works in bending.

The prime designer needs to give provisions for the load case 'span erection' at the conceptual design stage. The location and magnitude of loading onto the already erected deck and into the piers depends on how the decks are being built (systems 1 to 3) and the type of BDE being used (underslung or overhead launching gantry).

Arrangements at the rear support of a BDE are shown in Figure 15.25 and arrangements at the front support of the Launching Gantry (LG) are shown in Figure 15.26.

15.4.4.3 Fully precast spans

The fully precast spans are cast as simply supported girders. Generally they remain simply supported for the final stage, which would make the use of girders with double T cross sections suitable. However box and U-girders are most commonly used when applying this method, with box girders for road and

Figure 15.25 Arrangement at rear support of the LG, according to the system and type of LG (VSL International)

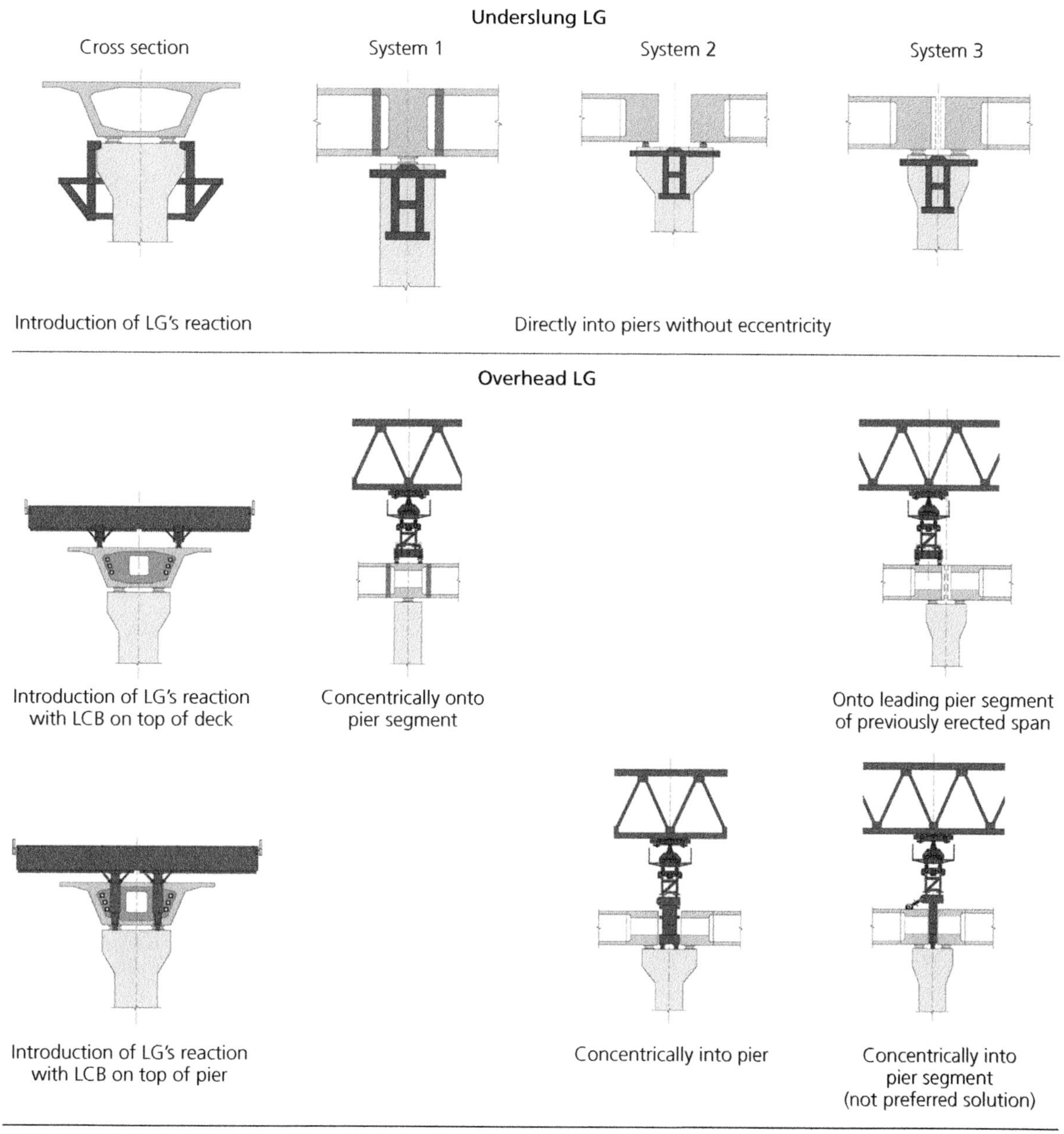

rail viaducts (both for high-speed trains and mass rapid transit/light rapid transit) and U-girders for mass rapid transit/light rapid transit. It is easier to establish longitudinal deck continuity over internal piers when using box girders than when using U-girders, since there is a lot of room available in the deck slab to locate the top prestessing tendons and inside the box to anchor the top and bottom PT. Girders are cast straight. Fully precast girders are either pre-tensioned and/or prestressed with internal tendons. In the case of box girders, use of external prestressing for a second-stage prestressing is

Figure 15.26 Arrangement at front support of the LG, according to the system and type of LG (VSL International)

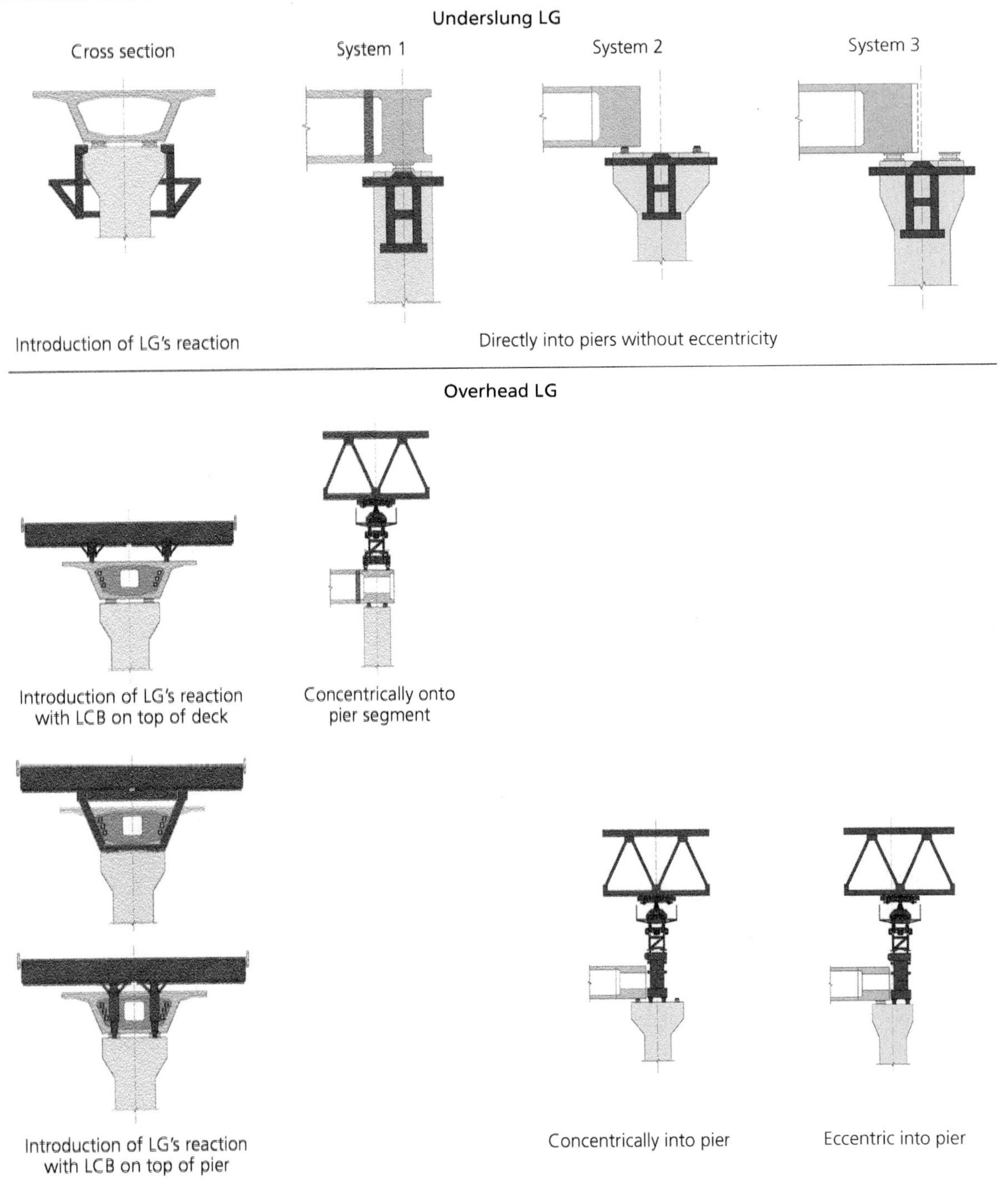

Figure 15.27 Fully precast box girder: isometric view and cross sections at midspan and through the end diaphragm (Meyer, 2004)

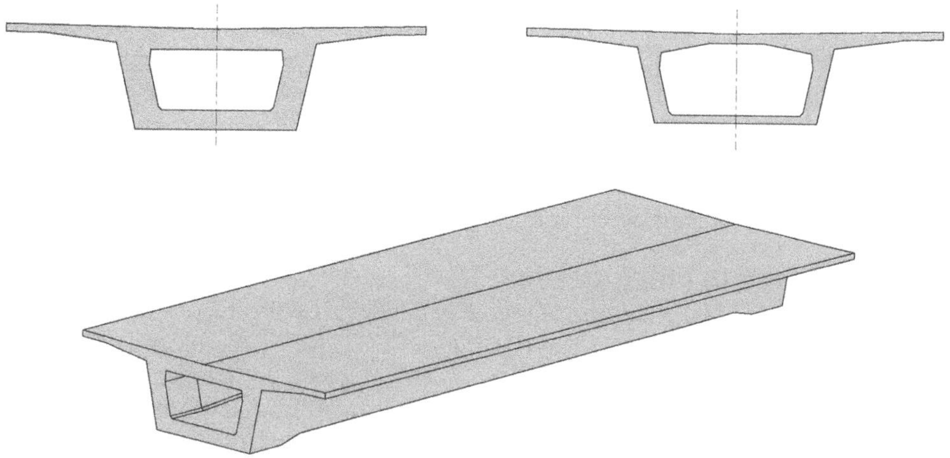

possible. A special category of girders is required for monorail systems. These girders have a rectangular cross section (solid or with lost void formers). The girder's top face is the monorail's running track. These girders therefore have to be cast to whatever the track geometry defines. Special formwork is needed for production of such girders.

To achieve economy when using the full precast span method, the cross section should be kept constant and the span length should be standardised to a few cases only.

Due to their size and weight, fully precast girders are mostly delivered to the erection front over the already erected part of the viaduct. This load case often becomes one of the controlling load cases in the design of the girders for the global analysis as well as for the deck slab analysis. This critical load case can be avoided in the case of parallel twin bridges as the load can be split between the two decks.

In order to minimise the number of casting cells for the production of box girders, the detailing of the box girder should ensure that the inner form can be a self-launching tunnel form. Hence the opening through the end diaphragms needs to be reasonably large, which results in a locally thickened top slab, bottom slab and webs. To increase the vertical dimension of the opening, the bottom slab can be thickened downwards rather than inwards. Refer to Figure 15.27.

Fully precast box girders are generally placed by overhead gantries. The pier head needs to be long enough to provide space for the front support of such equipment and piers and foundation need to be designed to be able to cope with the eccentrically introduced gantry reactions during the span placing.

15.5. Incrementally launched concrete, steel and composite bridge decks

The Incremental Launching Method (ILM) is used for construction of concrete, steel and composite bridge decks. The deck is broken down for construction transversely into sections of similar length, which are added to the already completed deck behind one of the abutments at a fixed location, either concreted in a formwork or assembled in an assembly area. The deck is pushed or pulled out of the formwork or assembly area and over the abutment onto the piers into its final position.

Provision of a launching nose stressed against the first deck segment can reduce the big bending moments in the deck near the front during launching. The deck bending moments in viaducts with only a few spans that are larger than the typical span length can be reduced by provision of temporary towers.

Special temporary bearings are required on the launching abutment and on every pier to move the deck over the piers towards its final position during the launching operation.

15.5.1 Concrete bridge decks built by ILM

The deck of an incrementally launched bridge is cast in a fixed form located on the bridge axis behind one of the abutments. It normally has a length of 1/4, 1/3 or 1/2 of the typical span length of the bridge. After the concrete of the deck segment N has achieved adequate strength and after the segment has been centrically prestressed against the previously cast segment N − 1, the deck is pushed or pulled forward (usually by means of hydraulic jacks) by the length of the casting bed. The casting bed is then free for casting of the next deck segment N + 1.

In order to be able to incrementally launch a bridge deck pulled or pushed out of a casting bed behind the launching abutment, the deck's soffit has to follow a straight line or a curved line with a constant radius in a plane, which is either flat or on the surface of a cylinder. For more details on geometry refer to Section II, Chapter 13. However generally not only the deck's soffit is kept constant, but the entire external cross-sectional geometry, since this greatly simplifies the construction and increases the economy of the incremental launching method.

Decks installed by the incremental launching method are centrically prestressed for the launching phase. The first-stage prestressing is supplemented for the final stage by continuity prestressing tendons, which typically run either internally parabolically profiled or externally with a trapezoidal profile. The box girder section is mostly used for ILM bridges. In a box girder, the ratio between bottom and top sectional core value is about 2, which minimises the amount of centrical prestressing. Furthermore, the second-stage prestressing can be easily curtailed along the bridge deck by anchoring tendons in blisters located inside the box. The top and bottom slab may have to be thickened locally to provide enough space for the first-stage prestressing tendons, which are typically straight tendons in the top and bottom slab of the box section and are coupled staggered at every alternate construction joint. The tables given in Figure 15.28, which considers the reduction of the critical negative and positive bending moments in the deck due to provision of a launching nose, can be used for sizing of the centrical prestressing. Economic savings are possible if the owner and the consultant can be convinced that there is no need to fully precompress the deck against longitudinal flexural fibre stresses, which occur during launching.

The segmentation for a box girder deck is preferably done in such a way that a tunnel form can be used as the inner form. Figure 15.29 shows how this can be achieved by positioning the construction joints at the front face of the pier diaphragms.

Special attention has to be given to the load introduction of the support reactions during the launching phase, which occurs anywhere along the webs of the box girder. The centre line of the temporary bearing has to be intersected with the centre line of the web and the bottom slab to avoid local bending

Figure 15.28 Minimum and maximum deck moment during incremental launching of a deck (Muller and Podolny, 1982)

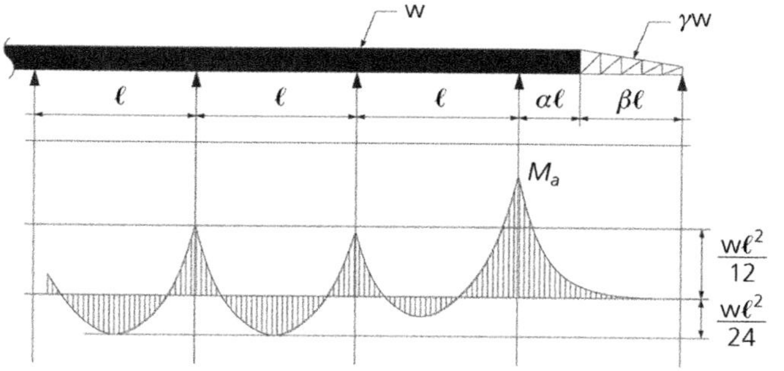

M_a for $\gamma = 0.10$ with multiplier $w\ell^2/12$

α	β	M_a
0.20	0.80	0.82
0.30	0.70	1.09
0.40	0.60	1.46
0.50	0.50	1.95
1.00	0.00	6.00

Critical minimum deck moment: $M_a = (w\ell^2/12)[6\alpha^2 + 6\gamma(1 - \alpha^2)]$

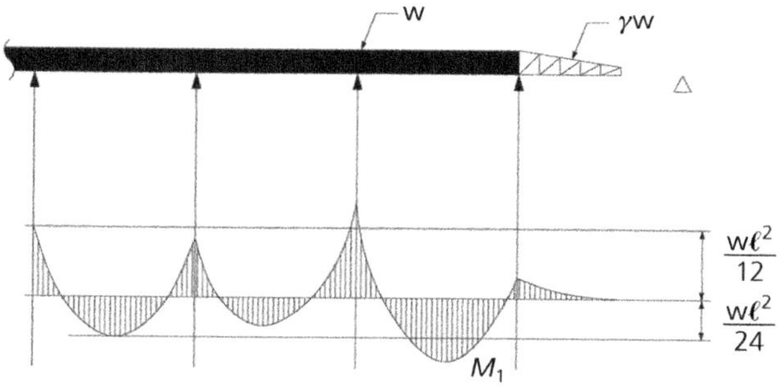

M_1 for $\gamma = 0.10$ with multiplier $w\ell^2/12$

α	β	M_1
0.20	0.80	0.74
0.30	0.70	0.79
0.40	0.60	0.83
0.50	0.50	0.86
1.00	0.00	0.93

Critical maximum deck moment: $M_1 = (w\ell^2/12)(0.933 - 2.96\gamma\beta^2)$

and punching shear in the bottom slab. The piers and the launching abutment need to be designed for the longitudinal forces occurring during the launching phase. These forces differ significantly depending on the method used to cast the deck's segments, the equipment used to push or pull the deck (location of equipment) and the type of sliding material or rollers used as part of the temporary bearings. It is possible to keep these forces internal by providing the movement equipment not just at the abutment but at all temporary bearing locations (for more details of this option refer to Section II, Chapter 13).

Figure 15.29 Segmentation of box girder deck so that tunnel forms can be used (VSL International)

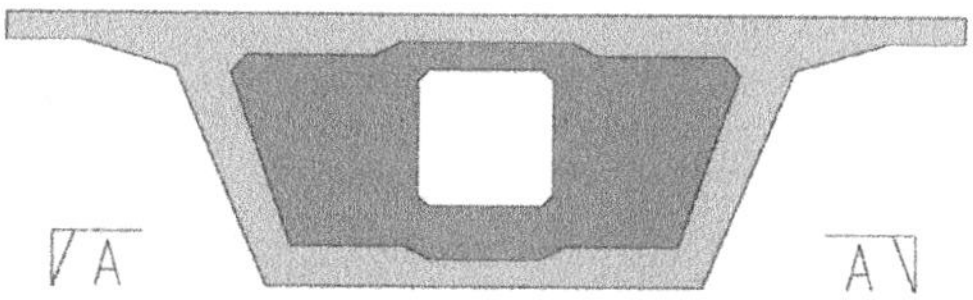

Cross section of deck

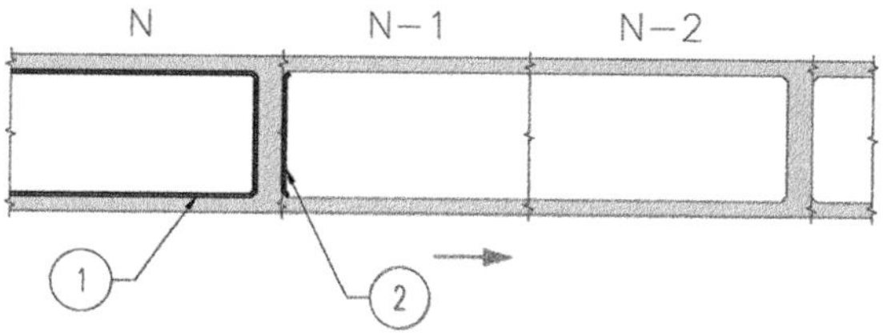

Horizontal section A–A through box indicating location of construction joints in relation to launching direction:

1. Inner form for casting of segment N (tunnel form)
2. Diaphragm form (front)

Figure 15.30 Deck cross section and plan of pier head (VSL International)

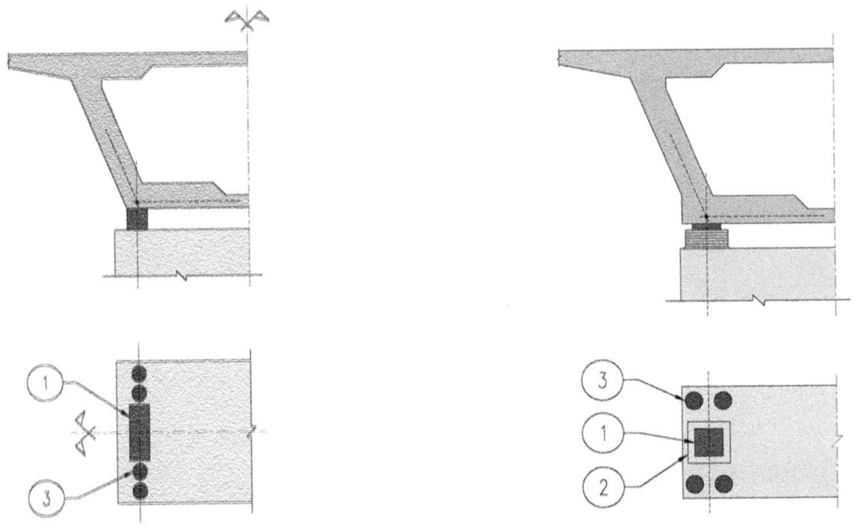

Case 1: Temporary bearings for launching

Case 2: Use of permanent bearings for launching: need to widen bottom slab to have permanent bearing centred below web

1. Temporary bearing/sliding plate 2. Permanent bearing 3. Hydraulic jacks

The pier heads need to be sized to provide enough space to position the temporary bearings and vertical jacks to be able to lift off the deck from the temporary bearings in case sliding plates get stuck. Figure 15.30 shows two different cases (1 and 2). In case 1 the deck is launched over temporary bearings. Temporary bearings can be positioned directly below the intersection of the web's and the bottom slab's working lines and can be made rectangular to fit within the width of the web's projection to the deck's soffit level. In case 2, the permanent bearings are used for the launching of the deck. Since permanent bearings are circular or square but not rectangular and have to be installed with decent edge distance against the web's outer face, their centre line can only be aligned below the intersection of the web's and the bottom slab's working lines if the bottom slab is widened.

15.5.2 Steel/composite bridges built by ILM

Steel has similar strength when subjected to tension and, if stability is not controlling, compression. Hence it is no surprise that the incremental launching method, where the deck has to cope at any cross section with positive and negative moments of comparable magnitude during launch, was used for steel bridges in the 19th century.

Composite bridge decks, which use steel and concrete for what they are best for, are, from a structural point of view, preferably built in two stages. The steel girders, which are good in tension and compression and are light, hence reduce the reactions into the temporary bearings and the longitudinal forces during launch, are installed first by incremental launching. The concrete deck slab is subsequently cast in situ with a travelling form. From a construction point of view, however, a solution whereby the deck slab is made up of precast planks and a cast-in-situ topping offers advantages, since it allows the bridge deck to be completed in a very short time after the launch. The precast planks are placed on top of the steel girders in the assembly area. However the planks over a length of about the

Figure 15.31 Steel girder with precast panels (Freyssinet International)

typical span length, at the leading end during the launch, are only installed after the completion of the launch. This reduces during launch the peak bending moments at the front of the girders (refer to Figure 15.31).

Alternatively to the precast planks, a 'lost' soffit in the form of steel planks can be installed on the girders in the assembly area (refer to Figure 15.32).

Structural composite action between the concrete deck slab and the steel girders by means of shear studs is established after the deck has been launched into its final position. This measure protects the concrete slab during launching from being subjected to longitudinal tension due to the deck's negative self-weight moments.

An alternative to precast deck slab planks is corrugated steel planks, which act not only as self-supporting formwork for the casting of the deck slab, but also structurally together with the concrete. Such steel planks can also provide temporary bracing between the steel girder's top flanges. To ensure the effectiveness of such a type of bracing, the stiffness of the panels and the quality of their connections to the plate girders has to be checked carefully. In some countries the use of corrugated steel planks may not be allowed.

The length of steel girders supporting a steel or composite deck built by ILM, and therefore the deck's transverse segmentation, is controlled by transport restrictions between the steel workshop and the bridge site. Segment lengths of up to 16 m are common (the stock-lengths of steel). Which lengths can be used for a specific project depends on the project's span length, local traffic regulations and on-site access conditions. Longer segments minimise the programme and the number of costly site splices in the main girders.

Figure 15.32 Steel girders with steel planks as 'lost' form for casting of the deck slab after completion of the launch (Freyssinet International)

The assembly of the deck behind one of the two abutments is, unlike with concrete decks, very often done on top of the finished embankment, since good access is required to deliver the steel elements on normal trailers. Hence after the deck has been longitudinally launched into its final position, the entire deck must be lowered by a considerable height down to the final level. It is important to ensure that pier head dimensions are big enough to allow for temporary support of the deck during launching at a raised level and for the vertical jacking system to lower down the deck to its final position (refer to Figure 15.33). Depending on the location of the temporary supports in the assembly area and between the assembly area and the abutment, there may be the need for a rear nose.

Even with steel being a very suitable material for incrementally launched bridge decks, the designer needs to address design and detailing issues which are specific to incremental launching of steel and composite bridge decks.

- During launching, there are moment reversals in every cross section of the deck's girders, which means that both flanges and the top/bottom parts of the webs need to be designed and detailed to be stable enough to also take longitudinal compression. The steel girder's top flange of a composite bridge deck can be relatively small in area for the final stage, since the concrete deck is basically the deck's top flange, but has to be big enough to take the considerable flexural stringer forces during launching.
- At the tip of the girder, a device is generally installed which allows the deck to be launched onto the temporary bearings when reaching a pier (by either vertical jacking or a front nose with a curved soffit). Whether or not this device is part of a real front nose – as is always used for concrete decks to minimise the peak moments during launch – depends on the girder's self-weight and on how much flexural capacity the girder has.

Figure 15.33 Elevated launching of the bridge deck (Freyssinet International)

- Introduction of temporary bearing reactions into the steel girder's bottom flange and the un-stiffened web has to be carefully analysed and flange and web have to be sized accordingly. Temporary bearings are mostly equipped with articulated rollers and not with sliding plates – as for concrete decks – to minimise the support width below the girder, so that the bearing stresses are either directly below or as close as possible to the web's footprint. They are always centred below the web. When determining the relevant support reactions, attention has to be paid to variation in support reactions during launch arising from non-uniform span lengths combined with steel pre-camber. This load introduction zone has to be checked for the combination of static forces from global action (moment and shear) and the reaction from the temporary bearing (patch loading). Eurocode EN1993-1-5 gives two approaches on how to analyse such a zone (effective length method and reduced stress method). Bolted splices require special attention to ensure that splice plates and bolts do not obstruct the temporary bearings and the web can still be stiffened, if needed, with vertical stiffeners. The width of bottom flanges is best kept constant to simplify the lateral guide system (refer to Figure 15.34). It is also important to check that potential local thickening of the main girder's bottom flanges at bearing location or bearing fixity provisions are compatible with the temporary bearings.
- Splices between girder segments: splices can be welded or bolted. Bolted splices are faster and can be installed by less-skilled workers. High-strength friction grip bolts have to be used for bolted connections. It needs to be verified that there is enough space to tighten such bolts.
- The connections between girders and noses (front and rear) and pulling beams normally require some local modifications to the deck's steelwork.
- For composite decks it is necessary to check whether a temporary horizontal bracing between the steel girder's top flanges is needed, since during launch the concrete slab is either not yet in place or structurally not yet connected.
- If for composite decks the deck slab is cast in situ after the steel girders have been launched into the final position, the self-weight of the travelling form has to be considered in the stage-by-

Figure 15.34 The bottom flange's splice plate, split into a pair to provide a corridor for temporary bearings and constant width of the bottom flange to simplify the lateral guide system (Freyssinet International)

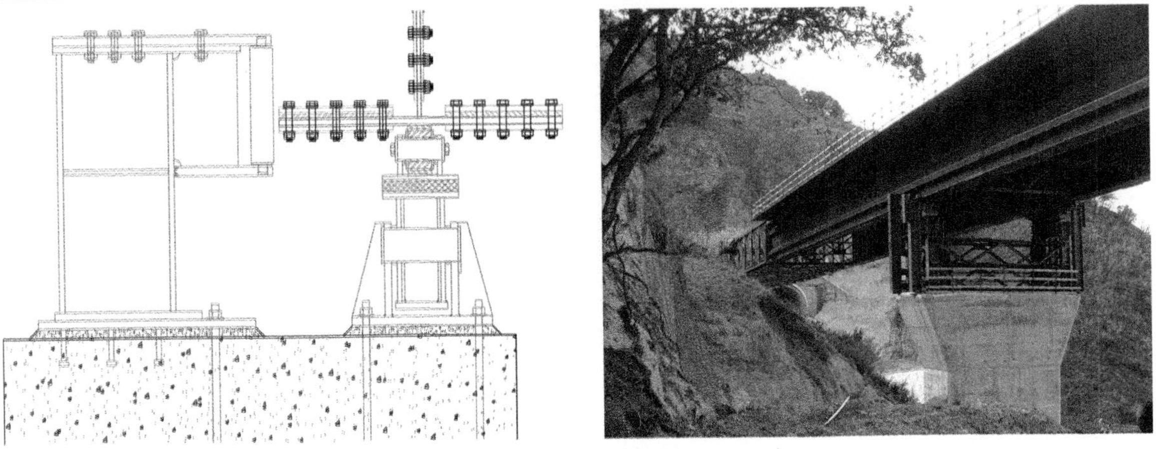

stage analysis of the deck. Travelling forms as well as travelling frames used to place precast panels sit on top of the steel girder's top flange. Shear studs therefore need to be arranged in such a way and have to be set out within certain tolerances to provide space for the supports of the travelling form (or travelling frame) during launching and casting (refer to Figure 15.35).

- After incremental launching the deck has to be transferred from the temporary bearings, which are often high above the final level, to the final bearings. The designer must give permissible differences in levels between the deck's support points during this operation. The location for temporary jacks needs to be defined with corresponding sizing of the pier cap and detailing of the main girder or the pier diaphragm (refer to Figure 15.36).
- Abutment and piers have to be checked for longitudinal forces introduced during launching.

Figure 15.35 Travelling form (Freyssinet International)

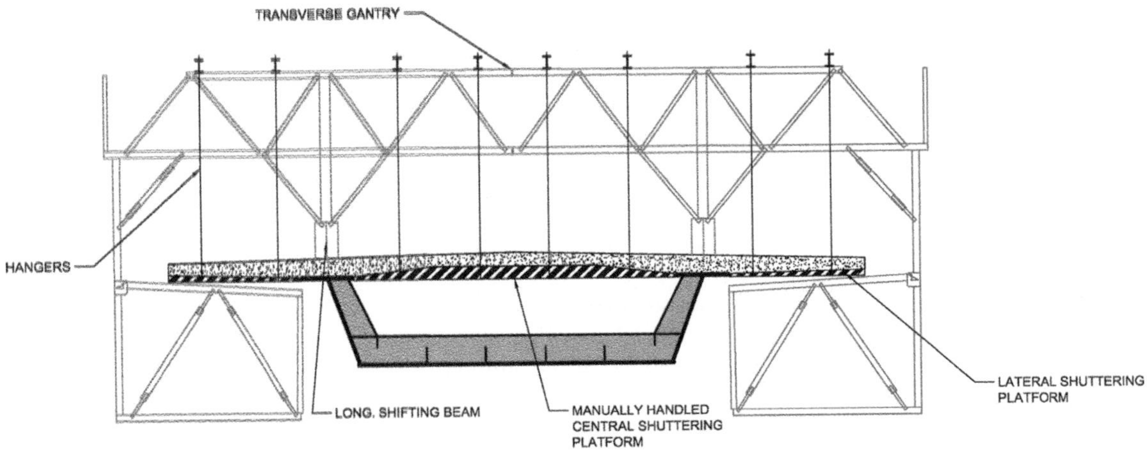

Figure 15.36 Jacking-up of deck for installation/replacement of permanent bearings (Freyssinet International)

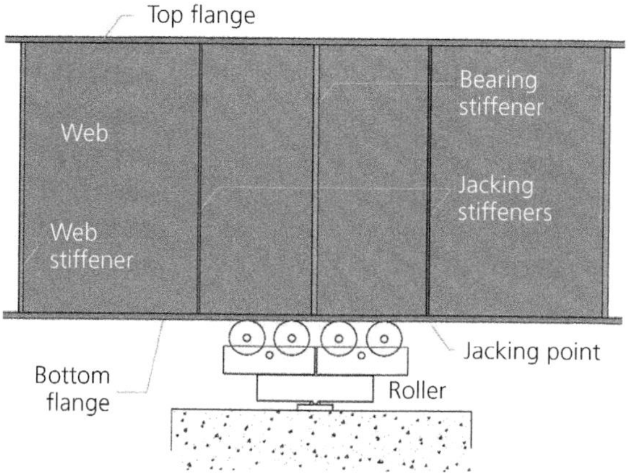

15.6. Stay-supported bridge decks made from cast-in-situ or precast concrete, steel or composite segments

The first major cable-stayed bridges were designed and built in Germany after the Second World War. Since then this bridge type has managed to penetrate more and more into span ranges, that used to be the clear domain of suspension bridges. A main reason for this development is that cable-stayed bridges are ideal to be built by balanced cantilever method and that parallel strand cables can be used as stays. A parallel strand stay can, regardless of its size, be installed and stressed strand-by-strand on site with light winches and mono-strand stressing jacks. This construction-friendly feature of state-of-the-art cable-stayed bridges compares favourably against suspension bridges, where extensive temporary installations are required to build the main cables in situ by aerial spinning or to hoist parallel wire cables, prefabricated over the full length of the bridge. Furthermore, for a suspension bridge, installation of the main cables and subsequently of the deck can start only after the pylons and the two anchor blocks have been completed. However, in the case of cable-stayed bridges, deck construction can start after completion or even only partial completion of the first pylon. This helps to make construction of cable-stayed bridges faster than the construction of suspension bridges.

Specialist bridge deck erection equipment is generally used to construct the decks of cable-stayed bridges. Decks can be made in steel or concrete or can be composite.

Stays often have to be stressed in stages so that static forces in stays and deck during erection do not become more controlling for the design of these elements than the static forces during the final stage of the bridge.

15.6.1 Bridge decks in concrete

Bridge decks in concrete built by BCM are either built in situ with travellers or, if precast, erected with cranes or lifting frames. Design and detailing of travellers and lifting frames is closely linked to the type of cross section used for the bridge deck, the number of planes of stays and where the stays are connected to the deck. Figure 15.37 shows examples of solid-slab, open and box cross sections. An open cross section is a section with a pair of main beams, cross beams and the deck slab.

Figure 15.37 Options for concrete cross sections of stay-suspended decks (VSL International)

Type of cross section		Number of planes of stays	Deck anchorage position	Cast in situ	Precast
Solid-slab section	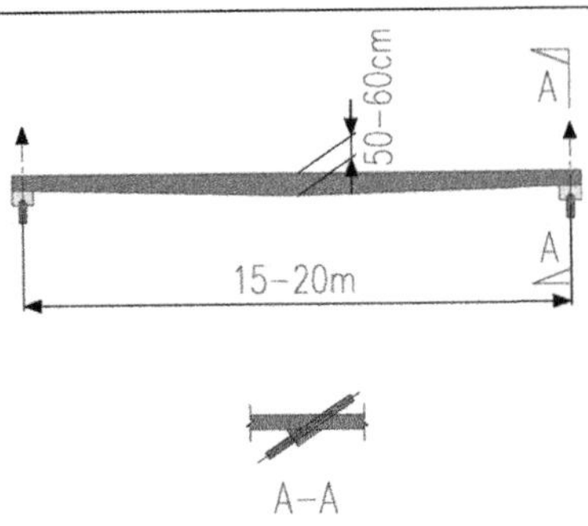	2	At edge of deck	X	N/A
'Open' section with main edge beams, cross beams and deck slab	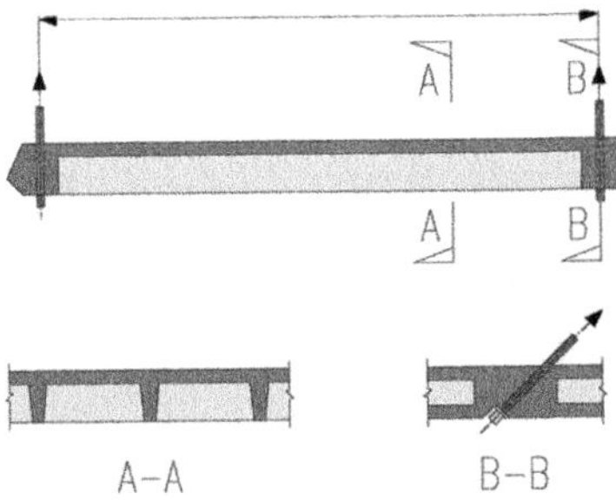	2	Through main beams (or outside of main beams at cross beams)	X	(X)
Single-cell box section with pair of internal struts	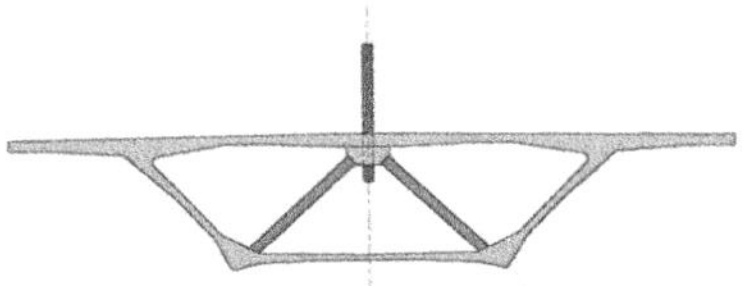	1	At centre of deck slab	X	X

When defining the concepts for travellers or lifting frames, the introduction of reactions during the segment casting or the segment erection and during the launching need to be carefully assessed in order to:

- be within the shear and flexural capacity of the cross section's element the BDE is standing on or is attached to
- permit the installation and stressing of the stays and the subsequent forward-launching of the BDE.

Furthermore, the overall structural impact of the BDE on the deck needs to be looked at with regard to its weight and the distance of its centre of gravity against the last-installed and stressed stays. Particularly for open cross sections, the segment erection is often a controlling load case for the bending in the deck and the tension in the stays. The main beams of an open section have no bottom flanges. The neutral axis due to bending may therefore be high up in the main beams and close to the deck slab, which can result in a non-ductile behaviour of the cross section. If this is a problem for a deck cast in situ with a traveller, the weight of the traveller and the static forces in the deck and the stays during casting of segment N can be reduced by connecting segment N's pair of stays to the traveller's tip and

Figure 15.38 Different casting options for decks with open section cast in situ with travellers (VSL International)

Main beam cast on front and cross beams and deck slab cast on rear part of underslung traveller

Full section cast at front of traveller (overhead or underslung)

stressing them partially prior to concreting. Alternatively, casting of the deck's full cross section can be carried out in two pours, staggered longitudinally rather than vertically. The pair of main beams is cast on the front part of the traveller, whereas the remaining part of the cross section (cross beams and deck slab) is cast on the rear part of the traveller. This method is especially effective and often superior to the first mentioned method when the decks are wide, since for wide decks the weights of the two pours are often of similar magnitude. For illustration refer to Figure 15.38. However, this method can be used only if stays are anchored at the centre of the main beams.

Box sections generally have more than enough flexural capacity to cope with the big negative bending moments during the segment erection.

15.6.2 Steel or composite bridge decks

For large and very large spans, the load case 'self-weight' becomes so dominant that the decks tend to have mostly composite or steel sections. Use of steel for the deck's main longitudinal members permits the segments to be two to four times longer than when using concrete as material for the deck. This helps to compensate for the two- to four-times-longer erection cycles of steel and composite decks when compared with precast segmental decks. The longer cycle is mainly due to the time needed for splicing the steel members and the casting of the stitches between the precast planks in the case of composite sections. As for concrete decks, the design and detailing of the lifting frames for steel and composite decks is also closely linked to the type of cross section used and the number of planes of stays and where the stays are connected to the deck. Figure 15.39 shows examples of decks with composite sections.

Steel or composite bridge decks built by BCM are erected either with cranes or lifting frames.

Cable-stayed bridges with two planes of vertical stays supporting a composite deck with open cross section enjoy huge popularity, since they are easy to build. The steel grillage is made up of a pair of main girders at or close to the edge of the deck, which are connected by steel cross beams typically at 3 to 4 m spacing. The concrete deck of a composite section is generally made up of precast panels, which are stitched against each other and connected to the supporting steel grillage by in-situ concrete. Precast panels can be installed on the steel grillage prior to its installation. Alternatively they can be installed after the steel grillage has been erected, connected to the tip of the deck, and a pair of stays

Figure 15.39 Options for composite cross sections of stay-suspended decks (VSL International)

Type of cross section		Number of planes of stays	Deck anchorage position
Open section with main edge beams, cross beams and deck slab	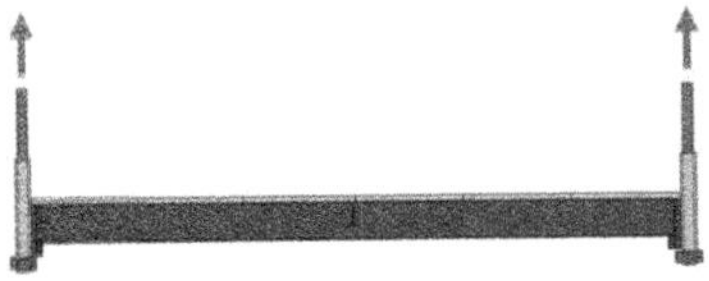	2	Outside of main beams
Triple-cell box section with pair of internal struts at stay location	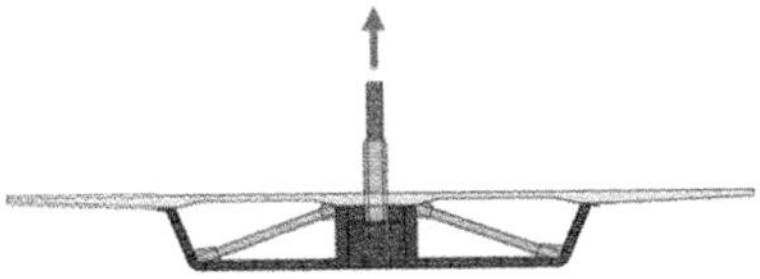	1	At centre of deck slab

has been installed and stressed. Whereas the first option helps to cut down the erection cycle, static forces in deck and stays are higher for the first than for the second option.

Steel and composite decks are much more flexible than concrete decks. This flexibility can make splicing of a segment to the tip of the deck a challenge, as is illustrated in Figure 15.40.

Such geometrical mismatches can be overcome relatively easily for open sections by load transfer of the segment's weight from the lifting frame to the deck and the installation of the splice bolts in several

Figure 15.40 Deflection issues during joining of steel deck segments when using lifting frames whose supports on deck are far away from the stay anchorage locations (VSL International)

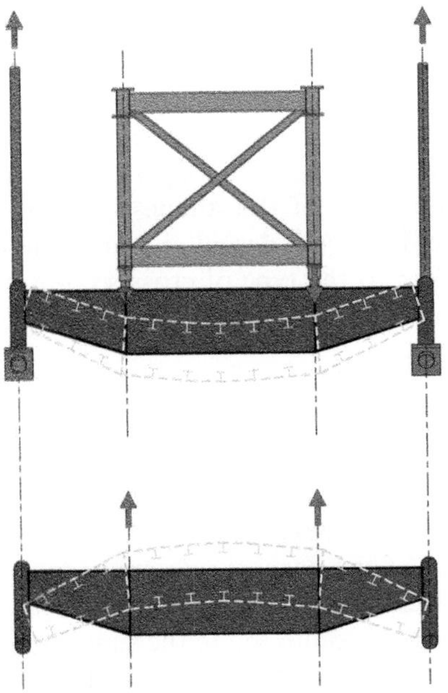

stages. However, the situation is significantly more tricky for box girders. A possible solution for such cases is to force the segment into the geometry of the deck at its tip by use of vertical props and stressed ties.

REFERENCES

Figures 15.1, 5.3, 15.5, 15.7–15.9, 15.11–15.19, 15.22–15.26, 15.29, 15.30, 15.37–15.40: VSL International.

Figures 15.2, 15.6 and 15.10: Meyer M (2009) 'State of the Art Bridge Construction – The Challenge to choose the right method', IABSE Symposium BKK.

Figure 15.4: AASHTO.

Figure 15.20: Menn C (1981), bridge construction lecture, ETH Zurich.

Figure 15.21: Ganz HR, Meyer M (1997) 'Segmentbauweise', *Schweizer Ingenieur und Architekt*, **26**: 4–9.

Figure 15.27: Meyer M (2004) 'Construction of the Taiwan High Speed Rail Contract 215', ASBI Convention, Tampa.

Figure 15.28: Muller J, Podolny W (1982) *Construction and Design of Prestressed Concrete Segmental Bridges*, John Wiley & Sons: 347.

Figures 15.31–15.36: Freyssinet International.

Bridge Deck Erection Equipment
ISBN 978-0-7277-6193-4

https://doi.org/10.1680/bdee.61934.357

Chapter 16
Equipment design

16.1. Introduction

This chapter describes the matters that should be considered in the design of the various types of BDE. Not all matters will be applicable to all such machines. The influence of the use of BDE on the permanent works design is discussed in Section I, Chapter 3, in other chapters relating to specific BDE types and in Section III, Chapter 15.

16.2. Philosophy

The most important consideration when designing any item of construction machinery is to keep it as simple as possible. There have been too many accidents on bridge construction projects created by operation procedures that were too complex for the site team to cope with. Repetitive processes add to the safety and efficiency of a machine, provided that effective steps are taken to avoid complacency of the workforce and any variations to the standard procedures are highlighted.

There is a balance between the cost and efficiency of a machine designed for the construction of one specific project, and one designed to cover a range of bridges. The danger of designing a multipurpose machine is that its cost, complexity and, usually, weight increases. However the benefit of being able to reuse it many times over can be huge as the initial design and construction cost can be spread more widely. The main influence on gantry cost is usually the need to cater for a wider range of bridge geometry than the minimum. Generally, plan curvature of less than 300 m starts to have a serious effect on cost, because of the difficulty in launching, placing of sections and the need to take account of the associated cross fall in the bridge deck. Large variations in span length can also be troublesome because the weight of a gantry suitable for building a long span may be too much for the foundation of a short-span column.

The concept of 'failing safe' is important in the design of BDE. Obviously the prime requirement is for things not to fail at all, and to make the operation of mechanical equipment components as close to 'fool-proof' as is possible. However it is a very useful addition to the safety of the BDE to give site operatives warning (for example by the provision of limit switches, or painted marks on the components) when critical conditions are approached. If a screw leg should not be extended beyond a certain height, locks or clearly visible markings should be used to either prevent the action or warn the operators.

Designing for structural safety is basically practising 'defensive design' where the experienced equipment designer will proactively incorporate elements that would anticipate and mitigate the risks of accidents on the project.

It is important for BDE design engineers to keep in mind that the complex equipment they are designing is ultimately going to be used by workers who are less familiar with the BDE but may try to push its use, on purpose or inadvertently, beyond its intended limits.

16.3. The design process

Experience has shown that the early establishment of the principles by which the BDE will operate is key to a successful design process. The common name for these principles is 'kinematics'. The usual form is a series of drawings that describe the stage-by-stage operation of the BDE during both deck construction and launch operations. The latter may not be a single process for a bridge in which the BDE has to perform different tasks depending on span length and deck geometry. The development of the kinematics is an iterative process as the design develops. It is essential that all people involved in the development of the BDE are kept informed about modified construction steps as the detail-designers will often need to refer to the kinematics to ensure their design can comply with the scheme's requirements.

The design should follow the process set out below, which relates to the flowchart in Section I, Chapter 3 (Figure 3.1).

Phase 1: Definition, and agreement between the stakeholders, of the performance specification, General Arrangement (GA) of the BDE, kinematics, design basis (loading, material, codes), permanent works interfaces. This is summarised in the Basic Methodology Report and the Equipment Documentation Report. This is an iterative process.

Phase 2: Detailed design, with fully developed kinematics, steel fabrication drawings, mechanical and electrical details and due allowance for access and maintenance for site personnel.

The result is the final detailed design, ready for final independent check and peer review, after which fabrication may start.

The stakeholders are the:

- permanent works designer
- main contractor
- construction engineer
- temporary works designer
- BDE designer
- BDE detail checker/peer reviewer
- suppliers of components
- assembly and commissioning team
- operators on the site.

Some of these stakeholders will be the same organisation in many cases.

16.4. Design codes

It is very important to make sure that the equipment is designed to the Codes and Standards that are recognised in the country of use. The proper implementation of a design QA/QC process is an essential step towards eliminating design errors.

Gantries are commonly designed to Eurocodes and these are accepted as being good for the task. Any internationally accepted Limit State code should be acceptable, although some codes are not as comprehensive as others. There is no single design code that covers all the rules necessary to design BDE, so a mixture of codes is inevitable. Caution should be taken when using rules from more than one code to ensure that a consistent and compatible design logic is adopted.

In many countries, reference is still made to FEM Rules for Design of Hoisting Appliances, even though the code was last revised in 1998 and is technically outdated. It still uses Working Stress Design even though Limit State Design is now accepted in most of the world. Furthermore, FEM does not address:

- quality control indications (e.g. minimum weld quality level)
- risk assessment, risk reduction and residual risk prescriptions.

The only area where FEM still has relevance is in regard to determining dynamic load factors. Hence whenever designing BDE, a combination of codes has to be used.

16.5. Access for personnel

Providing safe and effective access for project personnel to operate, inspect and maintain a BDE is essential to the safe operation of that equipment. The additional planning efforts and costs associated with providing proper access throughout the BDE will pay off in the long run by reducing the risk of accidents and by increasing productivity. The provision of adequate access needs to be considered at the basic design phase of the BDE since it will often be very difficult or even impossible to add adequate access if the matter is considered too late in the development process. The additional weight arising from the provision of safe access can be substantial and can affect the stability of the BDE during launch.

For example, it is not a good design practice to install a sheave that requires daily lubrication at the very front tip of an overhead gantry, where there is no easy access. The crew will be unlikely to follow the maintenance schedule if no access platform is provided for the mechanic to get to the sheave easily and safely. If it is really hard to provide safe access, maybe a different type of sheave should be specified.

In many cases, the BDE is used as the principal means of accessing the workplace. In such situations, additional care must be taken to provide safe access that will be used by the operational crews as well as inspectors and visitors.

The equipment should be designed to meet at least the safety regulations in effect on the project. If these regulations are lacking or substandard, it is the responsibility of the experienced BDE supplier and the main contractor to establish a set of minimum safety standards that will ensure the safe operation of the BDE as well as the safety of the workers and the general public. Although cost is a major consideration on the project, it should never be traded against the public and/or worker safety.

Some items, such as rigging safety factors, height of handrails and fall protection requirements, vary from one country to another and thus should be studied in detail when considering second-hand equipment. Minimum safety standards should always be complied with to create a safe working environment. The use of staircases is preferable to ladders for vertical access. A harness-free working environment is ideal.

16.6. Loading and load combinations

A prime difference between the design of BDE and that of normal permanent works is that for a BDE, the loads are genuinely known (and should be measured as part of the commissioning process) as opposed to the mostly 'notional' loads defined by the usual bridge design codes. This will almost certainly affect the load factors used in the design. Additional realistic allowance must be made for dynamic effects. Crane design codes give useful information on impact factors, although duty cycles for BDE can generally be assumed lower than for a long-term-use crane.

When assessing the loads for which a BDE has to be designed, all the practical effects that can arise during the project must be considered. For example, when lifting a precast unit, account has to be taken of the fact that it may not have fully 'dried' and thus the concrete density might be 3% higher than that assumed for the permanent works condition. Allowance should also be made for the fact that one cantilever of a 'balanced cantilever' deck will probably be 'newer' than that on the opposite side and thus there will be an out-of-balance effect in addition to anything caused by geometrical differences. Furthermore, a 'tolerance' allowance should be made of ±2% to account for variation in casting cell shape, some out-of-balance construction storage load (arising from equipment being stored on the deck) and the vertical component of wind load as required by the bridge design codes.

It is essential that accurate wind loading is available for the site. Local airports can be good sources of such data but if no suitable and reliable source can be found, it is essential that wind studies are commenced at the start of the project. It is rare that the assembly of an item of BDE starts within a year of the project commencement and thus it is usually possible to gather a year's data before they arrive. It is important to take readings at the correct height above the ground and the main contractor should be prepared to build a tower to achieve this. It should be noted that this task is separate from the monitoring of wind speed during BDE operation, an action that is also recommended.

It may be possible to reduce extreme wind-load values to allow for the relatively short duration of the construction operation, although this has to be a matter for engineering judgement in countries subject to typhoons. In such locations, the normal statistical approach to wind loading has less validity because a critical storm can hit a BDE even if its life on the project is short. Many BDE trusses are considered to be 'large' structures so the critical wind intensity will not hit the whole machine at any time. Some design codes allow for these effects. It will also be possible to avoid extreme combinations of wind and construction load by stipulating that site operations cease when the wind speed reaches a defined critical value (the level of which may depend upon the quality of weather prediction). Special braces or ties may be designed for installation when a storm is predicted. The use of a positive connection between BDE and the permanent works is seen as being good practice. In some developing countries where weather prediction is unreliable, the storm brace system may have to be permanent. The safety of site operators in high winds will usually govern when operations have to cease.

In practice it is unlikely that all possible load effects will combine at the same time and direction and a view may be taken on a compromise set of loads and factors to model a safe but reasonable situation on the site. There is a cost implication associated with this decision, to which the client may wish to be involved, as well as the certifying authorities in the region. Insurers may also have to be involved.

Lifting precast segments from a barge is often a matter for concern because of the possible impact on the lift gear if wave motion is significant. There are three options.

1. Limit the allowable wave movement and design for the appropriate impact force on the lift gear, and the rest of the machine. This adds a risk to the project as work may stop in times of bad weather.
2. Produce a scheme that can lift the precast section fast enough for it to completely clear the barge within the period of one wave. Subsequent lifts can be made more slowly. This was done successfully on Storebaelt Bridge, Denmark.

3 Produce a stillage design such that as soon as the segment is lifted from the barge, the support collapses and adequate clearance is achieved.

It is not common to include for seismic effects in BDE design criteria. It is generally accepted that, other than in countries which are very seismically active, BDE should not be designed for them. Even in countries where there is a reasonable chance of a machine experiencing an earthquake, reduced load factors would be acceptable, without 'service loading', some damage accepted, and simple restraints provided to prevent wholescale collapse of the structure. The responsible designer will discuss the consequences of such decisions with the main contractor, the project client and perhaps the project insurers.

The question 'What if?' is one that should always be at the forefront of the BDE designer's mind. For example, 'What if the BDE suffers an accidental lateral impact from a load suspended from a nearby crane?' Will the structure remain stable? If, for example, a support consists of a stack of unconnected stools, can the support be dislodged? This awareness of what can go wrong tests the designer's practical experience of such events. This can be as simple as the storage of materials on access walkways, something that can happen on poorly controlled projects. Does the designer place large notices prohibiting this, or does he design for a reasonable stored load? If, for example, stacks of shims are to be used in levelling supports as part of the BDE operation, the designer will not be popular if he prevents those shims from travelling with the machine, in which case their weight must be allowed for in the design.

16.7. Modelling

Steel design codes are written for design and detailing of structural steelwork in general and for selected applications, for example overhead cranes. They do not give guidance about sensible structural modelling, relevant loadings and load case combinations for the design and detailing of structural steelwork of BDE.

BDE design often demands a closer examination of structural details than 'normal' structures. This particularly relates to the articulation of the structure. Elements of BDE may well be more flexible than a 'normal' structure due to high stress levels in service, thus causing movements at supports that would otherwise be ignored in other structures. Simple finite-element computer modelling may fail to pick up these effects, resulting in a structure which performs in practice differently from that which a basic computer model might predict. It is therefore essential that the designer fully understands the way in which their structure will work to ensure secondary effects are properly accounted for. For example, a traverse beam sitting on stools on a bridge deck may appear a simple situation. However the beam will deflect in the vertical plane and the feet of the supporting stools will tend to move outwards. This will develop internal forces within the assembly for which connections have to be designed, unless suitable articulation is included that will limit the development of those forces. For illustration, refer to Figure 16.1.

Even the inclusion of elastomeric or stainless steel/PTFE sliding bearings will still create forces which have to be allowed for in the structure's detailing. The importance of maintaining a clear path for the necessary restraint for BDE is discussed later in this chapter.

Operational tolerances, such as the practical location and verticality of supports, are not taken into consideration in normal structural design codes. This is a specific feature of BDE design. There are many different possibilities (and combinations) for such effects and it is not practically possible to generalise them here. It is therefore necessary for the BDE's designer to review their design, and to take

Figure 16.1 An example of the importance of accurate structural modelling in BDE components (Tony Gee and Partners)

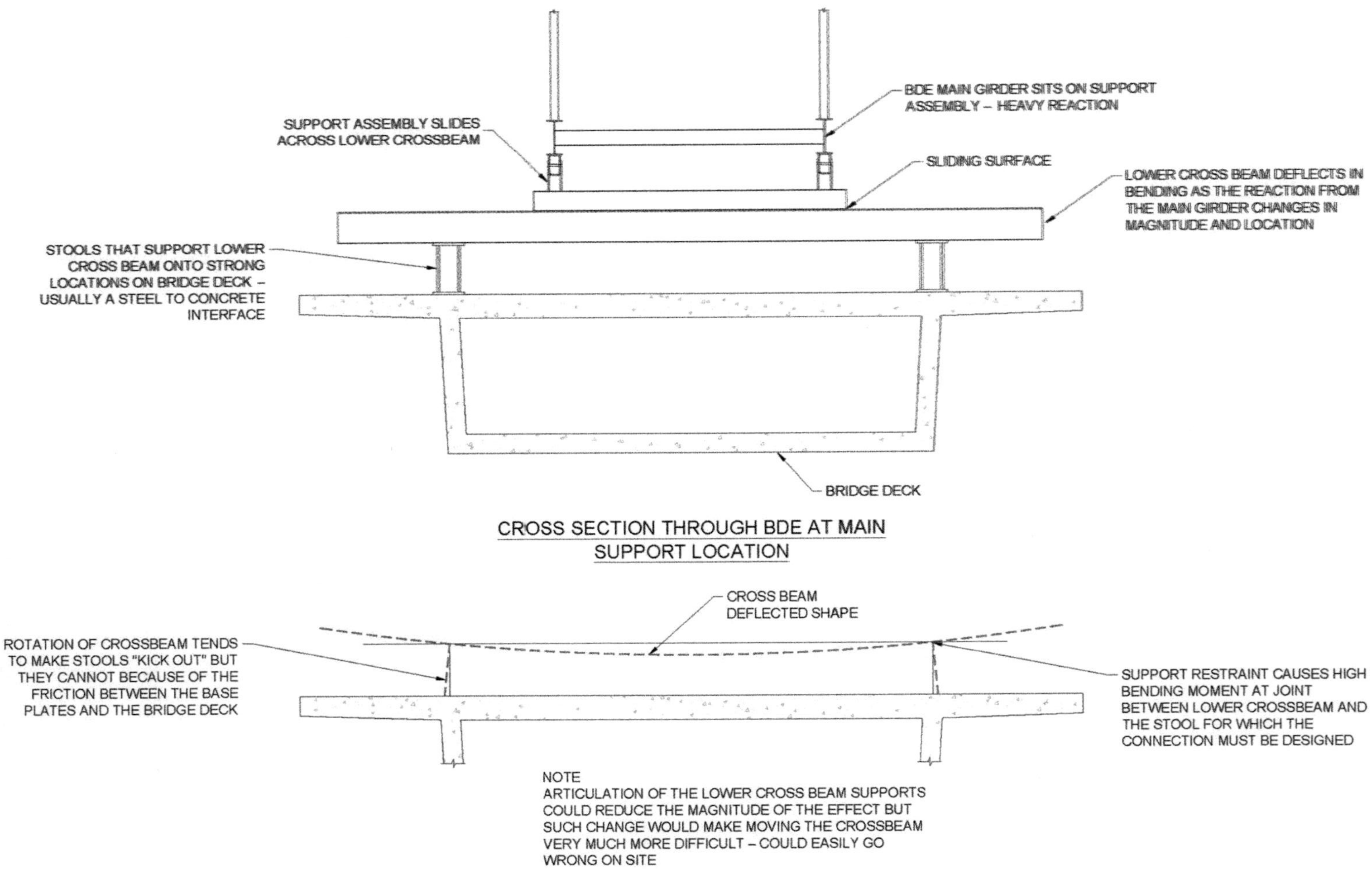

account of such effects. It is important that these features are listed and fully explained in the Design Basis statement and also that the limits of the designer's assumptions are listed in the operational tolerances for the working of the BDE. Whenever possible, features should be included in the design to prevent such tolerances from being exceeded.

16.8. Managing tolerances

Achieving acceptable geometry of the structure is perhaps the most important aspect of BDE design. Although the designer should aim for keeping all the stresses within the structure within acceptable limits, in practice the operator will be much more concerned about a 5 mm interference between two components of the BDE than he will about a 5% 'overstress'. Things have to fit together properly and it is that which makes the design of BDE so much more challenging than static structures.

It is frequently the case that the biggest problem in achieving an acceptable fit between the BDE and the permanent structure on which it sits is caused by inaccuracies in the construction of the permanent works. For illustration, refer to Figure 16.2.

It is essential that the BDE designer specifies the location of the BDE that is critical to his design, and then allows reasonable tolerance for the location of other supports and matters critical to the operation of the BDE.

The designer should always assume that the worst combination of tolerances can combine to stop something from moving. Experience shows that anything less than this will cause a problem eventually. If clearance is required between the BDE and the surrounding structure, 50 mm is an absolute minimum. As the design develops, the nominal clearances originally chosen will always decrease and less than 50 mm at the outset will be found to be too small at the end.

By the time the BDE elements are delivered to the site, there will be pressure to get the BDE operational. As amendments made on site are time-consuming (and therefore expensive), anything that can be done to eliminate 'teething trouble' on the site will be well worthwhile. Trial assembly of components at the factory should be considered as a minimum, and practice has shown that a full-blown trial assembly is worthwhile in almost every case.

16.9. Steel grade

The capacity/weight ratio of a gantry truss is of prime importance. That is one reason why most gantries are based on truss construction rather than box or I-section beams. If the BDE is too heavy, it may impose loads on the bridge deck or substructure that are greater than the permanent operational condition, thus demanding an upgrade to the permanent works design. The contractor will not usually be paid for the cost of such an upgrade so the situation must be avoided if at all possible, either by changing the operational method to avoid the critical load case, or by reducing the weight of the machine. Experience shows that the best BDE almost always use high-grade (355 MPa yield strength) steel rather than lower grades as the cost increase of the higher grade is very small (perhaps 5%) yet the steel capacity increases by around 30% depending upon the slenderness of the components in the structure (and, perhaps, fatigue considerations, as fatigue is independent of the grade of the steel).

Almost all BDE will be subject to some impact loading. Clearly this will be the case for machines that lift precast sections (even the smoothest-operating hydraulic winches will still impose some impact force on the supporting structure) but almost all machines are launched at some point of their

Figure 16.2 An example of the way in which a BDE has to cater for practical construction tolerances (Tony Gee and Partners)

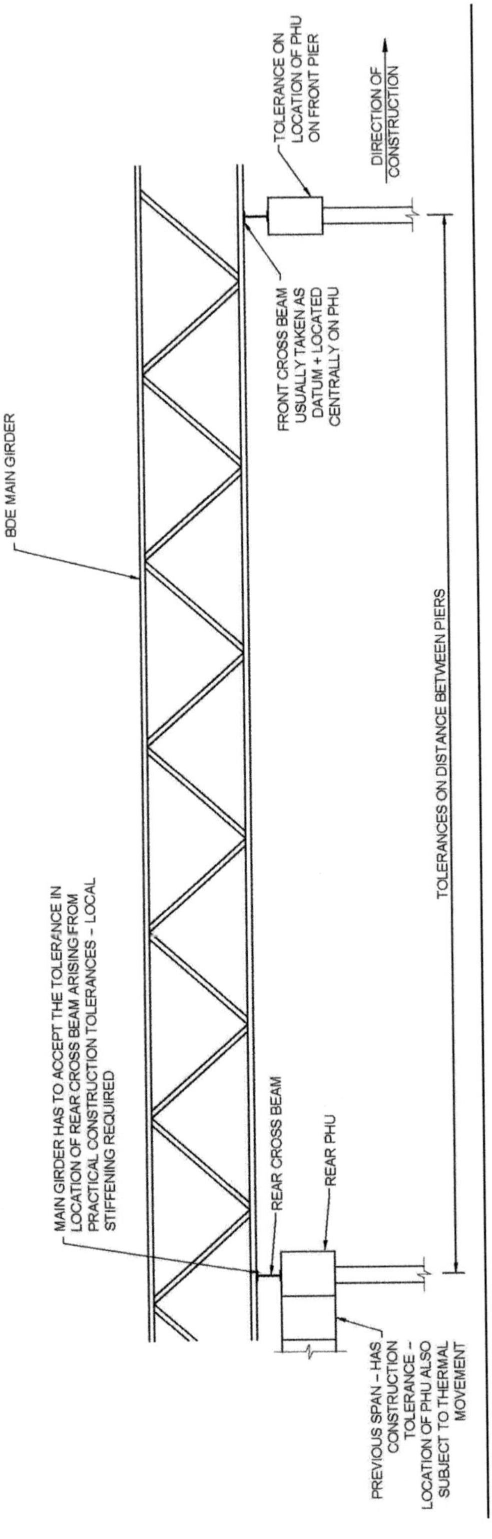

operating cycle and that act will inevitably create some dynamic effect (dirt on launch rollers, small steps in launch tracks) and the machine must allow for that. Unless the machine is certain never to have to work in anything other than a tropical climate, the steel will therefore require reasonable impact properties (Charpy value). In a cold climate, the specification for the impact toughness of the steel will be significant.

16.10. Stability and longitudinal fixity

Permanent works design codes do not have to consider matters such as stability of BDE during launch, so special consideration is required for that aspect of the design. There are various ways of doing that. Some prefer a low-factored restoring moment to be greater than a high-factored overturning moment. A minimum factor of 1.2 against overturning is commonly used, using Serviceability Limit State (SLS) factored loads. Others simply prefer the known location of a centre of gravity to be a reasonable distance on the stable side of the pivot point. Practical designers find the answers given by both methods to give a similar result.

It is clear that lateral and longitudinal stability of the BDE must be maintained throughout the construction of the bridge. This need will define the length of a gantry as it is clear that the centre of gravity of the whole must not reach the 'main' (central) pier until the leading edge of the machine reaches the next pier along. Similarly, the centre of gravity must have passed the central pier before the 'tail' of the gantry leaves the 'rear pier'. A distance of 2 m is generally considered reasonable but for light trusses (in particular) the site management must take great care to prevent the build-up of stored material on walkways if this value is not to be compromised. The location of the centre of gravity may be adjusted by moving a lift device, perhaps carrying a segment, to achieve the necessary stability. The correct implementation of this is obviously of vital importance to the safety of the structure. It is a method frequently used in gantry operation. A greater margin of safety on stability may be deemed wise on a site where major wind events can occur with little warning.

A check of the weight of the BDE, and a calculation of the real location of its centre of gravity, should be seen as an essential part of the commissioning process of any BDE. This is usually done by lifting the completed structure uniformly on hydraulic jacks and measuring the pressures. The task is particularly important after modifications have been made to any BDE.

It is essential that the BDE is properly fixed in position at all times. Leaving BDE sitting on rollers is **acceptable**. Even wind can have a big enough effect on a truss to cause it to move and accidents have occurred because of uncontrolled movements of a machine induced by wind. In fact lack of restraint is considered to be the largest single reason for BDE accidents.

16.11. Structural rigidity

The stiffness of an MSS-style BDE can be important as truss deflections will influence the shape of the deck, and pre-camber will be necessary to counter the effect. In practice, further difficulties can arise if the truss is too flexible because as the post-tensioning is stressed, and the deck starts to carry its own weight, the vertical reaction from the gantry could cause a continuous bridge deck to fail in 'hogging' (upward bending). Similarly, if a deck is poured in two stages, deflections of the MSS under the weight of the second pour will be imposed on the first (unstressed) pour and some designers may require that to be taken into account in the 'locked-in stresses' within the section. If the construction joint is at the bottom of the web, the bottom flange alone is so flexible that stresses can usually be ignored. If they are at the top of the web, the stresses are greater. This is a good reason for casting the whole section in a single pour if possible. Many designers allow the 'locked-in stress' to be ignored provided the second pour is made soon (say 48 hours) after the first one. An alternative is to lightly stress (using either temporary bars or by partial stressing of permanent tendons if they can be detailed suitably) the first pour to the old cantilever tip, so as to ensure no tension occurs at the construction joint.

Deflections of main girders used to erect precast segments are usually less critical as the support levels can be adjusted prior to fixing the segments together or to the substructure. However, excessive main girder deflections, either while supporting segments or launching from one span into the next, can affect the demands on the mechanical components. For more details on this subject refer to Section II, Chapter 6.

Main girders used for MSS-style machines will require connections that do not allow movements at splices (for example the use of HSFG bolts, welded connections, preloaded HT bars or pins in close tolerance holes). Cheaper forms of splices with limited rigidity, like non-preloaded bolted connections with normal tolerance holes, can in theory be used, for example, for the main girders of a BDE for which an exact shape is unimportant. The shape change needed to mobilise all bolts in a splice must, however, be possible and the effect of shape change has to be considered in the operation of the machine. The sudden physical and audible 'jolts' that occur as the machine changes shape are bad for the confidence of the site staff. Furthermore, it is common for BDE main girders to experience full reversal of bending moment between bridge-building and launch modes. This could cause hole-elongation over a series of spans and thus untorqued bolts working in shear are not recommended for use in major structural components. For illustration refer to Figure 16.3.

16.12. Matters specific to in-situ concrete BDE (MSS and form travellers)

There are some design aspects of BDE used for in-situ concrete casting that demand special attention. As with many construction engineering schemes, it is important to ensure that the structure is built in accordance with the prime design assumptions. This is particularly important when building in-situ concrete bridges that have continuous spans. The design will assume points of dead load contraflexure at about span/5 for internal spans and it is therefore convenient to take that point for the construction joint between concrete pours. That will mean that the MSS will typically support wet concrete for about 80% of the 'back-span' and 20% of the 'front-span'. If the 'end-spans' are typically 70% of the length of 'internal spans' this also works well for the efficiency of the MSS.

This is also good for the bending strength of the MSS because the negative moment at the front pier reduces the positive moment in the back-span. If the MSS designer takes advantage of this, they must make it very clear to the operator that concrete casting must start at the cantilever tip and work back towards the previously cast span to avoid overstressing the MSS main beams. This method has the added advantage of ensuring that bending and shear deflections in the MSS have taken effect before the wet concrete at the 'back end' of the new span connects to the cantilever tip of the previous span. The MSS is normally supported at the cantilever tip of the previous span. Hence there will be very little relative deflection at the construction joint. If the MSS is supported at the previous pier, jacking will be required to achieve 'zero shear deflection' at the cantilever tip.

'Tunnel forms', which are required when a section is cast in one pour, have to be removed from the 'old pour' and launched into the 'new pour' prior to the placing of the deck reinforcement in the new span. The speed at which this is done depends upon the permanent works design because of the size of 'windows' in the pier-head diaphragms, and the size and number of the PT anchorage blisters within the section, as these affect the difficulty in extracting the core forms.

Figure 16.3 Typical main-chord splices (Tony Gee and Partners)

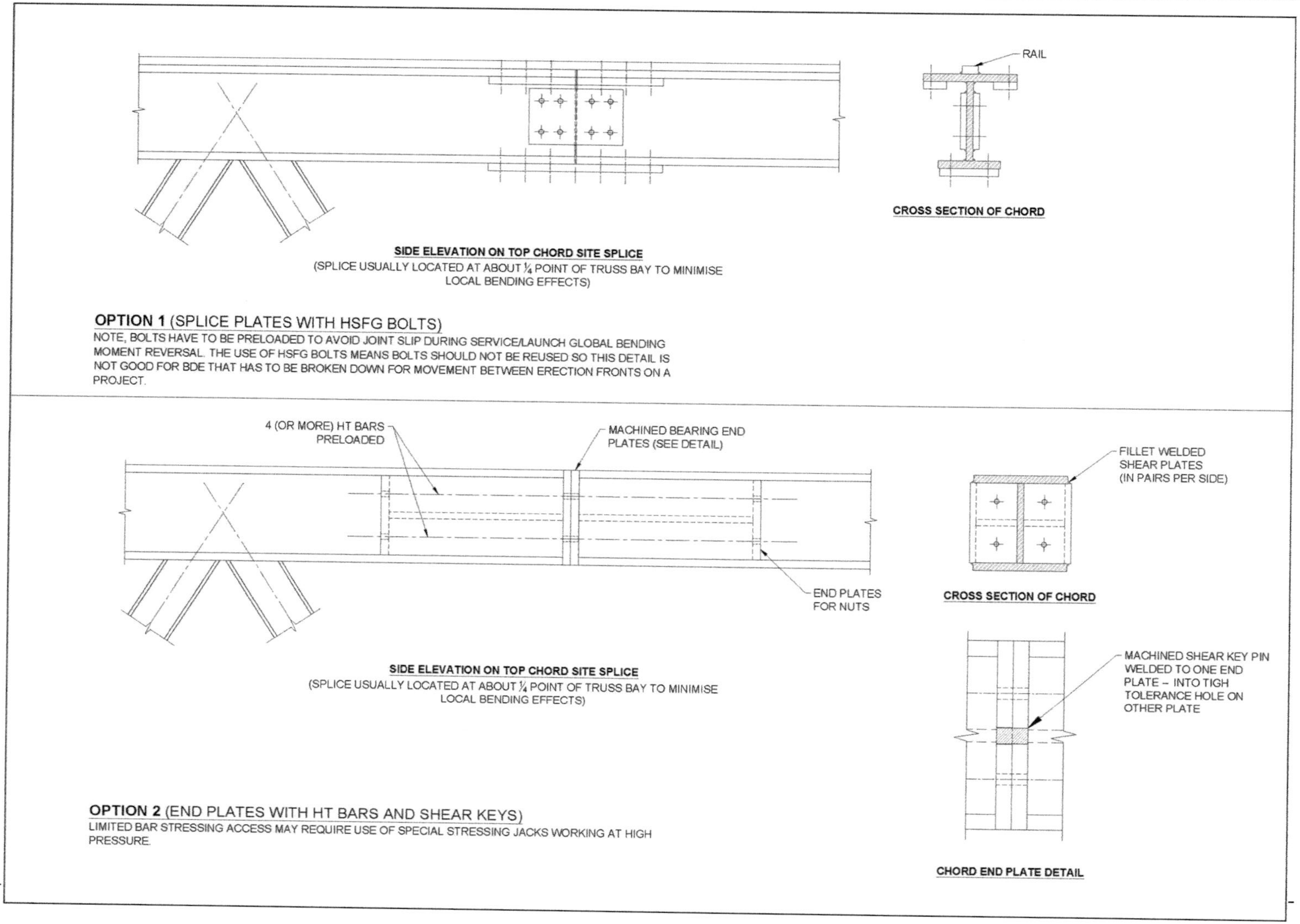

Form traveller operation also has to consider the interface between newly cast concrete and that of the previous pour. Hence, again, the concreting of the bottom slab starts at the tip and not at the construction joint. It is standard practice to cast segments cast with a form traveller in one pour. For more explanations about this subject, refer to Section II, Chapter 11.

16.13. Detailing

A rectangular cross-section truss is easier to fabricate, and the top chords have better lateral stability when in compression. The chord cross section is made of open-rolled sections. A pair of chords can be welded together at the fabrication shop, which enhances their robustness. In the case of a rectangular truss cross section, the gap between the chords can be used for storage of bracing members for transport. Also, it facilitates the access down the axis when the truss is shallow (less than 2 m between chords). Diaphragm braces are needed to maintain cross-sectional shape.

A triangular cross-section truss is more complex to fabricate but the cost increase can be partly justified by faster assembly on site, and the single top chord can be prevented from buckling by inclined diagonals, which increases the loads in them slightly. The cross-sectional area of a single chord will have to increase because of its increased slenderness. Diaphragm bracing is not needed to prevent the deformation of the section. Such a truss requires separate access to top chord level, although this is usually done from the trollies that run on the top chord.

Solid box-section beams tend to be heavier than trusses. Their design for the effects of localised roller loads from trollies (top flange) and when launching (bottom flange) is easier than for trusses. In hot countries, trusses are more user-friendly for site operatives. Solid box sections are sometimes used for gantries with vertical pinned articulation, necessary for low radius plan curvature structures. The stability of such structures is a matter for very careful design.

Good connection design should aim for as light and symmetrical welding as possible. The use of heavy plates should be minimised wherever possible. Complicated connections have sometimes been found to be a source of cracking. The reduction or elimination of complicated connections is therefore recommended.

If an existing machine is to be used for a new project, extensive and detailed Non-Destructive Testing (NDT) is strongly recommended prior to purchasing or commissioning the machine for the new project. Calculation of the potential 'fatigue' damage on an old machine is too complex and generally subject to error, due to limited available information about the machine's past use to make it worth considering. Experience shows that when cracks have been found they have generally appeared at heavily restrained connections.

It is important that the design prevents ponding of rainwater in components, because of the water's weight and also the potential for corrosion. To achieve this, either large drain holes must be provided or sections fully and properly sealed. This applies particularly to thin-walled tubular sections often used as brace members because of their high power/weight ratio and open-rolled section beams with stiffened webs in the horizontal plane.

16.14. Mechanical components

Before getting into detailed structural design of the main girders that are the core of the BDE, the designer should have a good idea of how the mechanical elements will fit around the BDE structure. Of particular interest are matters such as lateral guidance of trollies or main girders when launched over supports. The details used for flanged wheels, for example, are different from the use of discrete side-guide rollers.

A gantry designer has some difficult choices to make when specifying the mechanical components of a BDE. 'Normal' duty cycles for rollers, bearings, winches, electric motors and gearboxes, for example, will assume a longer life than that expected for the BDE. It is considered reasonable to take account of that when designing a BDE. By how much stresses can be increased is a matter of experience and specific guidance cannot be given here as the effect will vary depending on the circumstances. However the designer should remember that when a BDE fails to work properly, it is much more commonly due to a malfunction of a mechanical element than a structural one and thus money spent on the mechanical parts of the machine is usually well spent.

It is important to balance the strength of the structure with the capacity of the mechanical equipment. Common practice is that the mechanical components must not be capable of damaging the structure. In the case of hydraulic components, this may be limited by the pressure relief valves and the dangers of adjusting pressure relief valves under any circumstances must be made very clear in the Operation Manual. Controlled adjustments to the capacity of hydraulic components may be necessary during load tests but correct resetting afterwards is essential.

Special consideration is necessary when designing mechanical, hydraulic and electrical components. These methods are described in Section III, Chapter 17.

16.15. Design checking and peer review

The integration of the permanent bridge design and that of the BDE is important for the smooth running of the project. A good design sequence is regarded as the following.

1. Permanent bridge design
2. Permanent bridge design independent checking
3. BDE design
4. Check of the effect that BDE has on the permanent works and the stage-by-stage analysis/ design verification
5. BDE independent peer review
6. BDE independent detailed design check
7. 'Professional Engineer (PE) certification' to comply with local regulations

The practice of engaging an independent peer review and a detailed design check of the BDE is an essential step in a BDE design. It is important for the success of the project that the peer review is started early in the design development process.

Both the peer review and the detailed design check should be based on design assumptions, working drawings and Operation Manuals and not simply be an arithmetical check of the original calculations. Both must be carried out by a body familiar with BDE. The check and review can be under the supplier's scope of work, as long as they are performed by designers and experts not directly involved in the original design and detailing. Experience in the design of static structures alone is not considered adequate for BDE review.

A check must be made of the effect the gantry has on the permanent works. This is in addition to the design check the gantry designers should make themselves. This check can usually be made most efficiently by the permanent works designer.

The peer review will be carried out by a person experienced in design and operation of the relevant type of BDE. The reviewer takes a 'bird's-eye view' of the BDE and its interface with the permanent work. The reviewer will focus on support details, horizontal fixity and equilibrium during various stages of operation. The reviewer will carry out some detailed checks if, from their experience, they are deemed necessary.

The independent detailed design check comprises remodelling of static systems for selected controlling load cases and combinations and independent determination of static forces and respective capacity checks at governing sections.

The local PE certification may be required to satisfy local conditions but local knowledge of the country cannot be regarded as a substitute for expertise in the design and operation of BDE.

Too often the main contractor does not consider the matter of BDE early enough in their construction programme, resulting in inadequate time being given for the design and/or checking of the BDE. Designers (and suppliers) should resist attempts to rush these vitally important stages of the whole operation under pressure from the main contractor.

The check of the BDE cannot be regarded as complete until the BDE has been assembled on the site and checked to prove that it has been built and assembled in accordance with the design drawings.

16.16. Operation Manual and checklists

It is essential to regard the Operation Manual or Method Statement as an integral part of the design documentation for the machine. No temporary works design is complete without the Operation Manual. The Operation Manual is a very detailed document that describes the way in which the machine works and, most importantly, the limits on its function. The Operation Manual will be much easier to understand if simple illustrations are included.

In practice the Operation Manual will be converted into illustrated checklists for the operatives to work with in the field. It is good if the site staff develop the checklists but they must in turn be reviewed by the BDE designer for accuracy. For more detail of this, refer to Section I, Chapter 3.

The critical values of loads, allowable deflections and setting out tolerances must be included on the checklists, as must the actions to be taken by the site operatives if the measured values are outside the set limits. It is **absolutely essential** to make it very clear to the operatives that if anything unusual or unexpected occurs with the BDE operation, their first and vital duty is to make the system safe and to consult with the BDE designer for further instruction.

16.17. Commission and load and deflection test

Commissioning of BDE starts with a trial assembly at the fabrication shop. A factory trial assembly of the BDE is always time and money well spent and should be regarded as essential. Small design or detailing problems can be resolved much faster and more cheaply in the factory than on site and the BDE can be tested as far as is possible in amenable surroundings. The shape of the long elements of the BDE can be tested and measures taken to re-establish the shape quickly at the site assembly stage.

It is essential that the BDE's designer attends the final stages of commissioning (whether this is at the factory or on site), the load test and, whenever possible, the first deck erection and launch operation. The designer is the one best-placed to review the behaviour of the machine while it is tested for the first time and to review real performance compared with that predicted. Also in attendance will be the site engineer and the 'pilot' and 'co-pilot' of the operating team (refer to Section I, Chapter 3) as the commissioning is seen as an essential part of their training. These key people will be responsible for the safe onward use of the machine.

It is at this time that the Operation Manual (and checklist) is first tested. It is possible that some improvements can be made either to the design or to the way in which the machine is to be operated and any such changes will be much better and more quickly dealt with if the designer is on site and available immediately.

Prior to starting any mechanical testing, it is common practice to make a visual check of all the elements of the BDE, both for completeness (it can be quite hard for someone unfamiliar with the way in which it works to identify that something is missing) and also for the damage that is sometimes inflicted between departure from the factory and the completion of site assembly. The visual check is particularly important for mechanical components, which tend to be more fragile than structural ones.

Legislation in many countries demands that the BDE is load-tested prior to starting work. In addition to the load test, it is recommended practice to perform the following tests, even if legislation does not demand it:

- measurement of the BDE's self-weight and position of its centre of gravity
- deflection tests during operation.

This is fully described in Section I, Chapter 3, 3.5, 'Operations'.

All-welded structures tend to be slightly stiffer than calculation would suggest. This is generally attributed to the bulk of the connections that the mathematical model of the structure cannot replicate well. Structures connected by HSFG bolts act similarly to those welded. Structures assembled with untorqued bolts will not act elastically due to bolt slip at connections, unless there is no stress reversal in the BDE's operation, and that would be unusual. Thin-plated structures (not commonly used in BDE) tend to act with a stiffness equivalent to their gross section properties, in other words with shear lag ineffective on the section properties.

It is common to start commissioning individual parts of the BDE on their own as they become ready for service. For example, function testing of the lift devices does not have to wait until the whole BDE is ready. However the whole assembly is needed prior to undertaking the load test.

The BDE designer will define the principal acceptable limits on measured values and give instruction to those doing the work about what has to be done if the values are exceeded.

It is often impractical for the designer to describe the full set of acceptance criteria for a single item of a BDE because even the simplest BDE has to perform a variety of tasks. Thus the review of the practical performance and what must be done if the measurements are at, or close to, the expected limits is a matter for the designer to review. This is also one of the reasons why the designer should always be present during the commissioning of the BDE, and on hand thereafter should performance values appear to change to the point where the site operatives are concerned.

Deflection testing the gantry for the service and launch conditions is covered in Section I, Chapter 3. Any attempt to 'overload' the gantry for a launch condition could adversely affect the BDE's overall stability.

It is commonly the case that if a BDE is dismantled and moved to another part of the site for reuse, another set of load tests will be required.

16.18. Quality documentation

Items of BDE frequently pass from one project to another. It is rarely the case that the BDE's use on the later project is identical to that of the first project. In any case, it is important that a review of the design for the real application in the subsequent project is properly assessed. For that reason, the material certificates, NDT results, commissioning results, design change records and actual duty cycles should all be collected and retained because, without these, it will be very hard to make a proper assessment of the BDE's fitness for a later project. Its value will, therefore, diminish, possibly to the point where it cannot be used again.

16.19. Modifications

Experience shows that reuse of an existing gantry on a new project can rarely be achieved without any modification. Adaptation of an existing BDE to a new project almost never comes without cost. But the work can still be worthwhile if it saves cost or reduces lead time compared with a new BDE item. A proper assessment is needed before an accurate cost can be calculated and a decision made.

It is essential that the designer fully understands the way in which a gantry was assumed to work originally before planning any changes to the machine or to the way in which it is operated. Previous experience of gantry design is considered essential before attempting anything of this type. This is discussed more in Section I, Chapter 3.

Modifications must never be made on the site without reference back to the design team. Accidents have happened because this advice has been ignored. It is hard for a fresh design team to take on all the possible nuances of a BDE designed and developed by others. Site based temporary-works designers should only be allowed to modify or add to an existing item of a BDE in extreme cases and only with full coordination with the BDE designer. This applies even down to changes arising from, for example, a minor geometric change discovered during the commissioning process.

The correct way to deal with modifications to BDE is for the site team to issue a Temporary Work Modification Request (TWMR) back to the BDE designers. The effect of the change can be fully checked through and only once approval has been received should the change be implemented. It is clear from this that the speed of response from the BDE designer is of great importance.

Should cracks be found anywhere in the structure of a BDE, it is important to review other similar locations of the structure to see whether there is a fundamental weakness in the design or fabrication of the structure. When repairing any cracks it is important to ensure the root of the crack is identified and the weld repair made to fully obliterate the original defect. Consideration should be given to replacing elements which show extensive cracking.

REFERENCES

Figures 16.1–16.3: Tony Gee and Partners

Bridge Deck Erection Equipment
ISBN 978-0-7277-6193-4

https://doi.org/10.1680/bdee.61934.373

Chapter 17
Mechanical, hydraulic and electrical components

17.1. Introduction

This chapter gives an overview of the mechanical, hydraulic and electrical components used on a BDE. There are significant differences between a BDE and more conventional machines such as portal, crawler, mobile and tower cranes. This is because a BDE is:

- used for a relatively small number of operations on a specific project
- generally used only on one project or, if reused, only on a few projects typically during one to two years per project
- operated at low speeds.

A BDE is usually self-launching. It therefore needs to be moved (and of course also fixed) horizontally, both longitudinally and transversely (**horizontal movements**). During the launch operation it is generally necessary to raise and lower the BDE at its supports.

If the BDE is used to erect preassembled or precast bridge deck elements, it has upper cross beams or bogies, equipped with lifting and lowering devices, which can be moved along the BDE's main structure (**{main} vertical and horizontal movements**).

It is essential that the BDE's support conditions between various elements of a BDE and between the BDE and the permanent work remain as demanded by the design at all stages of its operation. Mechanical and hydraulic elements of a BDE – such as **spherical bearings, screw jacks/hydraulic jacks with safety rings, turnbuckles and pins** – play an important role in achieving this objective. They are crucial to ensure stability of the BDE at all times and to ensure there is no overloading of BDE elements and the permanent works, and also to avoid unplanned excessive restraint effects and second-order effects during operation. They help to:

- centre loadings
- obtain the geometrical setting-out of support members in plan and elevation
- maintain longitudinal and lateral fixity for the whole BDE and its various elements to prevent accidental movement.

High-tensile bars, which can be but are not always pre-loaded, are widely used, for example to tie down a BDE to the permanent structure, for attachment of preassembled and precast bridge deck elements to the lifting devices, and for suspension of precast segments from overhead gantries.

This chapter will address:

- vertical movements between a part A and a part B
- horizontal movements between a part A and a part B
- spherical bearings, screw jacks/hydraulic jacks with safety rings, pins and turnbuckles
- high tensile bars
- power supply
- control systems.

17.2. Vertical movements

For vertical movements, the options are:

- winch
- hydraulic solid ram jack
- Strand Moving Unit (SMU)
- centre hole jack or pair of solid ram jacks with a high tensile bar as tension member (similar working principle as SMU)
- pair of solid ram jacks with steel section as tension member and pins inserted through holes in the tension member for load transfer (similar working principle as SMU)
- chain hoists
- tirfors.

Winches, hydraulic solid ram jacks and strand moving units are most commonly used for the main vertical movements in a BDE. Main elements of these equipment components are shown in Figure 17.1.

A comparison between these three equipment components is given in Table 17.1.

17.3. Horizontal movements

Two parts can be moved against each other horizontally by rolling or by sliding. Table 17.2 gives a summary of the main options for rolling and sliding, with indication of permissible contact pressure and friction coefficient.

For propulsion, the following systems can be used:

- directly driven wheels
- chain drives with fixed Chain Pulling Unit (CPU)
- chain drives with movable Chain Pulling Unit (CPU)
- long stroke jack
- capstan winch
- rack and pinion system
- SMU.

Directly driven wheels, chain drive with fixed chain pulling unit, chain drive with movable chain pulling unit, long stroke jack and capstan winch are most commonly used for horizontal propulsion of a BDE. The main elements of these systems are shown in Figure 17.2.

Figure 17.1 Main elements of winches, hydraulic solid ram jacks and strand lifting units (A and B indicate the two parts between which the movement takes place) (VSL International)

Winch	Hydraulic solid ram jack	Strand moving units

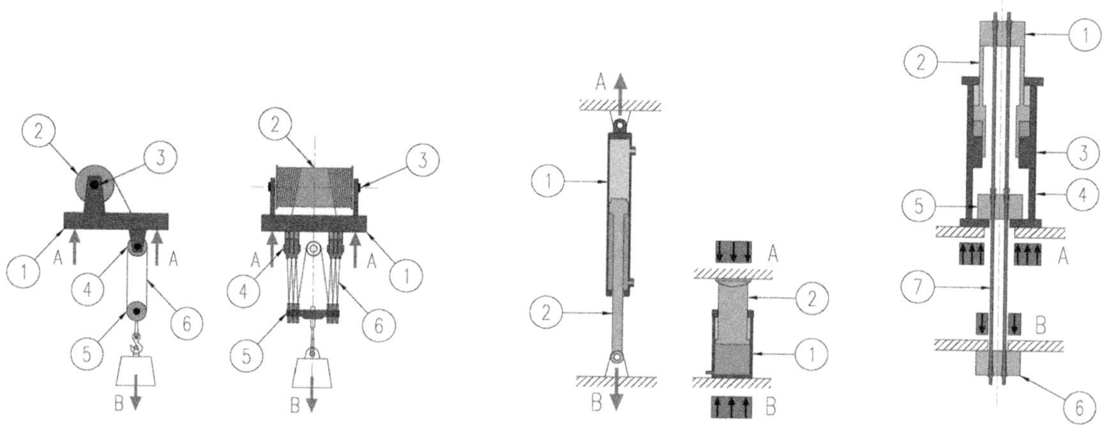

Winch	Hydraulic solid ram jack	Strand moving units
Winch sitting on A, lifting/lowering B:	Jack either hanging from A, lifting/lowering B or sitting on B, lifting/lowering A:	SMU sitting on A, lifting/lowering B:
1. Winch frame 2. Drum 3. Gear motor 4. Upper sheave block 5. Hook block 6. Rope	1. Jack body 2. Piston	1. Pulling head 2. Piston 3. Jack body 4. Chair 5. Lower anchorage 6. End anchorage 7. Strands

17.4. Spherical bearings, screw jacks/hydraulic jacks with safety rings, turnbuckles and pins

17.4.1 Spherical bearings

Spherical bearings are used to centre loading at support points. They cater for support areas not being at ninety degrees to the supported member's axis. Since they cannot transfer bending moments, they need to be treated as flexural hinges.

17.4.2 Screw jacks and hydraulic jacks with safety rings

Screw jacks and hydraulic jacks with safety rings allow for variation in vertical and horizontal distances between members and between a BDE and its support points subjected to compression loading. Their ends can be fitted with spherical bearings. If both ends have spherical bearings they are, from a mechanical point of view, pendulums, which can only transfer axial forces. It is important to check that provision of such pendulums does not jeopardise the overall stability of the BDE or its subsystems. The safety rings of hydraulic jacks ensure that, if engaged, the jack continues to transfer compression loading, even if the jack's seals start to leak.

17.4.3 Turnbuckles

Turnbuckles are axial support members which are adjustable in length. They compensate for construction and installation tolerances and provide a fully rigid connection either for tension or compression loading. They are particularly useful for fixing formwork panels, which have to be rotated for

Table 17.1 Comparison between winches, hydraulic solid ram jacks and strand lifting units

	Winches		Hydraulic jacks		Strand moving units	
Capacity	Load (in tension only)	Hook height	Load (in tension and/or compression)	Stroke	Load (only in tension)	Lifting/lowering height
	No. of falls × SWL of rope	<(Drum diameter × π × No. of rope layers on drum)/ No. of falls	Piston area × working pressure: piston areas for pulling and pushing are different	0 to ~5 m if in tension; 0 to ~500 mm if in compression	No. of strands × 100 kN per 0.6″ strand	<Length of strands
Speed (typical)	Upper speed of ~3 m/min and lower speed of ~0.3 m/min		Pump delivery/piston area: 0 to ~1 m/min		~0.25 m/min to 0.33 m/min loaded; 5 m/min to 10 m/min unloaded (if with motorised coilers)	
Risk of dropping load due to equipment failure and possible mitigation	■ Failure of gear: provision of double gear with individual gear to hold load on its own or provision of drum break ■ Failure of rope: no redundancy		■ Hose failures: provision of brake valves ■ Seal leakage: piston will fully retract (compression) or extend (tension). No special measures required if subjected to compression. If subjected to tension, piston's stop end to be designed to hold load after piston has been fully extended		■ Failure of a strand and/or slippage of its wedges respectively: redundancy for SMUs with more than three strands, since remaining strands are able to hold the load ■ Hose failures/seal leakage: load can always be held by lower anchorage	
Consumables during normal operation	N/A		N/A		Wedge grease	
Consumables during regular maintenance checks	Grease		N/A		Strands – change after 35 uses Wedges – Change after 2000 bites (~900 m of lifting/ lowering)	
Load reading	Need for load cell		Direct by reading jack's hydraulic pressure		Direct by reading jack's hydraulic pressure	
Control	Direct, remote or radio		Direct, remote or radio		Direct, remote or radio, but always need for operator at jack for greasing of wedges	
Power pack	Electrical power supply + electrical motor or hydraulic pump/hydraulic motor		Electrical power supply + hydraulic pump/hydraulic motor		Electrical power supply + hydraulic pump/hydraulic motor	

Table 17.2 Main options for rolling and sliding, with indication of permissible contact pressure and friction coefficient

	Permissible contact pressure: MPa	Friction coefficient: %
Rolling steel on steel		0.5
Rolling rubber tyre on concrete		1
Sliding steel on bronze	100	10–18
Sliding PTFE on stainless steel	12	0–5
Sliding fabric-reinforced polyester resin on stainless steel	35	5–12

stripping, during concreting and to stabilise supports of a BDE, whose position and/or length are variable.

17.4.4 Pins

Pins are a very widely used mechanical component in BDE. They allow the creation of hinged connections between steel members and between steel members and equipment components such as hydraulic jacks, wheels, etc. Pinned connections, as opposed to bolted connections, also help to minimise the time taken to assemble structural steelwork, such as the main girders of a BDE. Pins can be made from normal structural steel, but are often made from alloy steels with higher mechanical properties when reduction in size and weight are important considerations. Pins, unlike bolts, cannot be used to transfer shear between two plates only, that is through one shearing plane. In a pinned connection, at least two shearing planes are always required. Pins must always be mechanically secured by fitting cotter pins or retainer plates, so that they do not get dislodged due to vibration during operation.

17.5. High tensile bars

High tensile bars are widely used: for example to tie down a BDE to the permanent structure, for attachment of preassembled/precast bridge deck elements to the lifting devices, and for suspension of precast segments from overhead gantries. They can be, but are not always, pre-loaded. Table 17.3 gives guidelines about sizing of high tensile bars for various applications.

Reusability of stressbars and lifting bars is generally, even on very large projects, not controlled by fatigue strength. However, the biggest risk to the durability of stressbars and lifting bars is mechanical damage to the bars (deformed, not straight any more) and damage to threads due to handling of the bars. Hence for *every* reuse of stressbars and lifting bars, site operatives need to visually check the straightness of the bars and should confirm the integrity of the thread by threading a nut over part of the bar, where the nut will be located when the bar is activated.

Special attention needs to be given on site for the correct engagement of bars into couplers. Couplers of free-hanging individual bars should always have safety pins preventing the lower bar from being turned out by wind oscillation.

17.6. Power supply

The equipment components of a BDE are driven by electrical or hydraulic motors or directly by hydraulic oil (in the case of jacks). When hydraulic power is used, hydraulic pumps are required, which can be driven by electric motors or by diesel pump groups with combustion engines. Even when diesel

Figure 17.2 Main elements of horizontal propulsion systems: directly driven wheels, chain drive with fixed chain pulling unit, chain drive with movable chain pulling unit, long stroke jack and capstan winch (A and B indicate the two parts between which the movement takes place; F stands for force) (VSL International)

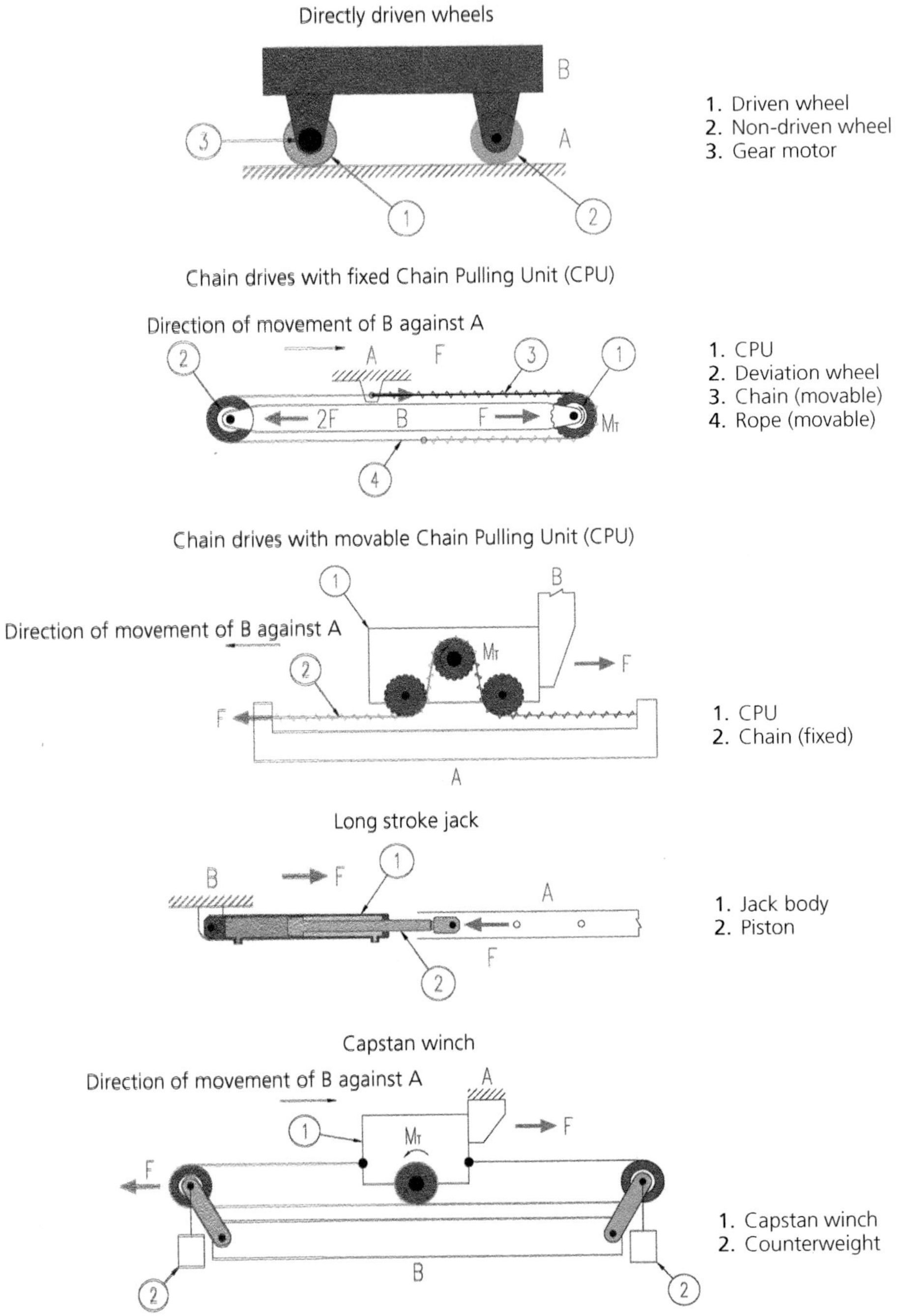

Table 17.3 Guidelines about sizing of high-tensile bars for various applications (P_o = stressing force after nut loss (anchored stressing force); P_u = ultimate capacity of HT bar; F_y = yield capacity of HT bar; μ = friction coefficient at contact area)

Application	Stressed with stressing force P_o (after nut loss)	Non-stressed	Detailing
External post-tensioning (temporary stressbars)	$P_o \leq 60\% \times P_u$	N/A	
Tie-down bars	$P_o \geq$ maximum service load; $P_o \leq 60\% \times P_u$	Maximum service load $\leq 50\% \times P_u$; need for redundancy in terms of number of bars	If not stressed, full articulation at both ends to ensure pure tension
Hanger bars	N/A	Maximum service load $\leq 50\% \times P_u$	Full articulation at both ends to ensure pure tension
Hanger bars for segment suspension in overhead gantries	Active build-up of hanger bar force with hydraulic jack(s) when transferring segment weight from winch trolley to hanger bars	Maximum service load $\leq 50\% \times P_u$; redundancy in terms of number of bars, when segments are kept suspended in hanger bars over live traffic	Full articulation at both ends to ensure pure tension
Lifting bars	Case I: Residual precompression in contact area is enough to transfer shear V from the segment-lifting beam to the element being lifted: $P_o \geq$ maximum service tension load + $1/\mu \times V$; $P_o \leq 60\% \times P_u$ and maximum service tension load in bar due to element being lifted $\leq 50\% \times F_y$. Case II: Residual precompression in contact area is not enough or cannot be guaranteed to be enough to transfer shear V from the segment lifting beam to the element being lifted: maximum service tension load in bar due to element being lifted $\leq 50\% \times F_y$ and equivalent stresses in bars considering normal and shear stresses (normal force, bending and shear) $\leq 1/1.5 \times F_y$	Maximum service tension load in bar due to element being lifted $\leq 50\% \times F_y$ and equivalent stresses in bar considering normal and shear stresses (normal force, bending and shear) $\leq 1/1.5 \times F_y$	If stressed (case II) or not stressed, need for provision of tight tolerance steel cone. Lifting bar is threaded into or is passing through this cone, which fits into a tight tolerance conical exit at top of lifting hole in segment (to transfer horizontal forces due to rotation of segment in space by bearing into the deck slab). Need to ensure that bottom bearing plate installed against soffit of deck slab can transfer part of bar force parallel to bearing plate by friction into concrete (friction plate, if angle of deck slab's soffit is too big, is better than pipe welded to bearing plate fitting into lifting hole, since attachment is absolutely rigid and effectiveness of fixity does not depend on hole accuracy and alignment at bottom of lifting hole; friction plate needs to have machined and hardened indentations)

Table 17.4 Technical data for some typical generators used to drive some types of BDEs

BDE type	Power: kVA	Dimensions W × L × H: m	Weight: kg (with fuel)	Fuel consumption: l/h
Lifting frame with 1000 kN capacity	42	2.3 × 1.1 × 1.8	1400	8
Portal crane with 800 kN capacity	100	2.6 × 1.1 × 1.9	1600	16
1000 kN capacity	150	3.1 × 1.2 × 2.2	2400	30
LG for segmental bridge construction, single winch	150	3.1 × 1.2 × 2.2	2400	30
LG for segmental bridge construction, double winch	250	3.1 × 1.2 × 2.2	2800	42

pump groups are used to drive the main hydraulic motors and components of a BDE, a small diesel-driven generator is installed for lighting and to supply electricity for auxiliary electrical tools, stressing equipment, etc. Electric power supply for the electric motors of a BDE comes from its own power generator set, driven by a diesel combustion engine.

The big advantage of using hydraulic jacks to move loads on a BDE is the fact that magnitude of the generated forces can be directly measured by reading the oil pressure and multiplying it with the active piston area. Furthermore, passive control systems can easily be incorporated, such as pressure-relief valves, which start to leak oil once the oil pumped into the jack reaches a certain preset pressure value, thus limiting the generated force. Gradual regulation of a jack's operating speed is achieved by the installing of proportional valves. Furthermore, a group of similar jacks can be hydraulically connected, so that each jack in the group takes a similar share of the applied load. Winches, chain pulling units and capstans can also be driven by hydraulic motors offering similar advantages as just explained for hydraulic jacks. Risks of loosing pressure in a hydraulic jack or motor in case a hydraulic hose gets damaged or bursts can be mitigated by the installation of break valves.

Noise generated by diesel pump groups and generators should not be underestimated, especially when a BDE works in urban areas at a high level between buildings. No one likes a big diesel engine working at night near their bedroom window. Super-silenced diesel engines or generators or additional quiet boxes are often the only solution to meet the required noise levels.

Technical data for some typical generators installed on BDEs are given in Table 17.4.

17.6.1 Location of generators

The location of the generator is controlled by the arrangement of the equipment components and by the objective to give the best possible accessibility for fuelling and maintenance, considering prevailing site conditions. Possible options for generator locations on gantries are shown in Table 17.5.

Portal cranes in temporary casting yards running on rails at ground level are often powered by on-board generators to avoid the need to install electric distribution lines along the rail tracks. Preferably the generator is then attached to the base of one of the portal crane's legs to simplify access for fuelling

Table 17.5 Possible arrangement of generators for gantries

Generator location	Advantages	Disadvantages	Comments
On trolley running on the erected deck behind the gantry, self-propelled or pulled by the BDE	■ Easy fuelling ■ Easy maintenance ■ No extra weight on BDE	■ The generator is an obstacle on the bridge deck, which needs to be managed ■ Need to plan generator movement during launching of the gantry	
At rear of the gantry's main girders	■ Generator moves with the BDE ■ No need to manage generator on the deck	■ Extra weight on BDE ■ Access is needed for fuelling and refilling	■ Fuel can be delivered over the deck ■ Use of rear winch trolley for servicing
Attached to outer side of one of the gantry's main girders	■ Generator moves with the BDE ■ No need to manage generator on the deck	■ Extra weight on BDE ■ Access is needed for fuelling and refilling	■ Fuel can only be delivered from the side ■ Need for auxiliary hoist for servicing

and maintenance. Sometimes this is not possible due to geometrical obstructions and the generator needs to be installed at a high level on the crane's main beam. In any case, access has to be planned and provided for fuelling and maintenance.

17.6.2 On-board power distribution

Starting from the generator, the electrical power is distributed to control switchboards and the electrical equipment components by electrical cables or electrical power rails. In the case of hydraulic power packs, the oil is circulating from the power pack to the hydraulic equipment components through pipelines (rigid metal pipes, flexible hoses). A main characteristic of a BDE is that its elements do not move as one unit but often against each other. The power distribution needs, therefore, to be arranged with flexible distribution systems both for vertical and horizontal movements.

Distribution of hydraulic oil from a 'fixed' pump to movable hydraulic components over an extended distance is not straightforward. Therefore hydraulic pumps, whether directly driven by a diesel engine or electrically driven, are generally travelling together with the hydraulic components they drive. However, reel systems or drag chains for flexible hydraulic hoses are available and are sometimes used (refer to Figure 17.3). For operation of segment support jacks on a gantry erecting segmental decks by the span-by-span method, fixed hydraulic lines with T junctions at the location of the segment support jacks can be used. This arrangement means that only the hydraulic control block, not the entire hydraulic pump, has to be moved along the gantry's main girder during activation of these jacks.

Distribution of electrical power from a 'fixed' generator or distribution board to movable electrical components is easier. For horizontally moving components this can be achieved either by festoon lines or by electrical power rails (refer to Figure 17.4). For vertically moving components, spring-loaded electrical power reels can be used. When the change of distance between the equipment components and the generator happens with significant time pauses and the power connection is not necessary for

Figure 17.3 Oil distribution by hose reel system (DEAL)

Figure 17.4 Horizontal flexible electrical distribution lines (DEAL)

Example of a festoon line installed on a lifting frame to supply power to the upper cross beam

Example of a power rail installed on a BCM gantry to supply power to the winch trolleys

the move itself, it is possible to arrange the power supply from different points along the equipment (socket/plug connection). The part to be moved is disconnected from the 'grid' before the start of the movement and is connected again after the new position has been reached.

17.7. Control system

The simplest way to control movement of an equipment component is to have an operator standing directly beside the respective equipment. In the case of hydraulic equipment, this operator opens and closes mechanically hydraulic valves at the hydraulic pump to control the movements.

Figure 17.5 Examples of control boards for remote-controlled equipment components (DEAL)

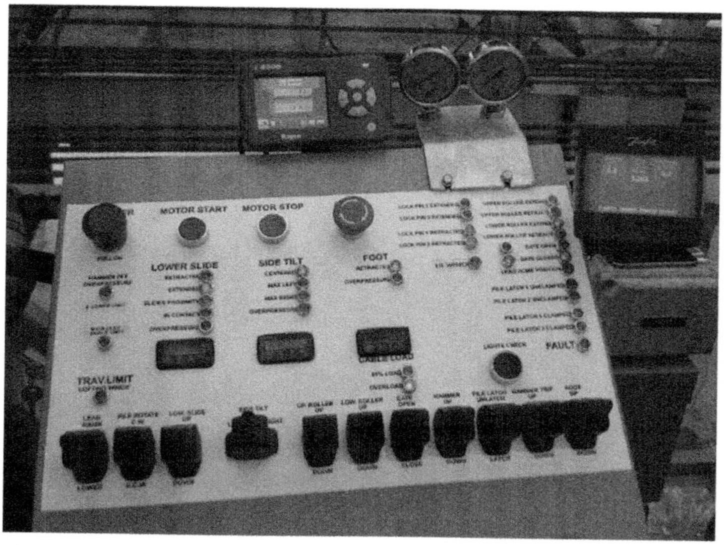

Centralised control board (inside or outside a cabin) for managing control and system diagnostic

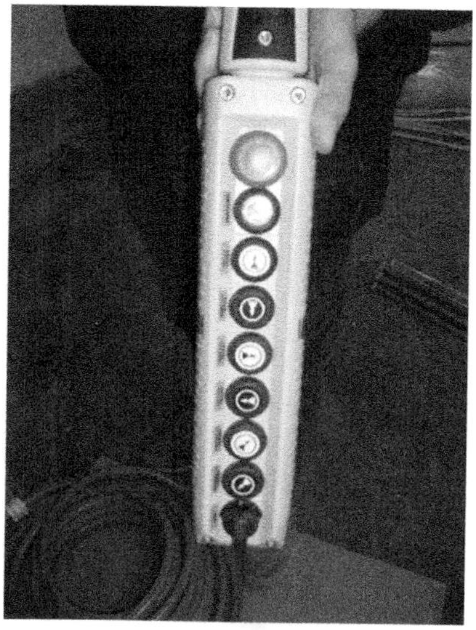

Push-button-type pendant: the simplest control board; applicable to control movement in a localised area; it gives the operator some possibility to adjust their position to have proper visibility

A characteristic of all BDE is, however, that they have numerous movable parts in a wide and complex area where visibility problems may be significant. For most operations, movements of several equipment components have to be synchronised and need therefore to be controlled centrically. The possibility of the leading operator having a good view of the ongoing movements is the first factor to consider in choosing the control system of the equipment. Another important point to increase safety of the operations is to have the main critical parameters under direct control or directly reported to the leading operator, for quick action in case of need.

The movement of hydraulic components is achieved by operating valves in control blocks generally attached to the pumps. In order to control such blocks by electrical signals, they need to be equipped with solenoid valves.

In BDE equipment, the use of a control cabin is limited to cases where there is good visibility. In any case, the operator is generally still assisted by a helper positioned at the location where the movement takes place and who is in constant contact with the operator by walkie-talkie.

The most used system for control of equipment components on BDEs is portable boards that can be more or less sophisticated according to the needs. They are either connected to the equipment by cable ('remote-controlled' equipment) or the connection is wireless ('radio controlled' equipment). For some examples of different options, refer to Figures 17.5 and 17.6.

As well as the active control systems described above, there is a need to build in monitoring and passive control systems such as:

Figure 17.6 Examples of control boards for radio-controlled equipment components (DEAL)

The board shown above is the most common solution; it gives the operator full possibility to adjust their position in order to have the proper visibility over the work area

This board is an evolution step ahead of the one shown on the left; the screen allows key data from field sensors (load cells, pressure sensors) and safety systems (warning alarms) to be viewed and assessed

- position-limit switches for winch operations (maximum top and bottom hook positions)
- load cells with digital load indicator and limit switches for winch operations
- pressure gauges on all hydraulic pumps, permitting the monitoring of the force generated in jacks driven by the pumps
- pressure-relief valves for hydraulic jacks to align the maximum force generated by the jacks with the structural steelwork design
- brake valves on hydraulic jacks to ensure that jacks keep pressure even if hoses connected to the jacks are damaged and start to leak.

REFERENCES

Figures 17.1 and 17.2: VSL International
Figures 17.3 to 17.6: DEAL

Bridge Deck Erection Equipment
ISBN 978-0-7277-6193-4

https://doi.org/10.1680/bdee.61934.385

Chapter 18
Lifting attachments

18.1. Introduction
This chapter considers the various ways in which the lifting gear that is a key component of most Bridge Deck Erection Equipment (BDE) can be connected to the precast concrete element to be handled. The advantages and disadvantages of each option are discussed.

18.2. Detailing of connections
There is no one 'best way' of making these connections. If there was, it would be used on all projects. Ten connection methods are illustrated in Figures 18.1 to 18.10. However there are certain aspects of the detailing of connections that are commonly accepted as pointing towards a good solution. These are as follows.

1. If the connection for a box girder is to be made using holes in the top flange, it is better that those holes are on the inner side of the webs as that makes the removal of the lifting attachments much easier. When deciding the attachment's location, it is important to consider the space for such holes, since they are commonly placed where the rebar and prestressing are congested. It is a good idea to make square box-outs slightly tapered (with the wider dimension at the top). That eases the removal of the form used to create the hole, and also provides a physical interlock between the surrounding concrete and the concrete used to fill the hole later. The alternative is to use a short length of galvanised corrugated duct, which can later be removed by 'unwinding' it from the segment. The corrugation gives a good bond with the segment and the Non-Shrink Grout (NSG).
2. If the connections are to be made using high tensile (HT) steel bars, they either have to be pre-loaded or the lifting attachments need to interlock with the segment to prevent movement of the bar, which can cause it to fail prematurely.
3. While nuts with spherical surfaces are very useful for absorbing casting tolerances and allowing correct alignment of the washer with the underside of the concrete surface, friction at the spherical interface will be too high for any further rotation to take place after the bar is under load. Thus further rotations will apply bending to the bar.
4. With regard to stress levels in stressed and non-stressed bars, refer to Section III, Chapter 17.
5. If bars are to be pre-loaded it is recommended that they are pre-loaded to a value of the nominal external applied force plus 20% (which includes an allowance for impact effects). This does not apply to method 4 in Table 18.1, for which the required pre-load is significantly higher.
6. A centre-hole jack is generally used for pre-loading. The pre-load may be kept in the bar either by tightening a nut or by keeping the jack on the bar. If the jack method is used, it should have a security ring so the integrity of the system does not rely on maintaining the hydraulic pressure in the jack.

Figure 18.1 Method 1 (Tony Gee and Partners)

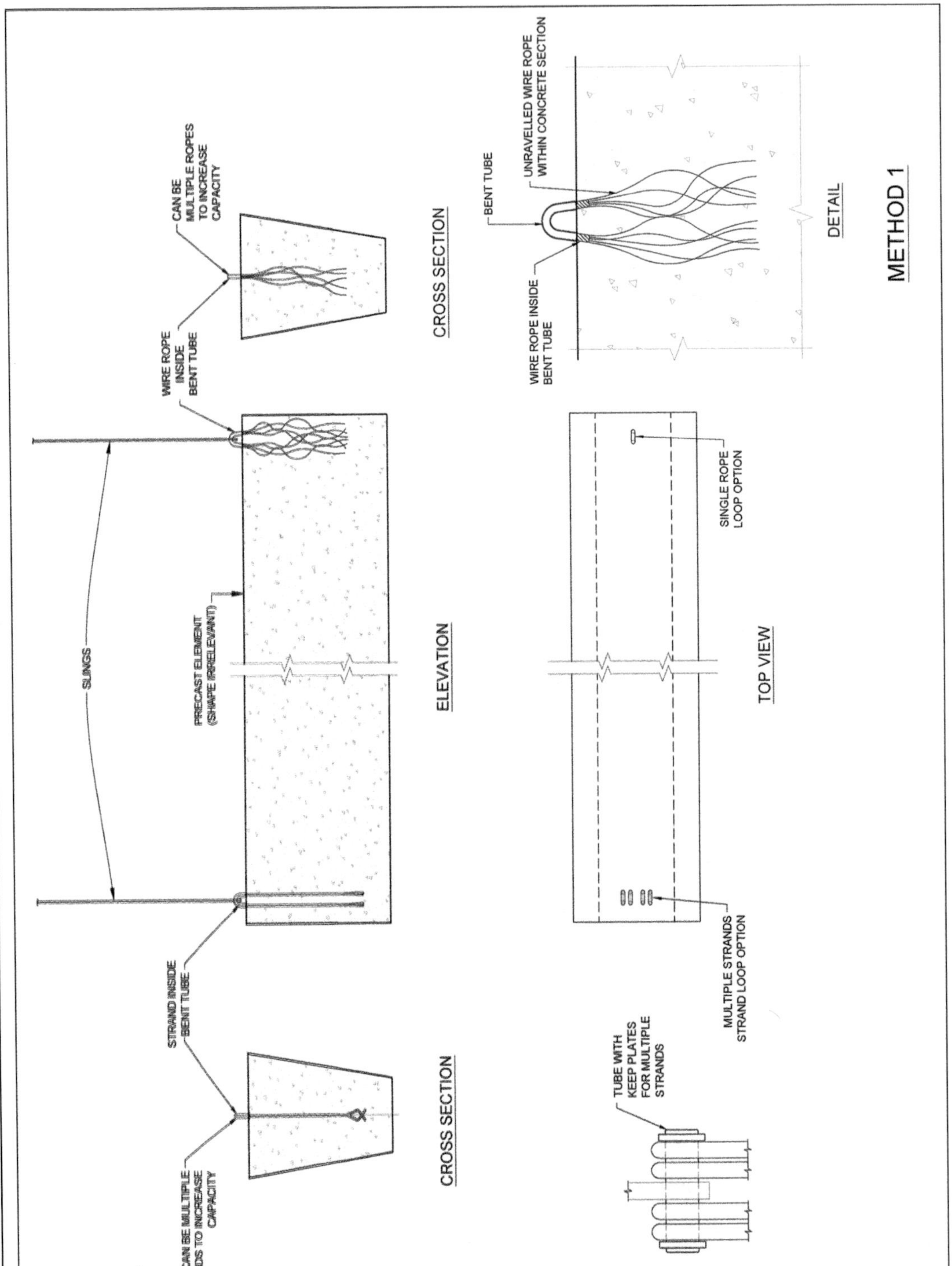

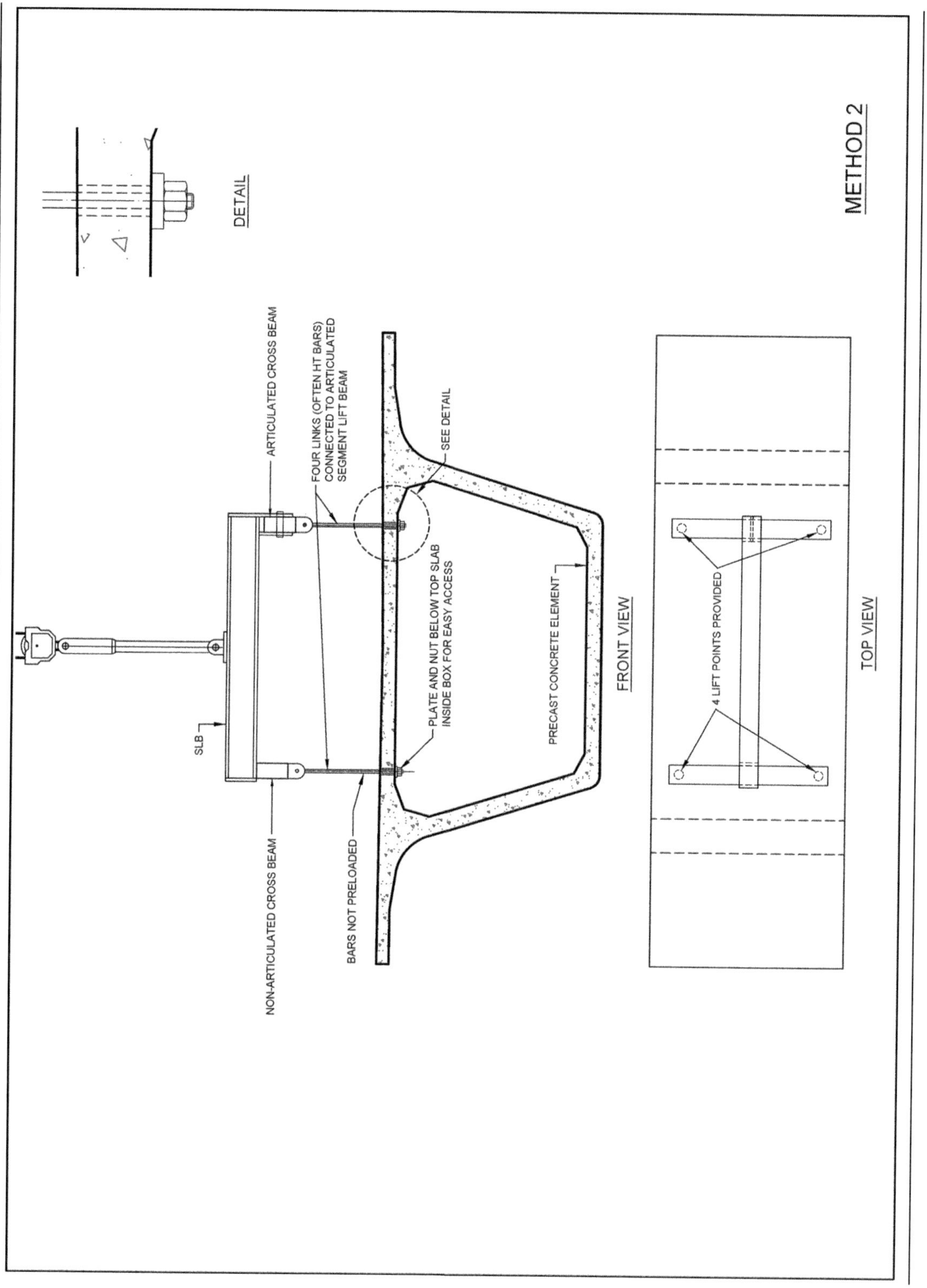

Figure 18.2 Method 2 (Tony Gee and Partners)

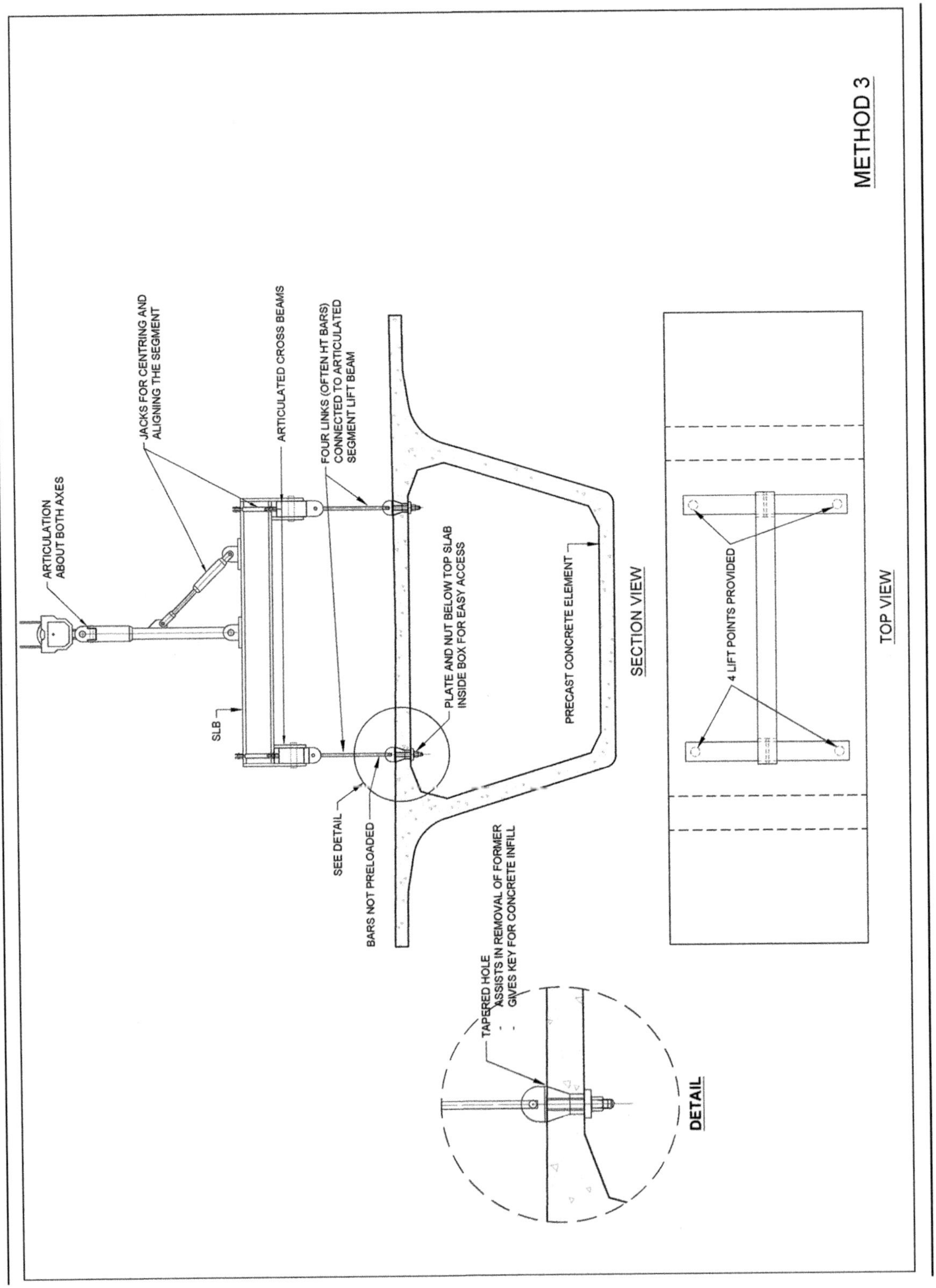

Figure 18.3 Method 3 (Tony Gee and Partners)

Figure 18.4 Method 4 (Tony Gee and Partners)

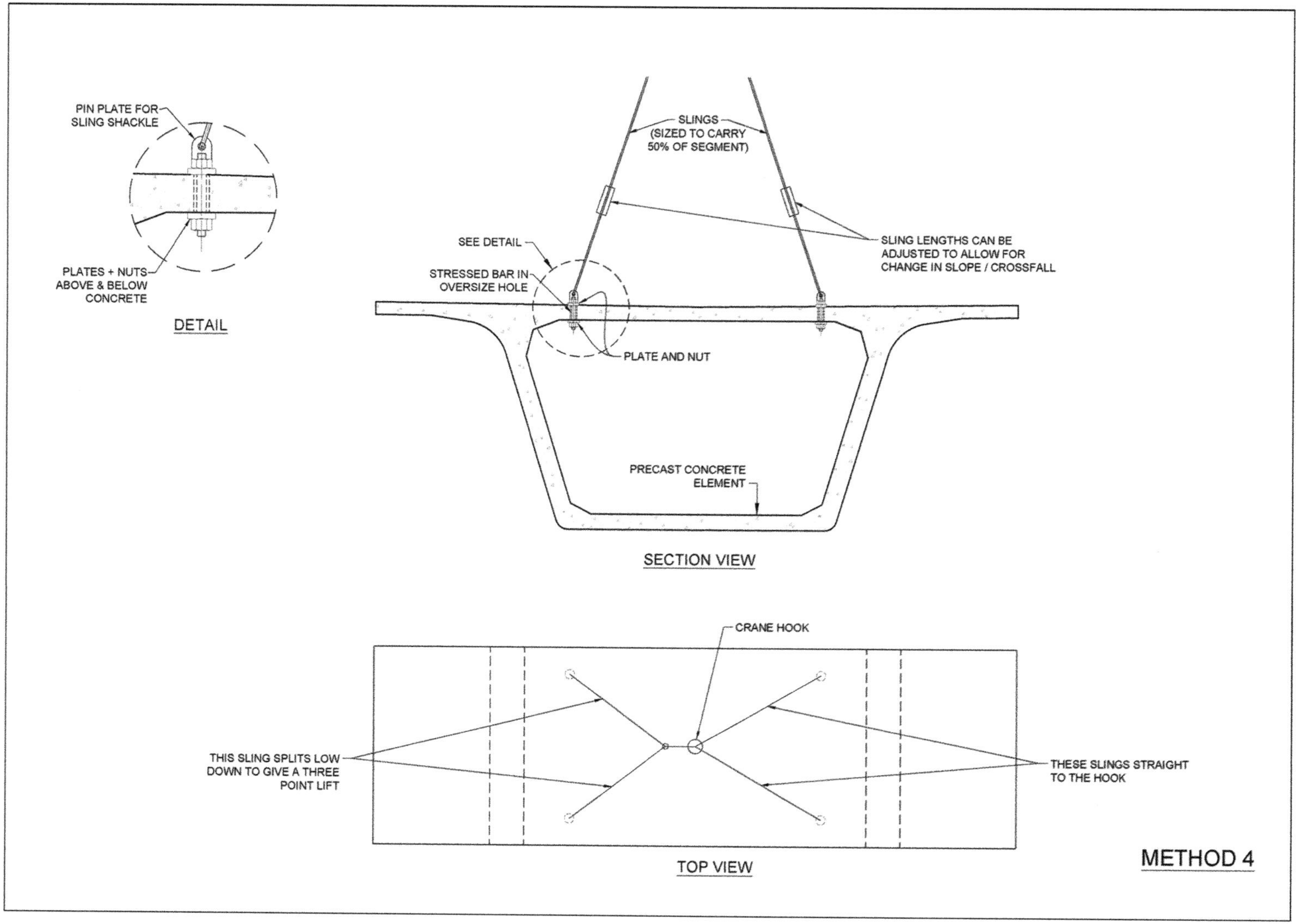

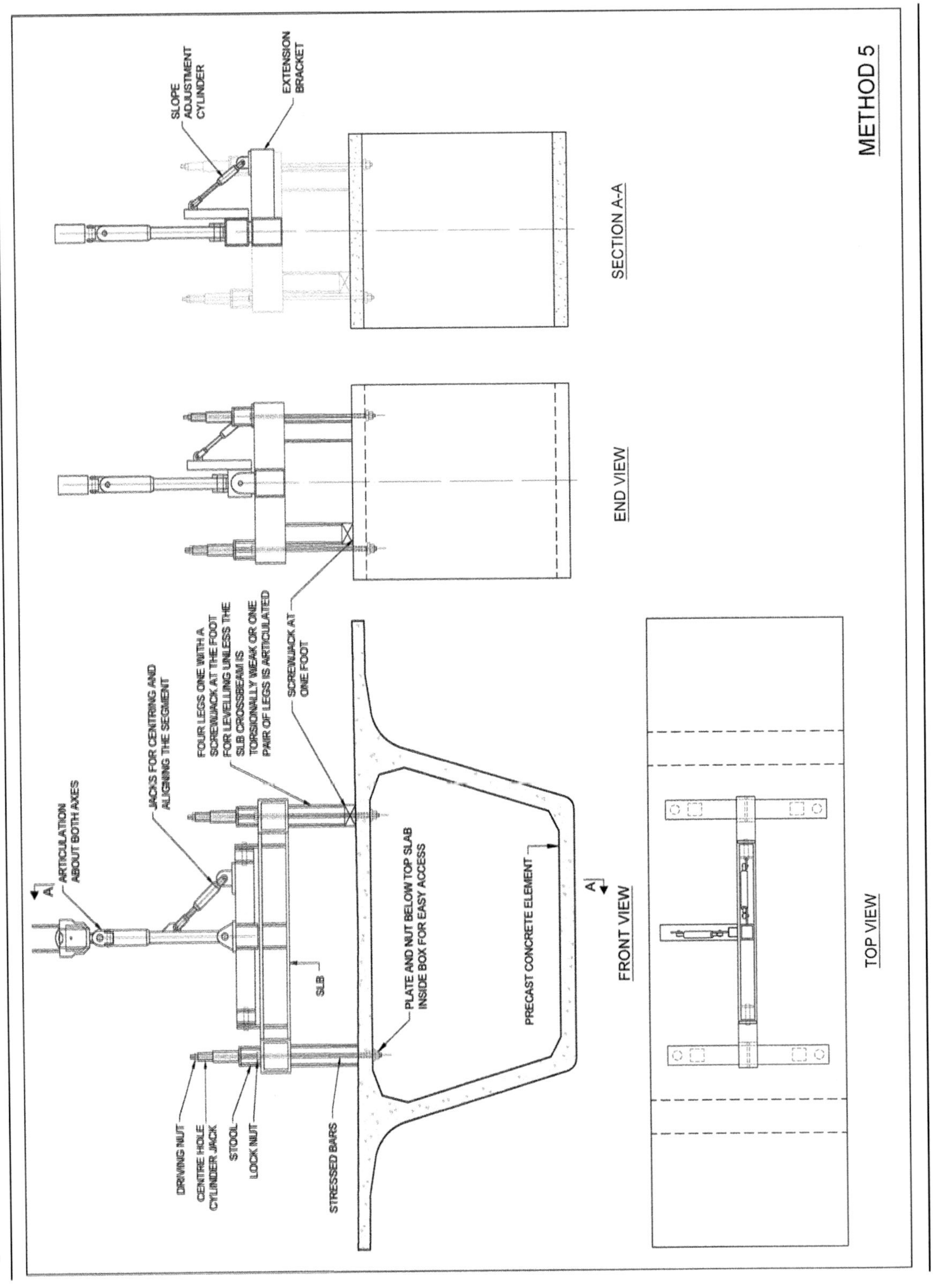

Figure 18.5 Method 5 (Tony Gee and Partners)

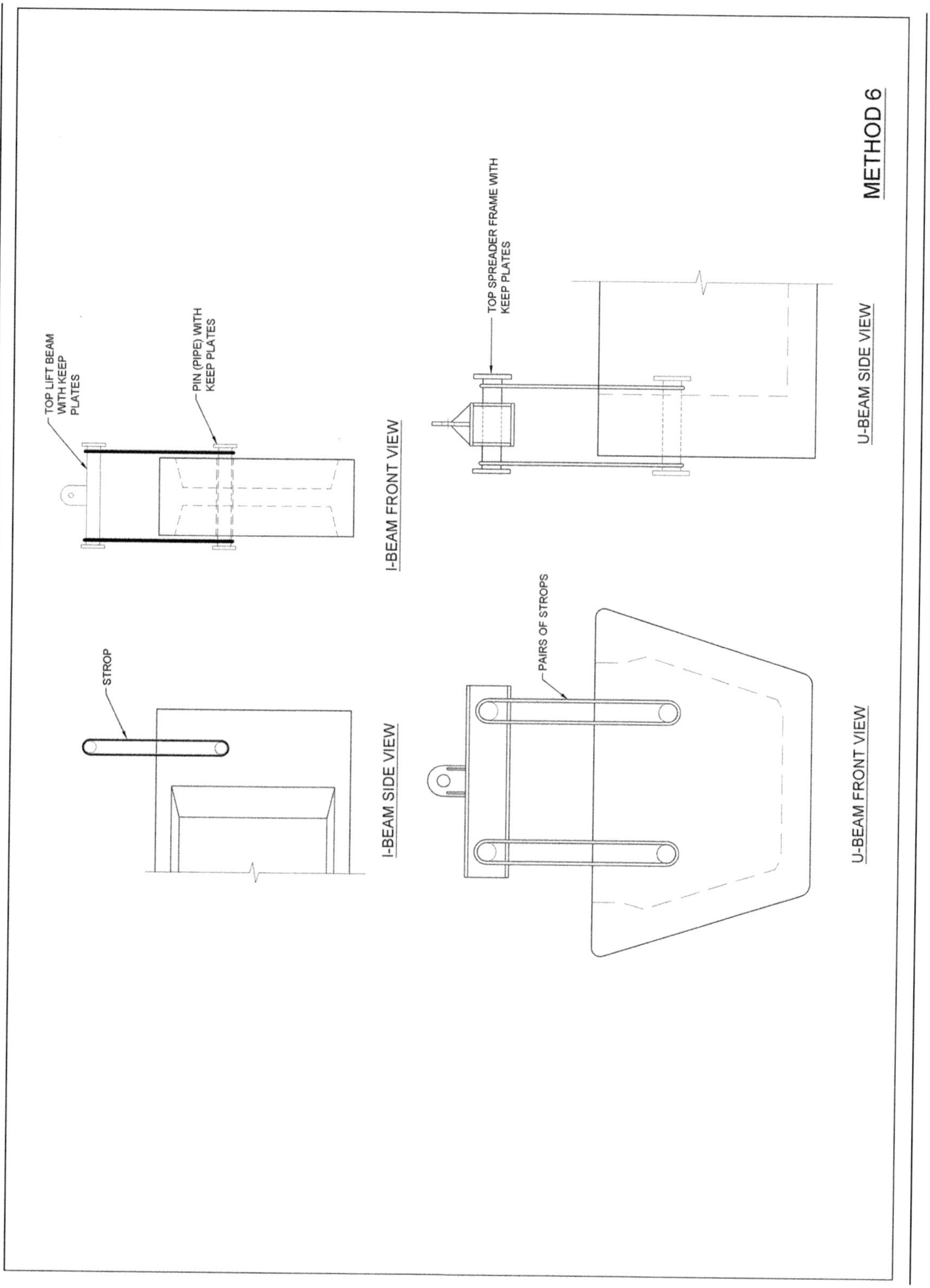

Figure 18.6 Method 6 (Tony Gee and Partners)

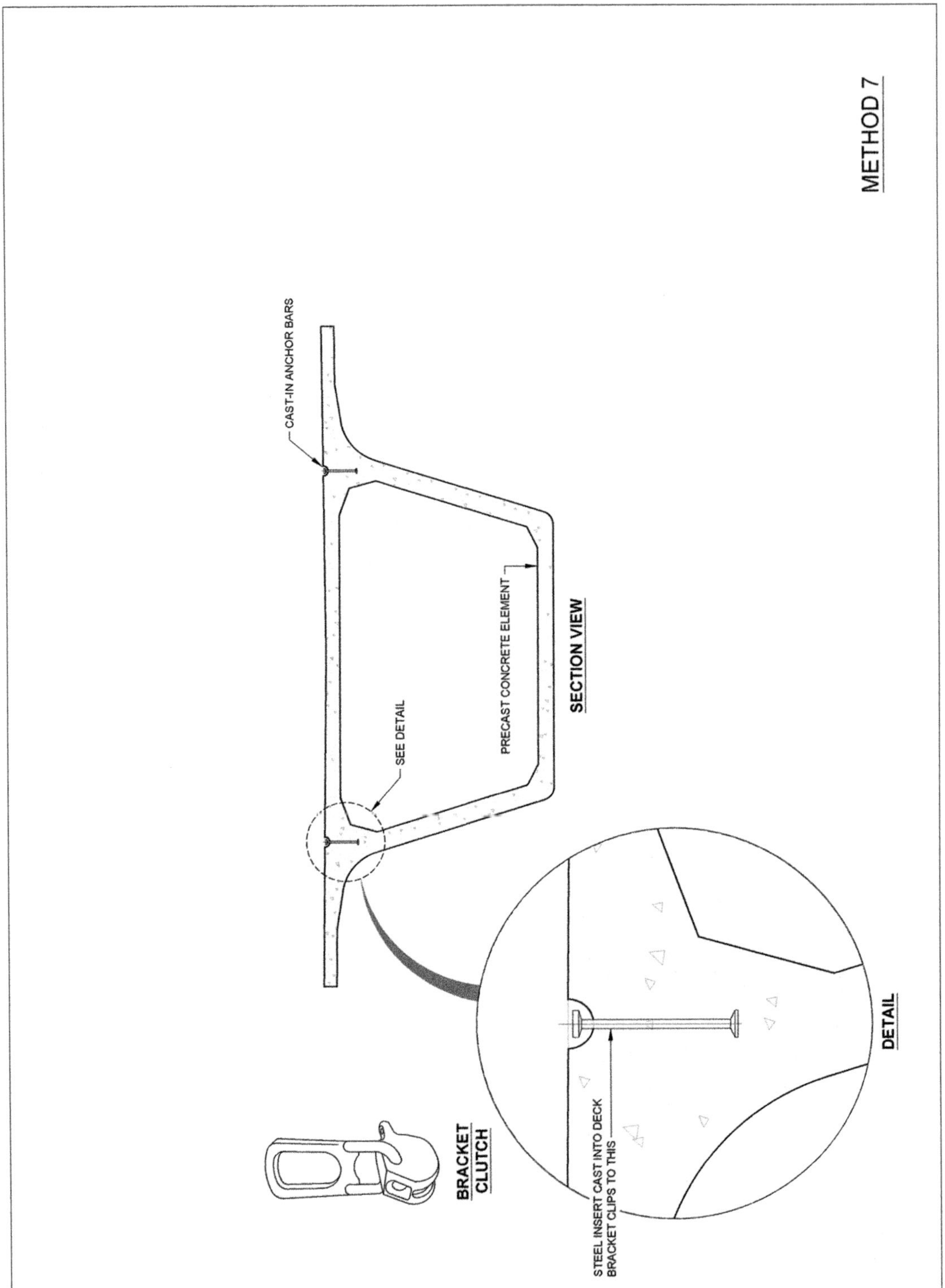

Figure 18.7 Method 7 (Tony Gee and Partners)

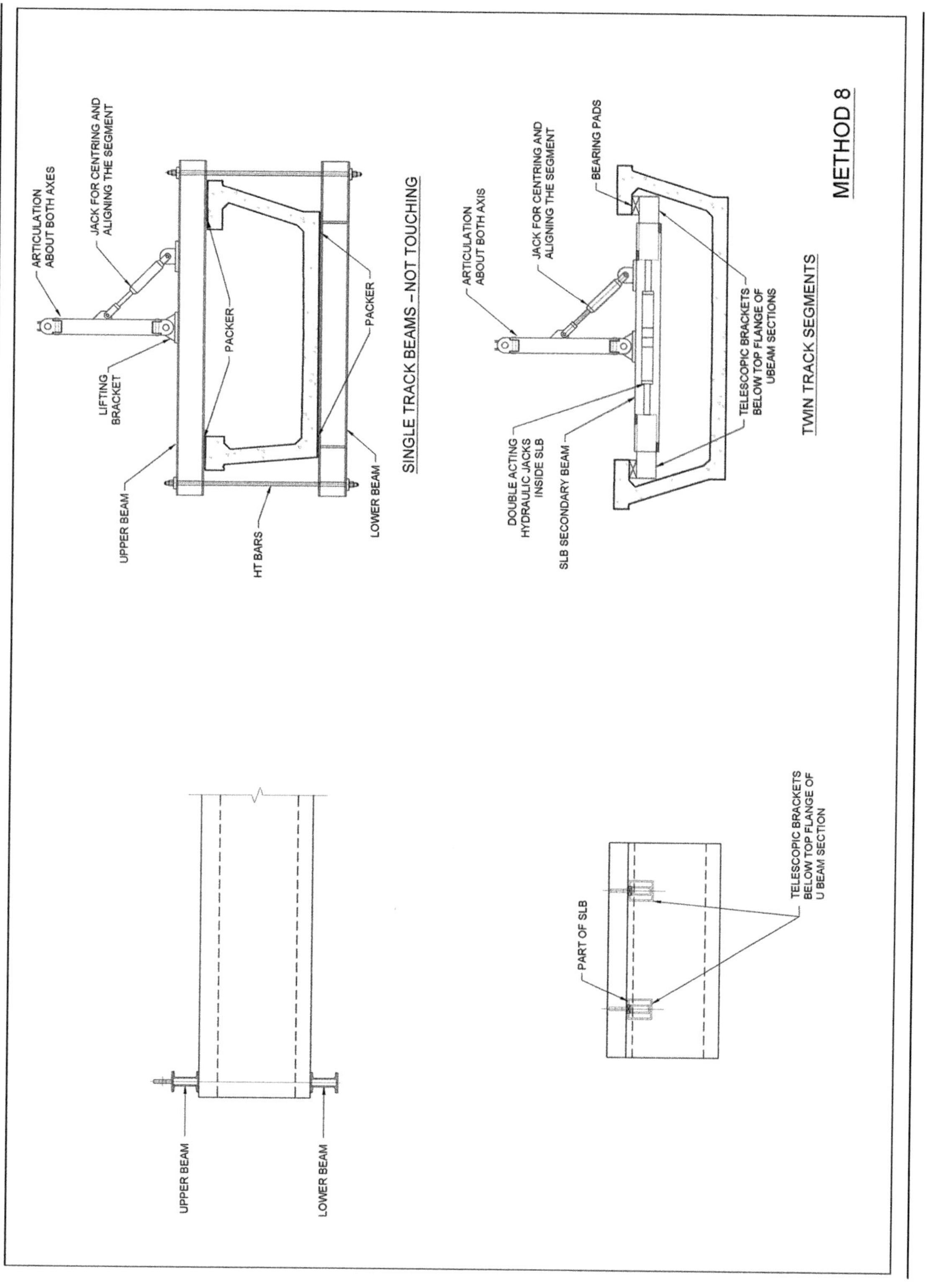

Figure 18.8 Method 8 (Tony Gee and Partners)

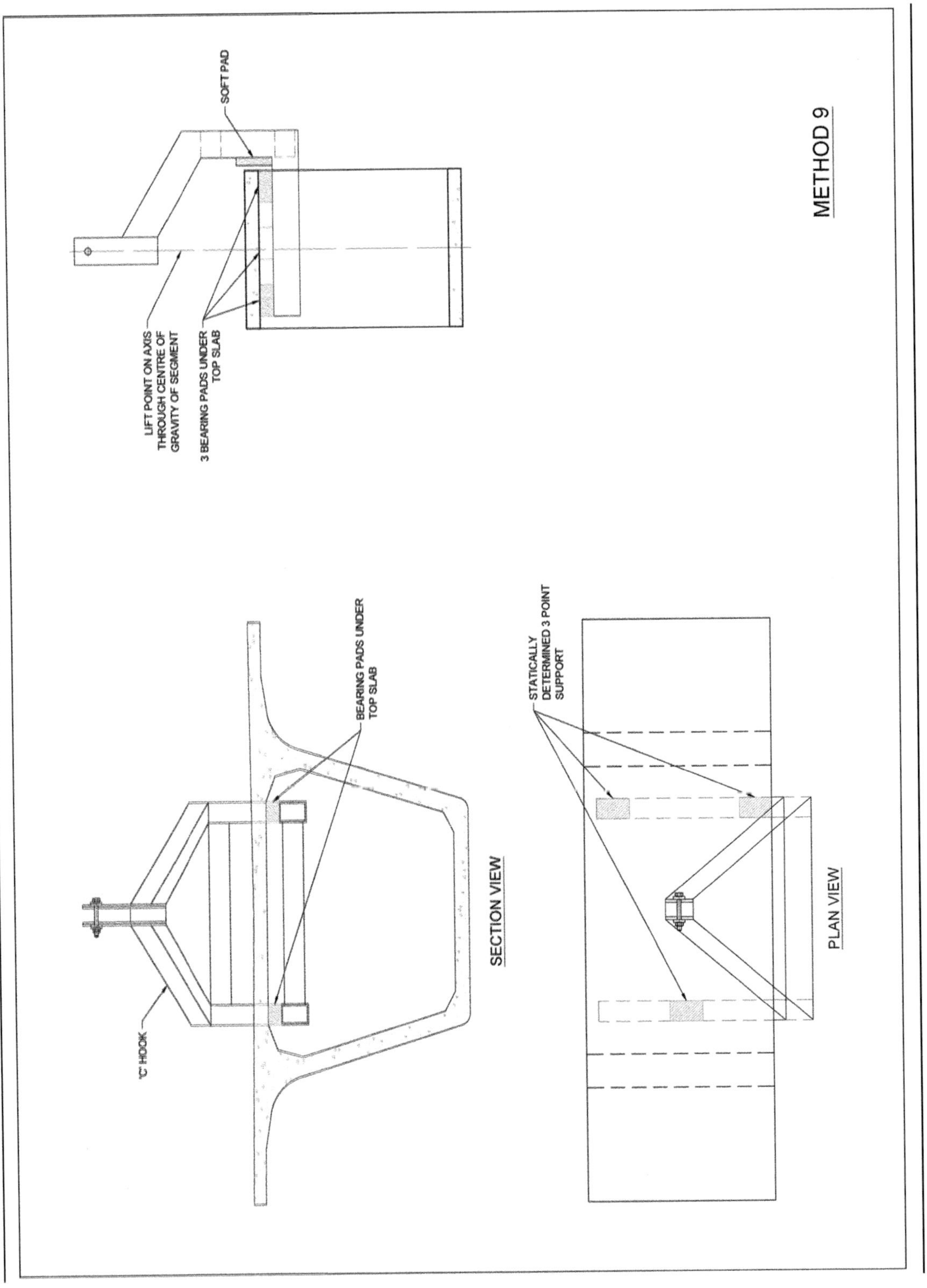

Figure 18.9 Method 9 (Tony Gee and Partners)

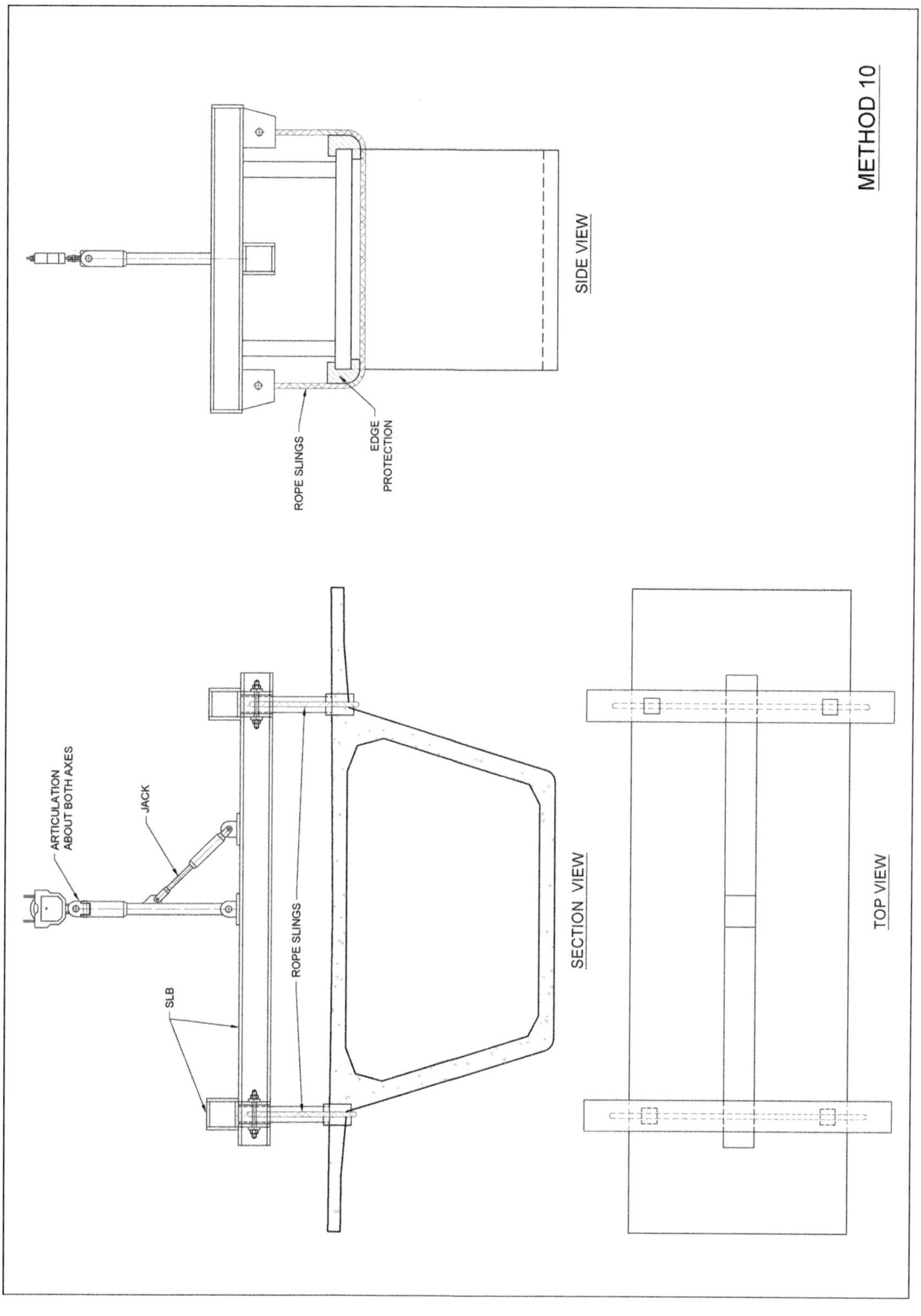

Figure 18.10 Method 10 (Tony Gee and Partners)

Table 18.1 Review of the options

No.	Description	Comments	Advantages	Disadvantages
1	One or more embedded rope or strand loops at each end of the beam for two-point lifting of precast beams. Single loops can connect to a crane hook or shackle. Multiple loops can connect to a bar with end keep plates	The 'traditional cheap' solution for narrow girders. If strands are used, they are usually fitted with a length of bent steel tube which the lifting sling bears on. The geometric accuracy of the system is low, which makes for simple installation. Embedment length of the rope/strand has to be checked. The centre of gravity of an asymmetric element must be calculated to enable the correct lift to be achieved	Low cost. Only simple slings are needed, which minimises the height between the lift point and the top of the concrete. The loss of cross-sectional area caused by the box-out (necessary to burn the rope off after final placing) is small. Most of the box-out will be within the 'cover' for the top layer of rebar so the effect on the permanent work design is minimised. Change to transverse cross fall is easily achieved as the beam is narrow	The system depends upon the rope being properly unravelled or the strand's wires opened up at the strand's end to look like an 'onion' (also called H bulb) and the embedded length within the concrete section to be long enough. Good quality control is needed in the casting operation, particularly if multiple loops are used. Once the loops are finally cut, it is impossible to reinstate the lift point. Localised cracking around the rope anchorages is to be expected. Any cracks are embedded by the NSG
2	Four unstressed bars passing through oversized holes in concrete to plate below, with segment lifting beam above	If the lifting bars are not to be pre-loaded, the use of an segment lifting beam is essential to avoid bending the bars due to inclination of the lifting slings. Refer to note 1 in subsection 18.2, above, for the formation of the holes	Cheap. Fast to install. Minimum loss of top flange section, although coordination with permanent work designer will still be needed. Section will have to be filled with NSG to restore global section to the bridge deck, but the volume is small	This system only works for precast elements with symmetrical sections and where there is to be no change in longitudinal or transverse cross fall between lift-off and placing. Generally only used in casting yards
3	Four unstressed bars passing through concrete with cone guidance with segment lifting beam above	The system is similar to option 2, but the interface between the male cone that is part of the segment attachment and the female cone shape cast into the section minimises bending on the bar	Fast to install and remove (no need for stressing and de-tensioning). Can be used where a change in longitudinal or transverse cross fall is required (as is almost always the case)	Geometry of lifting hole's cone needs to be accurate (use machined cone formers). Loss of permanent work cross section is greater, and the female cone shape may be harder to fit between permanent work rebar and tendons. Volume of NSG needed for section restoration is higher than for option 2. Balance system within bars is advised to keep the bar size down

4	Four stressed bars passing through oversized holes in concrete with no segment lifting beam above	Friction between segment attachment and the concrete prevents relative movement. Segment attachments are bigger than for option 5 because of the need to resist horizontal and eccentric loading due to gradient and super-elevation and the inclination of the lifting slings	No need for machined lifting-hole formers. Can be used where there is a change in cross fall and/or gradient by adjusting sling length	Increases height between the lift point and the top of concrete. The lack of the segment lifting beam increases pre-load demand on the bar. Volume of NSG needed for section restoration is more than for options 2 or 5, but less than for option 3. Slower to install and remove than options 2 or 3
5	Four stressed bars passing through oversized holes in concrete with segment lifting beam above	Friction between segment attachment and the concrete prevents relative movement	Minimises size of bar and pre-load demand, so easier to fit within permanent work. Cost of segment lifting beam is offset by low cost of work in each segment lifted. Can be used where there is a change in cross fall/gradient	Increases height between the lift point and the top of the concrete. Balance between bar slings (part of segment lifting beam) minimises loads. Slower to install and remove than options 2 or 3
6	Large holes through the end diaphragm	Commonly used system for U- or I-beams. Shear pin passes through the hole. Rope strops are wrapped around the pin (keep plates to hold the strops on). The upper end of each strop can connect either to a 'Ram's horn' crane hook or a short-lift beam	Cheap and simple to install and operate. In most projects there will be no need to backfill the pin hole because the permanent work design can accept the loss of section	Small quantity of extra rebar around the pin hole. Some projects may require the hole to be filled after beam placement
7	Proprietary cast-in segment attachments (various types)	Cast-in segment attachments (in recesses) to which vertical lift slings can be connected	Very fast to fix and remove lifting slings or the segment lifting beam. Can easily re-attach lifting gear until box-out is backfilled. Minimum demand for NSG. Should be easy to fit with permanent work rebar and tendons	Inserts are relatively expensive and are usually lost in each segment. Need to cut attachments and to fill recesses with NSG
8	Lifting without segment attachments by direct bearing of segment lifting beam against precast element	The purpose-designed segment lifting beam either sits below the precast element or hooks under a part of it (for example the inner flange of a U-beam)	Little or no change required to the permanent work. If the slope of the lift has to be adjusted and is significant, the friction between the lift and the precast element can be increased by friction pads. Very suitable for lifting of elements having U-section with top flanges	The segment lifting beam's removal may not be simple (need for 'telescopic' arms for U-sections)

Table 18.1 Continued

No.	Description	Comments	Advantages	Disadvantages
9	C hooks	The C hook allows the support of the precast segment through bearing against its top slab soffit. This method is of particular use in stockyards where box-section segments are stacked	Fastest option for installation and removal. No need for any people on top or inside the segment. No need for any inserts. If the slope of the lift has to be adjusted and is significant, the friction between the lift and the precast element can be increased by friction pads	Decent space corresponding to length of the C hook on one side of segment is required for installation and removal. Hence generally not suitable for segment erection. There is a danger that impact between the C hook and the match-cast face of the element may cause damage. Lack of positive location may cause the lift to be off-level and could cause distress to the segment
10	Soft slings (for lifting of segments which have wing slabs)	The edges of the segment have to be protected aga nst damage from the sling, which is simply wrapped around the wing slab close to the web of the box section	Works well for box-section segments where there is no danger of instability (centre of gravity well below the lift point). Difficult to install slings	Could damage the end faces of the element. Cannot be used for erecting segments when they are placed against anything else (so really only for underslung gantries or in stockyards)

7 The method of connection will have to be coordinated with the permanent works designer to control local stresses. Conversely, a competent permanent works designer will consider the way in which an element may be lifted.
8 Generally a precast element is connected to the Segment-Lifting Beam (SLB) at four points. To ensure that, for a symmetrical element, the forces at the attachment points during lifting are similar, the SLB has to be articulated to provide a statically determinate configuration. If no articulation (or no SLB) is provided, each sling and attachment point must be designed to carry half the weight of the element.
9 If only three slings are used, or if the element is asymmetric, it is best to attach the slings around the centre of gravity so as to equalise forces as much as possible. It will probably be necessary to incorporate a 'length adjustment device' in two of the slings to allow for construction tolerances, and the need to level the element prior to its joining with the rest of the structure.
10 The angle between the slings at a hook must not exceed 90 degrees. For large elements it will be necessary to use long slings, and this affects the height of the hook and the geometry of the whole lift scheme.

18.3. Review of the connection options

A comparison between the ten connection options is presented in Table 18.1.

REFERENCES

Figures 18.1–18.10: Tony Gee and Partners

Bridge Deck Erection Equipment
ISBN 978-0-7277-6193-4

https://doi.org/10.1680/bdee.61934.401

Chapter 19
Precasting of bridge decks built by BDE

19.1. Introduction

As explained in Section I, Chapter 2, bridge decks are sliced for construction into elements to rationalise the time- and labour-intensive falsework systems or to move as much activity as possible off site into factories and to allow use of highly mechanised erection equipment. This chapter is about precasting the bridge deck partially or fully in factories, which are either permanent facilities or specifically built for a project. Figure 19.1 shows how decks can be broken down into precast elements either longitudinally or transversely.

Longitudinal cuts create **precast girders** (with more than one precast girder per span), which have I, T or U sections.

Transverse cuts create **precast segments** as follows.

- Precast segments if there are more than one segment per span.
- Fully precast spans (Full Span Method/FSM) if there is only one segment per span.
- Segments for incrementally launched decks (Incremental Launching Method/ILM), which are cast in a stationary casting yard behind an abutment.

When cutting the bridge deck transversely, the precast elements generally have the section of the bridge deck's final cross section. Hence there is less site work involved when building precast bridge decks with segments than with girders.

The main advantages of using precast elements for bridge decks are as follows.

- The construction time is shorter, since the casting and the erection of elements can be overlapped and can be carried out with highly mechanised equipment.
- There is less impact on traffic and the environment, since the concreting is not done at the final location and the deck erection takes less time.
- The elements can be mass-produced under factory conditions.
- On-site activities are reduced.
- It is easier to manage quality in a factory than on site.

The main disadvantages of using precast elements for bridge decks are as follows.

- The permanent work quantities are generally slightly higher.
- The standardisation, which is necessary to make precasting economically viable, limits geometrical variations.

Figure 19.1 The creation of precast deck elements by longitudinal or transverse cuts (Meyer 2009)

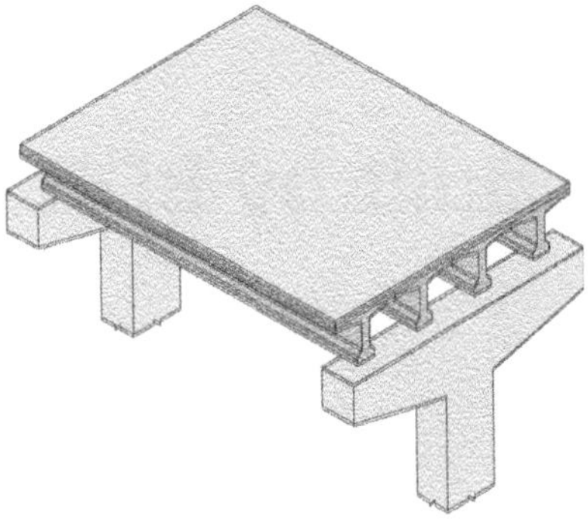

Precast deck elements created by breaking the bridge deck longitudinally into girders

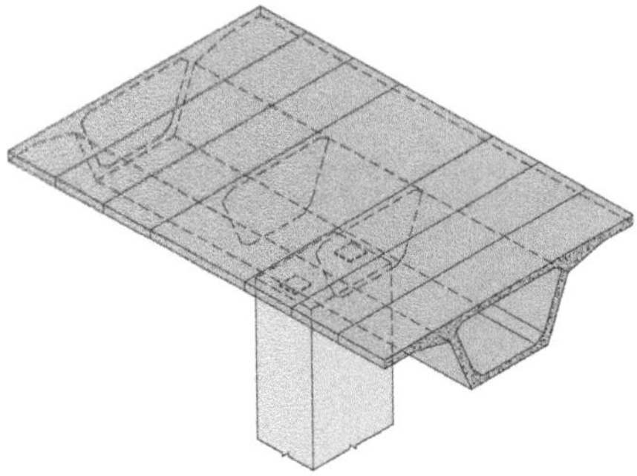

Precast deck elements created by breaking the bridge deck transversely into segments

As explained in Section III, Chapter 15, prestressing is a key structural element in bridges with concrete decks. Table 19.1 gives guidelines about which type of prestressing (pre-tensioning, internal or external post-tensioning) can be used with which type of precast elements.

This chapter covers production aspects for precast girders, precast bridge deck segments, fully precast spans and for incrementally launched bridge decks. Each subsection will address formwork concepts, permanent work aspects, including geometry control if applicable, fabrication of reinforcement bar cages, placing and curing of concrete, cycle times and risks. Finally there is a subsection about handling of precast elements.

19.2. Production of precast girders

19.2.1 Formwork concepts

Different formwork concepts are used for production of precast girders depending on whether for first-stage prestressing they are pre-tensioned or post-tensioned.

The production of pre-tensioned girders requires the set-up of a pre-tensioning bed. The formwork and the bed need to be designed and detailed as one unit. The main elements of a typical bed and formwork for production of pre-tensioned U-girders are shown in Figure 19.2.

When stressing the pre-tensioning strands in a bed, the strands are anchored in the bulkheads. The bulkhead transfers the prestressing forces either into compression props (option II in Figure 19.3) or into vertical posts, which are anchored in 'dead men' (option I in Figure 19.3).

In option I the bulkhead bears against a pair of vertical posts anchored in a 'dead man'. The compression force from the prestressing is centrically introduced into the bed's base slab. The 'dead man' counteracts by its own gravity the overturning moment due to the vertical eccentricity between the working line of the pre-tensioning strands and the centre line of the base slab.

The bulkhead of option II bears against a pair of compression props extending over the full length of the bed. Since the compression force from the prestressing can normally not be introduced into the compression props without a vertical eccentricity, props deflect during stressing and bounce back

Table 19.1 Type of prestressing and its application for different precast elements

Location in bridge deck	Type of precast element	Pre-tensioning	Post-tensioning	
		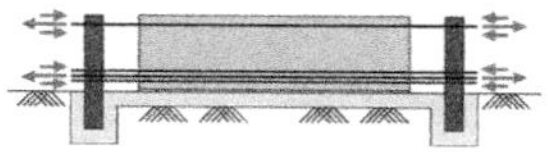	Internal 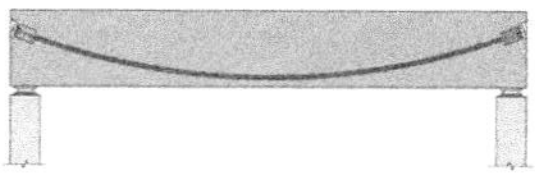	External 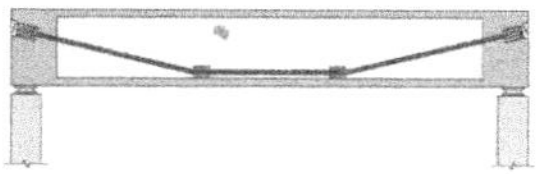
Transverse	Segmental bridge decks	(X)	X	
Longitudinal	Precast girders	X	X	
	BCM (Cantilever PT)		X	
	FSM (first-stage PT)	X	X	
	FSM (second-stage PT)		X	X
	Precast segmental		X	X
	Decks by ILM (first-stage PT)		X	
	Decks by ILM (second-stage PT)		X	X

Figure 19.2 Main elements of a typical bed and formwork for production of pre-tensioned U-girders (VSL International)

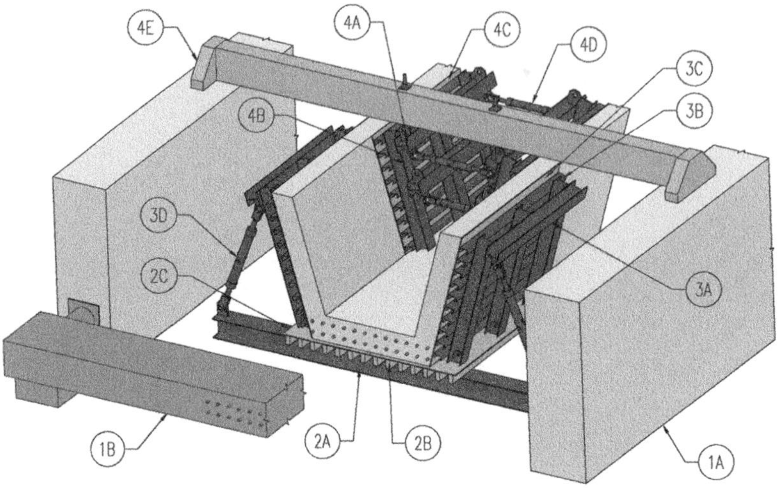

1. Bed
1A. Compression prop
1B. Bulkhead with solid ram jacks for load transfer
2. Bottom form
2A. Waler
2B. Runner
2C. Form skin
3. Outer side form
3A. Waler
3B. Runner
3C. Form skin
3D. Prop
4. Inner side form
4A. Waler
4B. Runner
4C. Form skin
4D. Props
4E. Suspension system
5. Stop end: not shown for clarity

during load transfer. Hence for option II the girder's base slab is generally structurally separated from the compression props.

Option I is selected when the working line of the pre-tensioning strands is close to the soffit of the precast element and for very long beds, where more than one element is cast in the same stage. Option II is chosen when the working line of the pre-tensioning strands is high above the soffit of the precast element and/or for short beds, where only one or a few elements are cast at the same stage.

The formwork for post-tensioned precast girders consists, like the formwork for pre-tensioned girders, of soffit form, external side forms, internal side form (for U-section only) and stop ends. Since pre-stressing is applied against the cast girders, there is no need for a pre-tensioning bed. Therefore many more options are available for the layout of casting yards which produce post-tensioned girders when compared with yards which produce pre-tensioned girders. When producing post-tensioned girders it is, for example, possible to have extra sets of soffit forms. Soffit forms are simple to build and relatively cheap when compared with the side forms and the stop ends, which are more complicated. With the provision of extra sets of soffit forms, the cycle time for the side forms and the stop ends can

Figure 19.3 Pre-tensioning beds – options for anchoring pre-tensioning forces (VSL International)

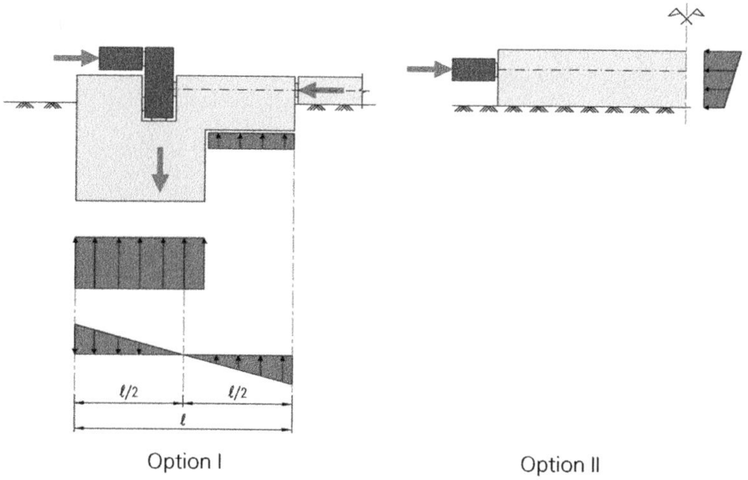

be reduced. The construction stages for precasting of a post-tensioned U-girder are shown in Figure 19.4.

Whether pre- or post-tensioned, particular attention needs to be given to the elastic deformation of the precast girder due to prestressing (refer to Figure 19.5). It is recommended to place the soffit form at the ends of the girder on elastomeric pads, which do not provide a rigid support for the girder during the load transfer. This minimises the risk that the girder's ends are getting damaged.

The type of cross section of a girder has an impact on production aspects. The external side forms of U-girders can be fixed. This is different to I- and T-girders, which have bottom flanges. For I- and T-girders, the external side forms have to be moved laterally out of place for lowering of the rebar cage and for the lifting of the completed girders. Furthermore, air bubbles can move without obstruction upwards along the form's skin when vibrating the concrete of a U-girder.

Figure 19.4 Construction stages for precasting of a post-tensioned U-girder (VSL International)

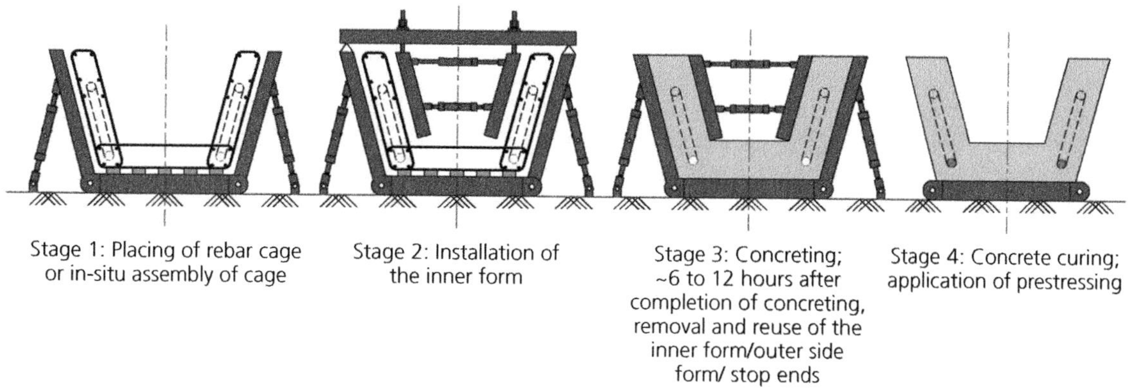

Figure 19.5 Detailing of soffit form at end of girders to cope with elastic deformation due to prestressing (VSL International)

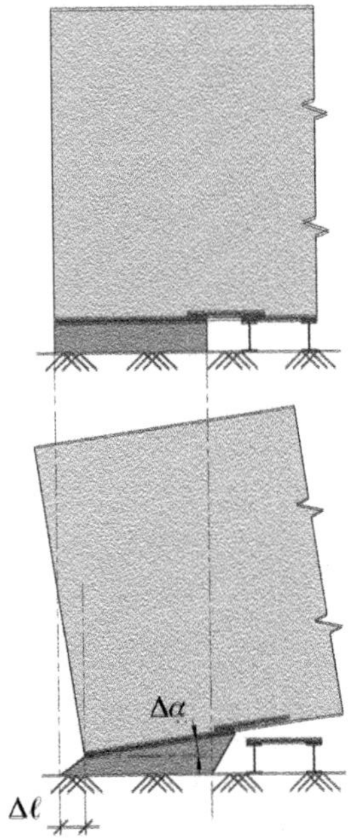

19.2.2 Permanent work aspects

The prestressing of precast girders is designed to introduce an internal stress state which favourably superposes with the flexural stresses from bending moments due external loading. The prestressing is designed to counterbalance a certain percentage of the bending moments due to external loading. Whereas it is straightforward to achieve a design bending moment in a simply supported girder when using post-tensioning by detailing a curved tendon profile, this is not possible when using pre-tensioning, unless special measures are taken. These measures are either to un-bond strands towards the ends of the girder in steps (option I) or to deviate some of the strands to achieve a trapezoidal profile (option II). The bond between strands and concrete is prevented in option I by encasing the strands in individual plastic tubes. Over the un-bonded length the strands are not directly protected against corrosion. However the plastic tubes are sealed off at the internal end by masking tape and at the outer end by the epoxy paste typically used to corrosion-protect the pre-tensioning strands at the end of the girder. Option II requires not only special devices for the deviation of the strands, which are partially lost in the girder, but also provisions in the casting bed to deviate the strands during stressing until the load transfer of the prestressing force from the bed to the girder has been accomplished. Such provisions are particularly cumbersome if girders have different lengths and their strands are deviated at different locations in the bed. For illustration refer to Figure 19.6.

As explained in the introduction to this chapter, geometrical variations need to be minimised to improve the economic viability of precasting. If girders of different depth are needed, the cross sections of these girders should be developed out of the same basic geometry, so that rebar jigs and forms can be reused as much as possible. Rotated stop ends, vertically to compensate for gradient and horizontally to cope with skew support lines, are more easily achieved with pre-tensioned girders than with post-tensioned girders, where the tendon's anchorages are fixed against the stop ends.

Figure 19.6 Design bending moment and moment due to PT for post-tensioned and pretensioned girders (VSL International)

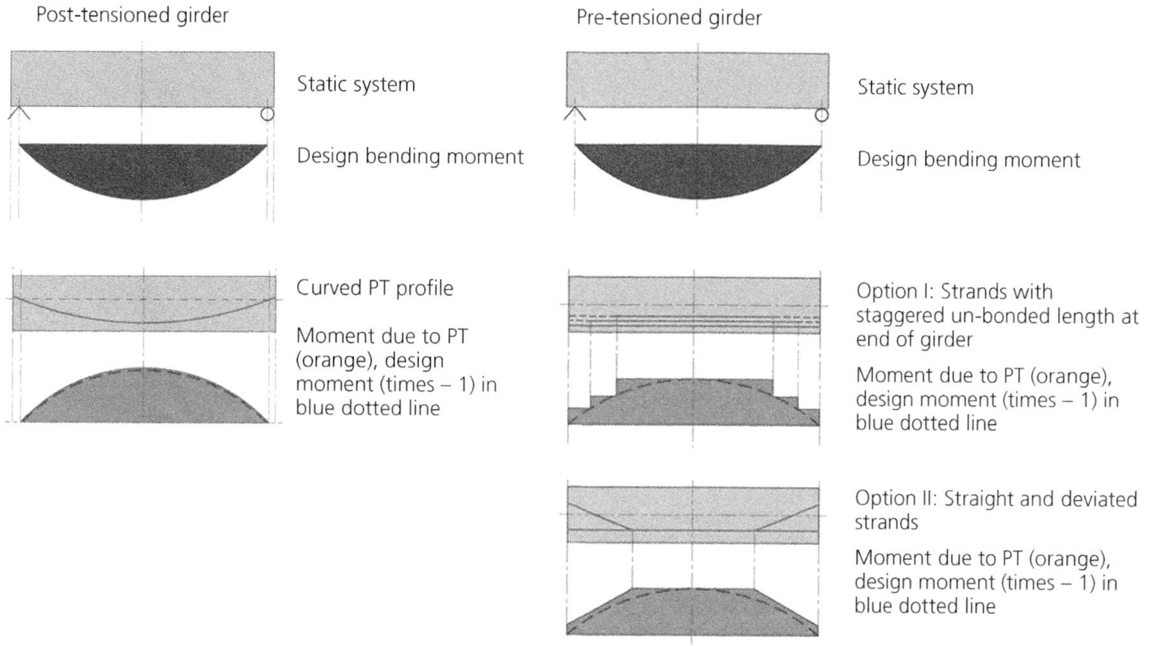

Design and detailing the starter bars at the stop ends (for deck continuity) and for the end and intermediate diaphragms is controlled as much by static forces as by constructability issues. If it is not possible to bend and fix starter bars during production against the form or let them protrude through form skin made up, for example, from expanded metal, rebars should be coupled. For the longitudinal reinforcement of diaphragms, it is possible to cast in sleeves through which these rebars can be slotted.

For girders supported directly on permanent mechanical bearings, typically box-outs have to be provided for the bearing's shear studs, which are formed during production, preferably by steel forms attached to the girder's soffit form. Geometry of these box-outs needs to be sized big enough to cope with variable gradient (the girder follows the gradient, but the bearing is installed truly horizontal). These box-outs also have to keep clear of reinforcement and, most importantly, of the pre-tensioning strands, which, unlike reinforcement, cannot be locally diverted.

19.2.3 Fabrication of reinforcement bar cages (rebar cages)

For pre-tensioned girders, rebar cages are basically always prefabricated to maximise usage of the beds. What has proven to be very efficient is to preassemble all the transverse reinforcement in slices on a template (Figure 19.7). As explained above, when producing post-tensioned girders an option is to provide more soffit forms than side forms and stop ends. For such set-ups, tying reinforcement directly on the soffit form is an option.

19.2.4 Placing and curing of concrete

The different methods used to place the concrete are

Figure 19.7 Preassembly of transverse reinforcement in slices on a template (VSL International)

- pouring of the concrete into the beams directly from ready-mix trucks driving along the girder's formwork
- use of a portal crane to place the concrete delivered by a ready-mix truck with a bucket
- pumping of the concrete with a stationary placing boom into the girders.

To reduce the cycle times, especially when the girders are produced in pre-tensioning beds, the concrete curing is often accelerated by hot-air or steam curing.

19.2.5 Cycle time

The various activities in the casting moulds for the production of post- and pre-tensioned girders are listed in Table 19.2.

For production of a 30-m-long, 700 kN girder, activities 1 to 5 and 7 to 11 in Table 19.2 typically take about 12 hours. Curing may vary from about 12 hours if girders are steam-cured to 36 hours if girders are naturally cured.

19.2.6 Risks

The list of risks given in Table 19.3 is by no means comprehensive, but addresses risks which are particular to precasting of post- or pre-tensioned girders. For a specific project, a main contractor needs to carry out a full risk assessment of the operations performed in the casting yard, considering the actual features of the yard, the formwork, the plant and the detailed construction method adopted.

19.3. Production of precast segments

This subsection deals with the production of bridge decks made up of match-cast precast segments. There are two methods used for the production of match-cast segments, the short-line method and the long-line method.

Table 19.2 Activities in casting mould for the production of post- and pre-tensioned girders

	Activity	Post-tensioned girder	Pre-tensioned girder
1	Cleaning of formwork, application of form oil	X	X
2	Lowering of prefabricated rebar cage together with strands and stop ends	X	X
3	Coupling of strands and stressing one by one with mono-strand jacks		X
4	Closing of side forms	X	X
5	Placing of concrete	X	X
6	Curing	X	X
7	Removal of stop ends/anchorage recess formers, installation of PT anchor blocks	X	
8	Opening of side forms	X	X
9	Load transfer by moving bulkhead with solid ram jacks		X
10	Stressing of PT tendons with multi-strand jacks	X	
11	Cutting of strands and removal of stop ends		X
12	Lifting of girder out of casting mould	X	X

19.3.1 Formwork concepts

The main elements of a short-line cell are shown in Figure 19.8.

The main steps when using a short-line cell for production of match-cast precast segments are given in Figure 19.9.

The main elements of a long-line cell are illustrated in Figure 19.10.

The main steps when using a long-line cell for production of match-cast precast segments are shown in Figure 19.11.

For span-by-span decks of system 3 type construction (for explanation, refer to Section III, Chapter 15), the pier segments are often match-cast against the first and last field segment in a short line cell, with all the field segments produced in a long-line cell.

Special attention needs to be given to the design and detailing of the bulkheads. Smooth segment end faces which fully match when joining the segments in the field are of upmost importance to avoid hard spots and achieve the stress distribution across the joints that the designer has assumed in the design. Smooth segment end faces are also important to get leak tightness for PT ducts crossing the segment joint, a precondition to be able to effectively grout the tendons. The need for regular splicing of the bulkheads should be avoided as much as possible. After some reuses, steps can form along the joint lines where concrete may spall off when breaking the bond between the new and the conjugate segment. Splices in bulkheads can be avoided altogether or minimised significantly by fabricating dedicated sets of bulkhead form panels for different cross sections.

Table 19.4 gives a comparison between segment production using the long-line or the short-line method.

Table 19.3 Risk assessment for the production of precast girders

Activity	Risk	Mitigating measures	Post-tensioned girder	Pre-tensioned girder
Pouring concrete at top of girder	Vertical access and work at height: climbing up and down, work performed at height (at 2 to 3 m above ground)	Access staircase, working platform along side form	X	X
Fixing girder to lifting devices for removal from casting mould				
Lowering of rebar cage and closing of side forms	Cage does not fit into casting mould	Use of template when preassembling rebar cage	X	X
	Cage distorts excessively during handling	Use of rigid spreader frame with multiple suspension points	X	X
	Damaged and dislocated concrete spacers	Use of engineered spacers efficiently tied to rebars	X	X
Tendon ducts	Crushing during concreting	Install strands prior to concreting	X	
Stressing of strands	Failing strands/wedges/couplers with subsequent release of built-up energy in bulkheads and jacks	Exclusion zone behind stressing jacks		X
Stressing of tendons	Failing strands/wedges with subsequent release of built-up energy in anchorages and jacks		X	
Movement of portal cranes	Interference with people and goods crossing portal crane's rail path	Warning lights and acoustic signals	X	X

19.3.2 Permanent work aspects

When detailing the cross section of a bridge deck being produced as match-cast segments in a factory, the consultant needs to not only satisfy the structural requirements, but also consider the production aspects.

When using box-girder cross sections, the minimum depth of the segment should not be less than 2.2 m. There needs to be sufficient vertical clearance inside the segment during production for the cantilever beams of the inner form traveller and for the workers. The outer geometry of the segments, which have constant depth, shall be kept constant for a match-cast run and standardised for a project

Figure 19.8 The main elements of a short-line cell (access not shown) (VSL International)

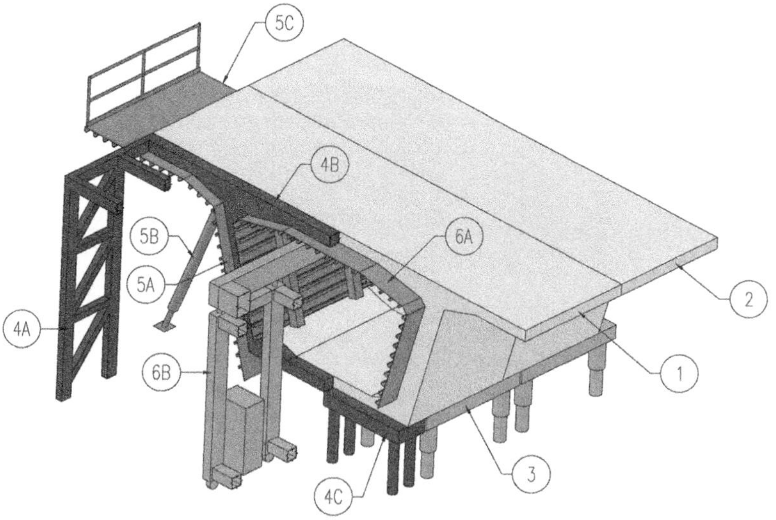

1. New segment
2. Conjugate segment
3. Soffit Form (SF): 3 per cell
4. Bulkhead (BH)
4A. Support frame
4B. Form panels
4C. Stationary part of soffit form
5. Outer Side Form (OSF): in pairs
5A. Form panels
5B. Struts
5C. Access/support for wing slab's stop end
6. Inner Form (IF)
6A. Form panels
6B. Traveller
7. Transfer cart: not shown for clarity

as much as possible. For each different external geometry, a dedicated cell will be required. Variation in the wing slab length of a box girder is easily accommodated if the variation takes place where the deck slab has constant depth. Variation in segment depth by extending the typical cross section downwards often occurs for bridge decks built by Balanced Cantilever Method (BCM). Such variations complicate the formwork significantly. The 'vertical' length of the inner form changes linearly over the segment length. Each segment requires a different bulkhead. The outer forms, however, are not affected. They just need to be deep enough for the tallest segment, since they overlap with the soffit form. Matters get even more complex when the webs are not vertical but inclined. With inclined webs, the width of the bottom slab varies linearly over the length of the segment.

Whereas aesthetic aspects play a part for the outer geometry, the same is not the case for the inner geometry of a box girder. The inner geometry of a box girder segment is rather complex, with a lot of variations (refer to Figure 19.12). The inner geometry shall therefore be optimised to the best compromise between structural efficiency and constructability. Since match-cast moulds are expensive, the time a segment stays in a cell has to be as short as possible. Therefore the full rebar cage of match-cast segments has to be prefabricated outside of the cell. In addition, the inner form for the field segments should be designed in such a way that it can be driven through the bulkhead into the rebar cage and out of the segment in one operation. To be able to achieve this, the inner geometry of a segment needs to be detailed accordingly, that is

- no ribs/beams below the deck slab
- no intermediate diaphragms (but, for example, provision of a pair of internal diagonal struts for load introduction from deck stay anchorages located at the centre of a deck)
- preferably no ribs/beams across the bottom slab
- change in the depth of webs and bottom slab in steps, near to the bulkhead if depth is reducing and near the conjugate face if depth is increasing
- the overall width of top blisters at the corner between the webs and the deck slab in a section should always be less than 1/4 of the clear distance between the webs minus the depth of the

Figure 19.9 Main steps when using a short-line cell for production of match-cast precast segments (for the components of a short-line cell refer to figure 19.8) (VSL International)

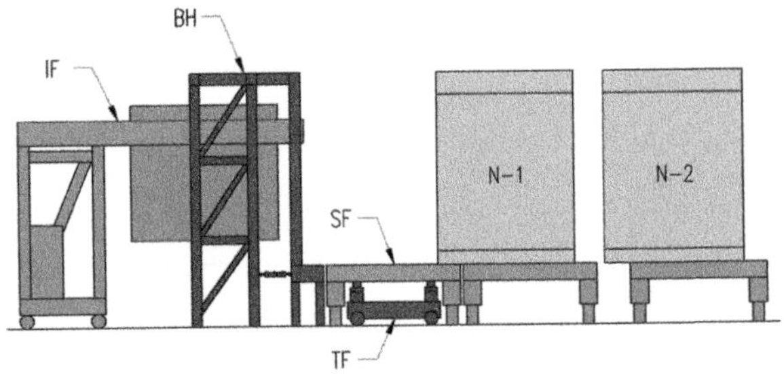

- Segment N – 2 in parking position ready for removal into storage yard
- Conjugate segment N – 1 positioned for casting of segment N
- Soffit form in place and adjusted
- Outer side forms closed

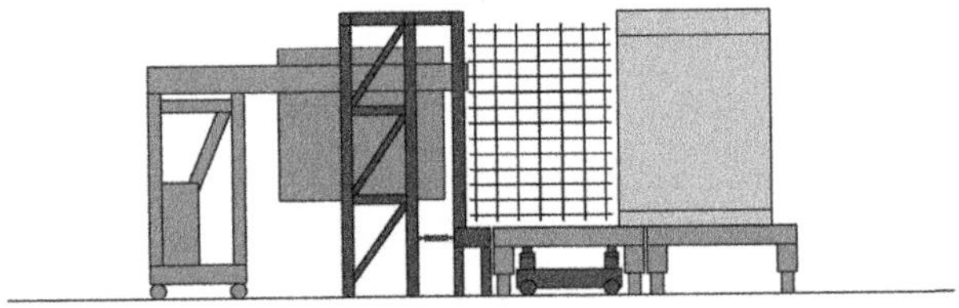

- Lowering of prefabricated rebar cage of segment N

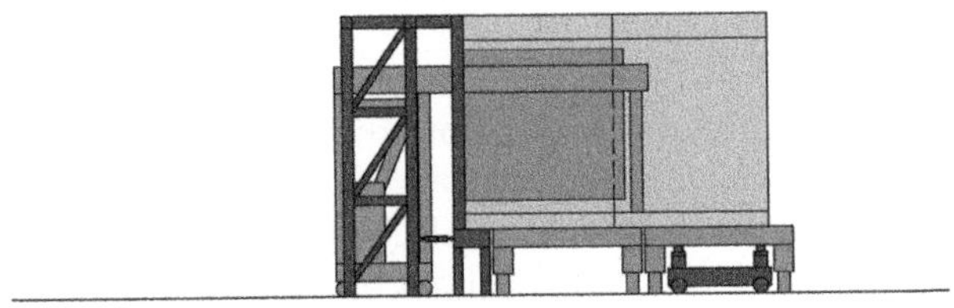

- Moving form traveller with inner form into casting position
- Casting and curing of segment N

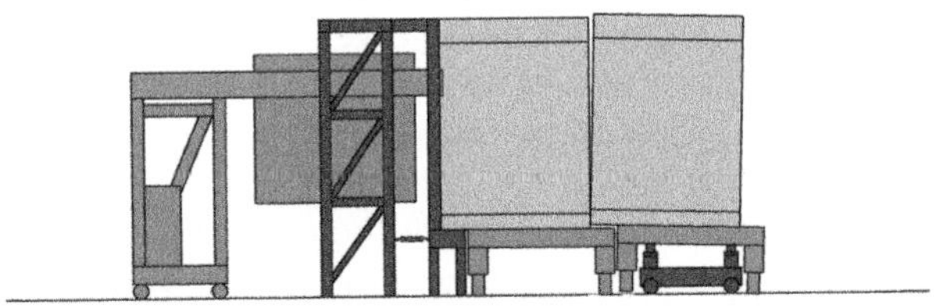

- Stripping outer side forms
- Stripping internal form
- Move traveller with inner form out of segment into parking position
- Breaking of bond between conjugate segment N – 1 and new segment N by activating transfer cart below conjugate segment's soffit form

formwork (this allows the entire side and top blister formwork to move sufficiently transversely to clear the top blister when driven into the rebar cage or out of the segment)

- end faces and top/bottom faces of blisters formed by formwork should be tapered between 1 : 10 to 1 : 5 to allow easy stripping by rotation and/or horizontal translation of the inner side form.

19.3.3 Geometry control

The geometry control starts with defining how the deck is broken down into segments by transverse cuts. The starting point is to establish the geometrical relationship in the deck's cross section between the road's or rail's Setting Out Line (SOL) and the segment's centre line. The casting and the erection methods need to be fixed to know the location of the wet joints, the casting and erection direction and the type of casting cells used (short line; long line with pier segment by short line, etc.). Assuming that the deck's cross section is made up of one box girder only, with zero transverse offset between its centre line and the SOL, the centre lines of the segments are in plan a series of tangents intersecting with the SOL at the segment joints. The bulkhead is located at these intersection points and its orientation is

Figure 19.10 The main elements of a long-line cell (access not shown) (VSL International)

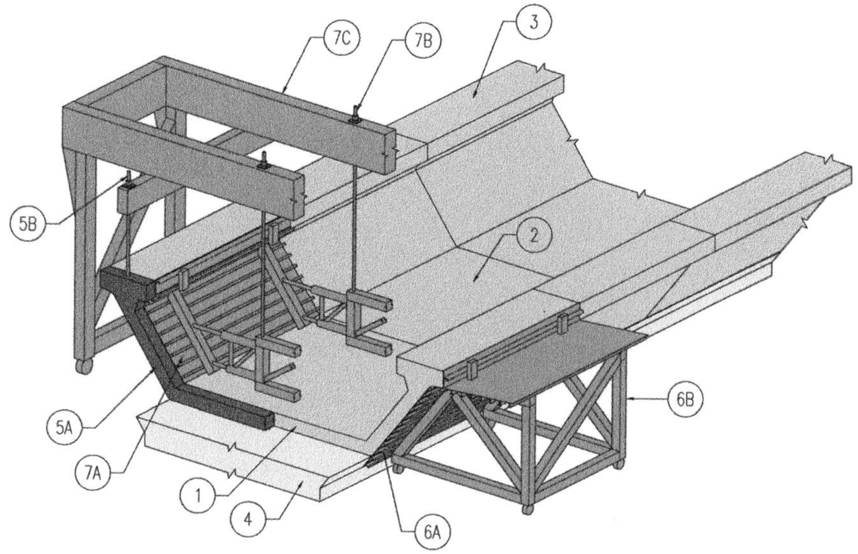

1. New segment (segment N)
2. Conjugate segment (segment N – 1)
3. Segment N – 2
4. Soffit Form (SF): fixed
5. Bulkhead (BH)
5A. Form panels
5B. Suspension system
6. Outer Side Form (OSF): in pairs
6A. Form panels
6B. Traveller
7. Inner Form (IF)
7A. Form panels
7B. Suspension system
7C. Traveller

perpendicular to the segment centre line. A match-cast run generally starts with a segment whose end faces are parallel (permanent and temporary bulkheads are perpendicular to the segment's centre line). For a deck built by BCM, the starting segment for a match-cast run is the pier segment (Figure 19.13).

For a simply supported deck built span by span, the starting segment for a match-cast run cannot be the pier segment, since the gap between the pair of pier segments on a pier generally needs to be constant and perpendicular to the SOL's tangent. Hence one of the first field segments in a span is cast as the starting segment with parallel bulkheads (Figure 19.14).

Segmentation in elevation is controlled by the need to have the pier segment, the pair of pier segments (if the pier table is split into two segments) or the pair of expansion joint segments (over an internal pier) centred at soffit level over the pier.

Short-line casting allows production of segments for decks having any kind of geometrical road or rail alignment in plan and elevation and variation in cross fall, since the conjugate segment can be orientated in the casting cell to any geometrical position against the segment to be cast. This is illustrated in Figures 19.15 and 19.16.

In order to be able to cast segmental decks, the theoretical setting-out geometry for each segment has to be calculated. The casting cell needs to be designed and detailed to allow for all the geometrical variations of the segments with regard to its geometry, taking into consideration

- geometrical adjustment capacity of the transfer cart (vertical; rotation about all three axis) to position the conjugate segment correctly in relation to the segment to be cast
- length of the soffit form and length of its overlap with the conjugate segment (no clash with the soffit form of the conjugate segment for any combination of segment length between the new and the conjugate segment and/or for curvature in plan)

Figure 19.11 Main steps when using a long-line cell for production of match-cast precast segments (for the components of a long-line cell refer to figure 19.10) (VSL International)

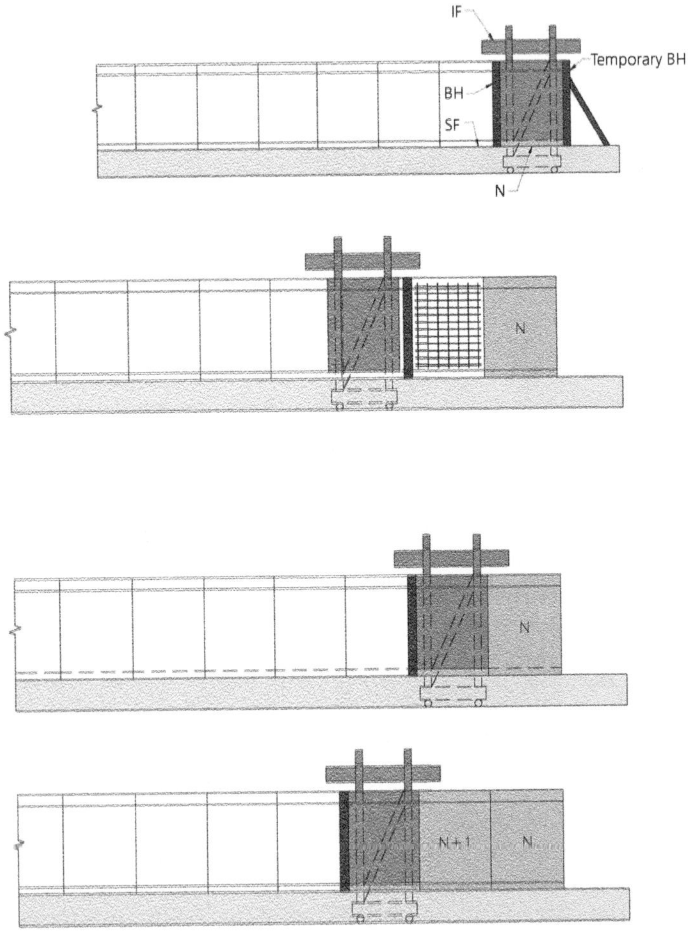

- Casting of first segment N with temporary bulkhead

- Stripping of outer side form, inner form and bulkhead
- Moving bulkhead with inner form traveller to casting position of segment N + 1; parking of bulkhead
- Moving and parking of inner form traveller in front of casting position for segment N + 1
- Moving outer side form travellers into casting position for segment N + 1
- Lowering of prefabricated rebar cage of segment N + 1

- Moving inner form traveller into casting position for segment N + 1
- Casting and curing of segment N + 1

- Match-casting of segment N + 2 against segment N + 1

- length of the outer side form and length of its overlap with the bulkhead and the conjugate segment, which is different on the inner and outer side for decks curved in plan (for tight curvature in plan the outer side forms can be detailed to be longitudinally movable)
- length of the inner form and length of its overlap with the conjugate segment (no clashes with internal blisters for any combination of segment length between the new and the conjugate segment and/or for curvature in plan).

The setting-out dimensions have to be transferred into the coordinates of the casting cell's local coordinate system. Since only the geometry of the deck really matters for the bridge's final geometry, which needs to be achieved contractually, control points are all placed on the deck slab. Six control points – with four located over the webs to omit influences due to local deflection of the deck for level control and two located at the segment's centre line for transverse offset control – are still used widely; even so, with the technical development of total station theodolites, four points are sufficient (refer to Figure 19.17).

Table 19.4 Comparison between segment production using either the long-line or the short-line method

	Long line	Short line
Alignment in plan and elevation	Straight	No restriction
Cranage	Heavy-duty gantry over production bay	No need for heavy-duty gantry over production bay
Geometry control	Simple: defined by setting out of the bed and by the level of the movable bulkhead	Complex: need to modify setting-out geometry for casting of segment N on basis of as-cast geometry of previously cast segments 1 to N – 1
Ideal applications	■ U-girder (always straight) ■ Haunched box girder with straight alignment	For any type of girder with variable alignment

Control points need to be robust and to provide a clear punch mark so that there is no ambiguity about where to place the measuring tool. Accurate setting out of the control points against the bulkhead and the downstream control points of the conjugate segment is preferred, since there is no need to record the setting errors of the control points. Such recording is necessary when ordinary bolts are used for the level control points, which are pushed manually into the wet concrete only after completion of the concreting.

After a new segment is cast but before removal of the conjugate segment, the two segments are surveyed to record the as-cast geometry. The as-cast geometry of a segment is always different from the theoretical geometry of the segment by the so-called casting errors, because

- it is impossible (and not at all necessary) to position the conjugate segment with zero tolerance

Figure 19.12 Isometric view of a box-girder segment showing the complex inner geometry and the geometry of the match face (adapted from FDOT's Segmental Manual)

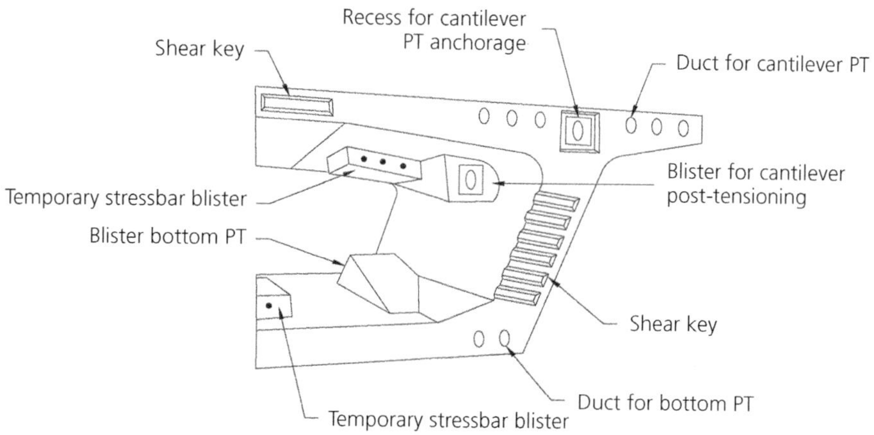

Figure 19.13 Example of segmentation in plan for a bridge deck built by BCM with casting direction from centre line of pier towards centre of span; pier segment is cast in a special pier cell as a square block (VSL International)

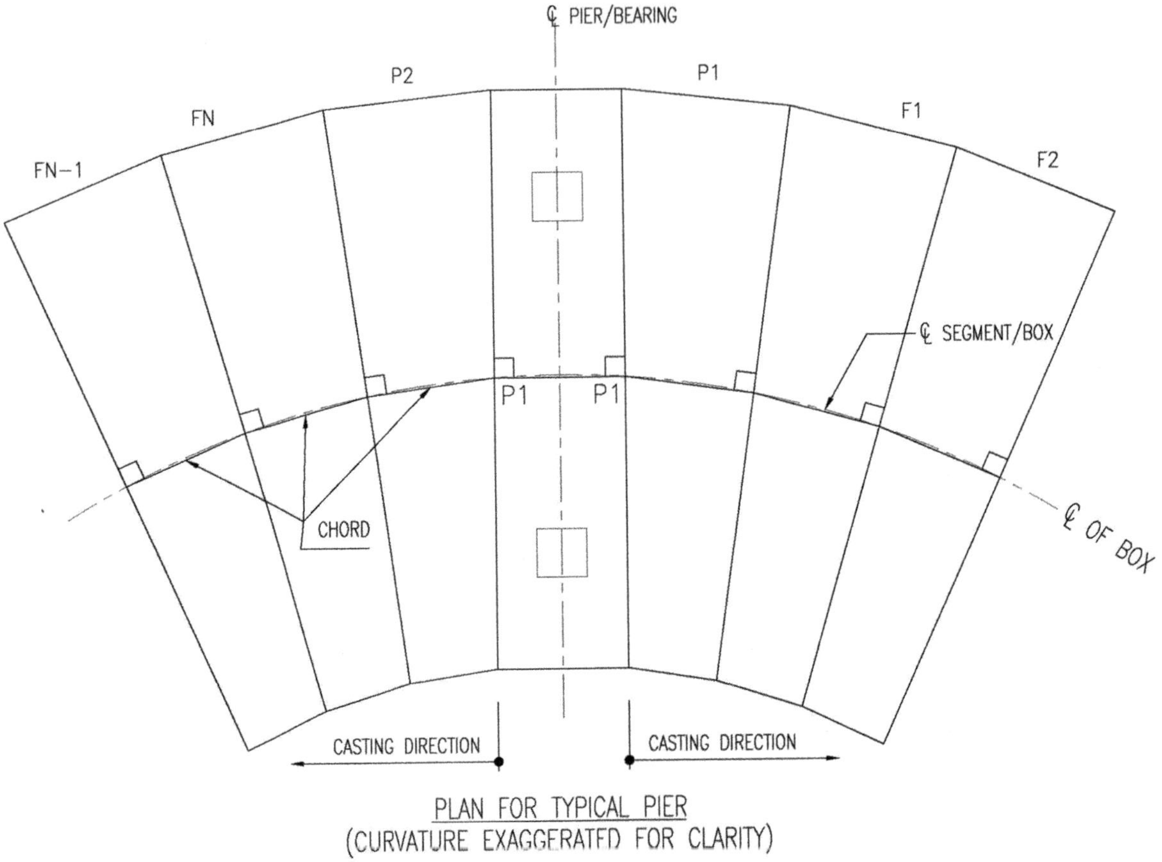

- during the concreting, the cells' formwork – which is not indefinitely rigid due to tolerances in pin holes etc and is elastically deforming – and the conjugate segment move slightly.

Casting errors are unavoidable and are not bad per se (even though they are called errors), as long as they are not ignored by the site team but their magnitude and direction is duly recorded. This allows modification of the theoretical setting-out position of the next segment to be match-cast.

The geometry control for segmental construction is all about mastering the relative geometrical relationship between the match-cast segments. Hence settlement in casting yards is acceptable as long as the casting cell is on a rigid foundation slab, which if it settles moves as a solid body. If the geometry control is done by measuring the global coordinates of the control points with a total station theodolite, one survey tower may be able to cover several cells. Its position is not very critical (verticality, relationship against the cell centre line), since whenever in cell N a conjugate and/or new segment need to be surveyed, the position of the instrument is determined against the actual position of cell N's bulkhead at deck slab level (reference control points on the bulkhead). The only thing which then really matters is the rigidity of the survey tower. The survey tower has to be rigid enough so that it does

Figure 19.14 Example of segmentation in plan for a simply supported bridge deck, where end faces of spans have to be parallel and orthogonal to SOL's tangent at centre of pier: the field segment F1 to FN are cast in one match-cast run in a field cell; F1 segment is cast as a square block; pier segment P1 is match-cast against field segment F1 and pier segment P2 is match-cast against field segment FN in a special pier segment cell (VSL International)

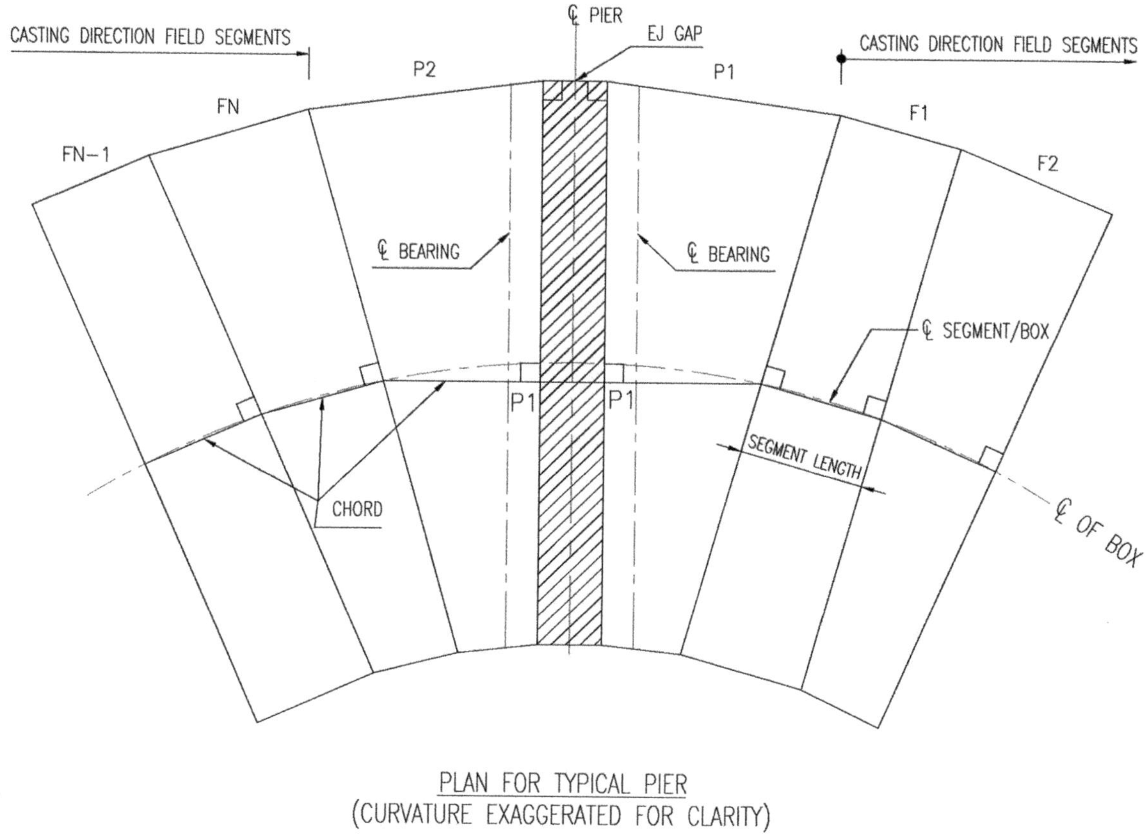

not vibrate due to activities in the yard and/or due to wind during the time the measurements for the conjugate and/or the new segment are taken in a particular cell.

The segments can be cast by the classical long-line method, where the soffit form is straight and fixed, when the segments produced in one casting run are straight in plan and elevation. Of course it is possible to use long-line cells also for decks with modest variation in geometry if the long line's soffit form is designed and detailed in such a way that its geometry can be modified for different casting runs. Whatever method is used, the deck's geometry is defined by the geometry of the soffit form. Even when casting segments by the long-line method, it is still necessary to use control points at the deck level to control the accumulated length of the segments cast in one run and to record the as-cast geometry of the deck. The as-cast geometry of the deck is needed for setting out the segments in the field.

19.3.4 Fabrication of reinforcement bar cages (rebar cages)

As mentioned above, the prefabrication of the rebar cages is a prerequisite to minimise the cell production cycles and therefore to reduce the number of cells required on a project. The rebar cages

Figure 19.15 Short-line cell: orientation of conjugate segment in plan to achieve a horizontal curvature of the deck and in longitudinal elevation to achieve a vertical curvature of the deck (adapted from FDOT's Segmental Manual)

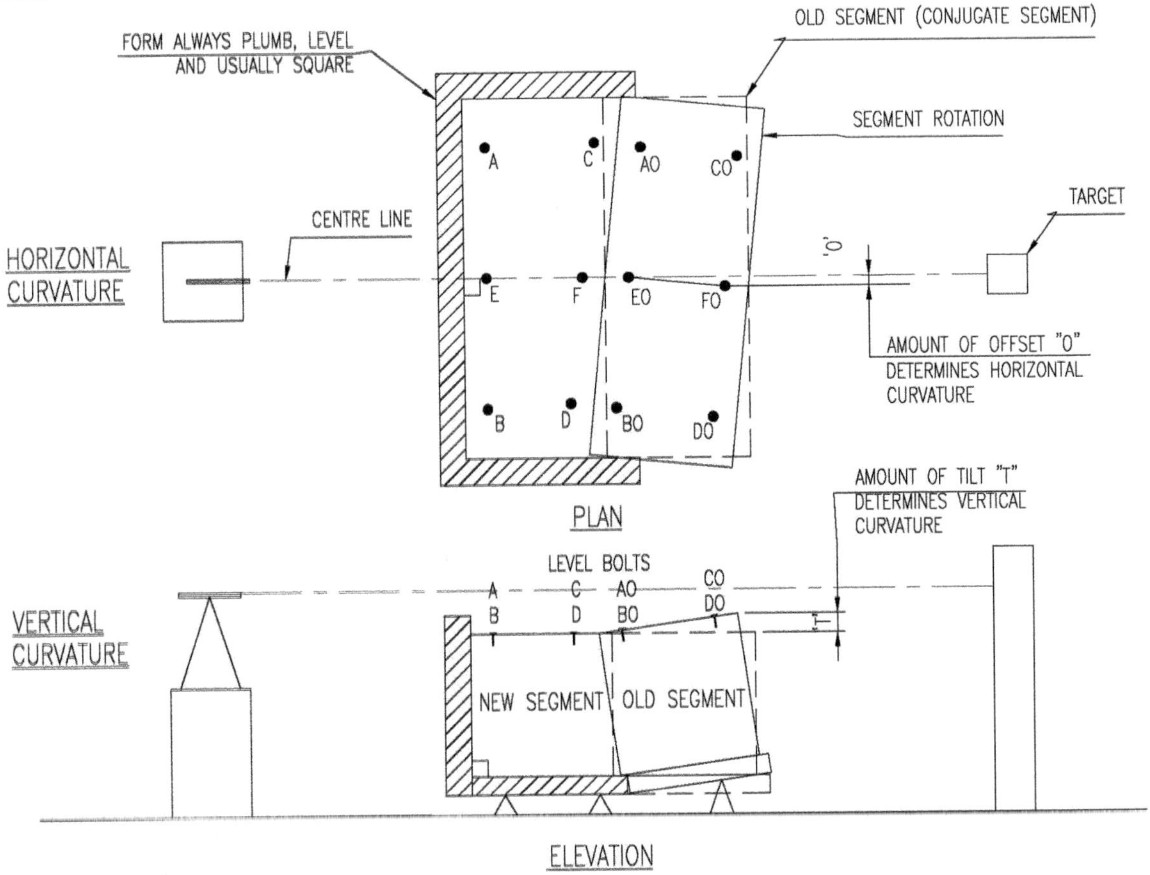

are fabricated in jigs to ensure they are fabricated within tolerances and fit into the cell with the specified edge distances to the segment's inner and outer faces to achieve the required concrete cover. The hardware of the ducts for the internal PT is preferably detailed so that the ducts can be positioned correctly and tied firmly into the rebar cage already in the jig. Once the rebar cage has been lowered into the cell, the longitudinal gaps between the ducts and the bulkhead and the ducts in the conjugate segment have to be bridged. In addition, the ducts need to be fixed at the segment's faces so that they do not move during the concreting. This can be achieved by the use of inflatable rubber hoses and by duct locators at the bulkhead. The inflatable hoses also stiffen the ducts so that they are not squashed during the concreting. Surprisingly, it is only recently that the coupling of the internal ducts across match-cast joints with segmental duct couplers has started to become a requirement. Without couplers ducts are not leak tight. The protection level as defined by *fib*, the International Federation for Concrete, is only Protection Level 1 (PL1). For illustration refer to Figure 19.18.

A very effective method to produce accurate rebar cages is the slice method (Figure 19.19). This method can only be used if the specifications allow tag welding between rebars. The rebar cage is

Figure 19.16 Short-line cell: rotation of the conjugate segment to achieve a change in cross fall of the deck over the length of the segment (adapted from FDOT's Segmental Manual)

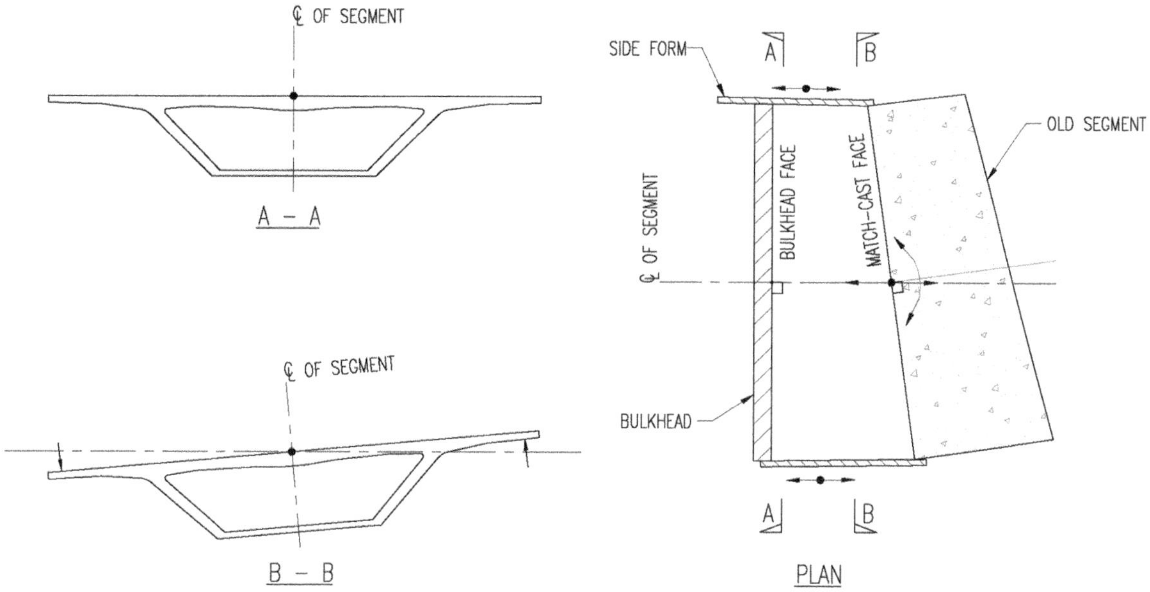

broken down into transverse and longitudinal rebars. The transverse rebar is detailed in such a way that each set of transverse reinforcement in the deck and bottom slab can be combined together with a set of the web's 'vertical' reinforcement into a transverse rebar slice. These rebar slices are pre-assembled on horizontal templates and lifted and spread out in a simple rebar jig. The rebar cage is completed in this jig by the installation of the longitudinal reinforcement and the PT ducts.

19.3.5 Placing and curing of concrete

The different methods used to place the concrete are

- use of a portal crane to place the concrete delivered by a ready-mix truck with a bucket
- pumping of the concrete with a mobile concrete pump into the segments
- pouring of the concrete into the segments directly from a ready-mix truck with/without a conveyor belt.

To reduce the cycle times, especially in cold countries, concrete curing is often accelerated by hot-air or steam curing.

19.3.6 Cycle time

A typical casting cycle for a field segment of constant depth and a weight of 700 kN is shown in Table 19.5. It is important to note that the cycle shown is for illustration only and that actual cycles need to be worked out for individual projects considering project-specific features and constraints.

19.3.7 Risks

The list of risks given in Table 19.6 is by no means comprehensive, but addresses risks which are particular to precasting of segments in short- or long-line moulds. For a specific project, a main contractor

Figure 19.17 Short line: traditional geometry control in the casting yard with 6 control points (4 for levels, 2 for transverse offset) and a set of survey tower and target per casting cell; the bulkhead with the all-important reference control points is not shown for clarity; also not indicated is the fact that the length of the segment is controlled by measuring distances between corresponding control points over the webs (adapted from FDOT's Segmental Manual)

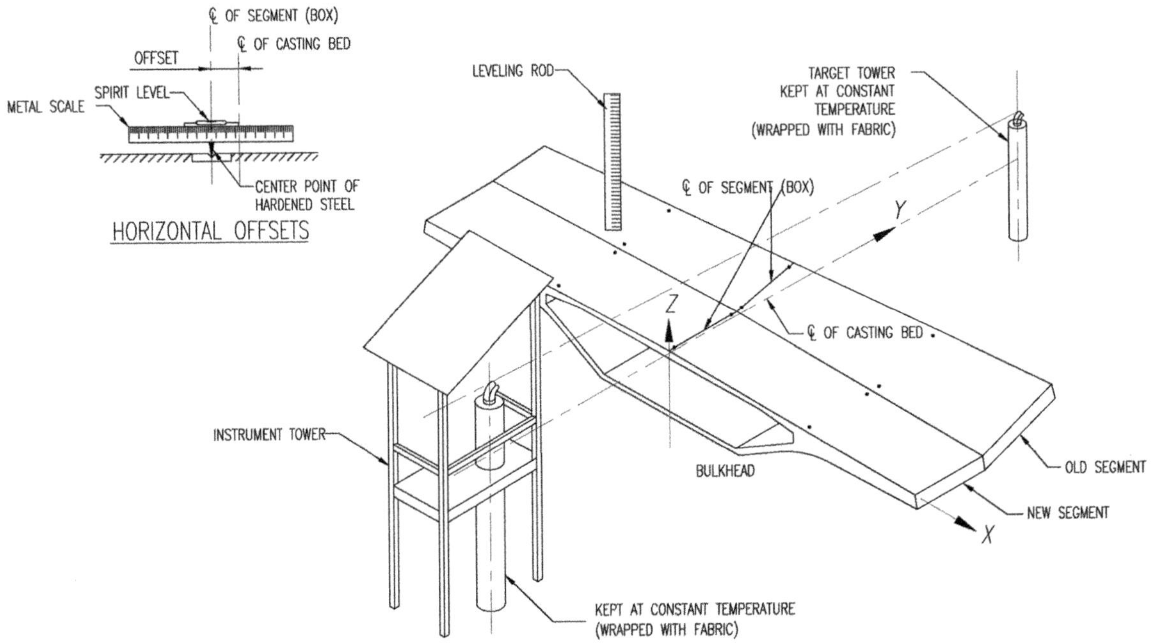

Figure 19.18 Inflatable rubber hose, duct locator and segmental duct coupler (VSL International)

Inflatable rubber hose

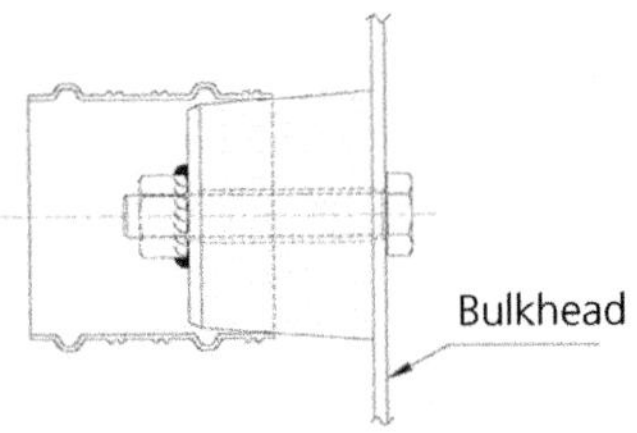

Typical duct locator

Duct locator at bulkhead

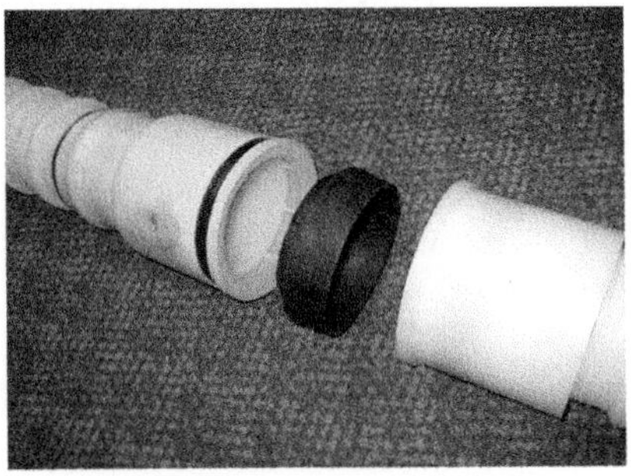

Segmental duct coupler at match face

needs to carry out a full risk assessment of the operations performed in the casting yard, considering the actual features of the yard, the formwork, the plant and the detailed construction method adopted.

19.4. Production of fully precast spans (Full Span Method/FSM)

This subsection covers production of fully precast spans (PS) for deck construction by the Full Span Method (FSM). PS have between 5 to 15 m wide deck slabs, are between 15 m to 70 m long and have

Figure 19.19 Preassembly of the rebar slices on a template (VSL International)

weights between 1000 kN and 15 000 kN. Even the reinforcement bar cages (rebar cages) of such girders, which need to be prefabricated to cut down on the number of cells, weigh between 50 kN and 1000 kN. One of the technical challenges when producing PS is to handle the heavy rebar cages from the jig to the formwork and the PS from the formwork to the storage yard and to the loading bay for transport to the erection front. Two types of formwork concepts are used for the production of pre-tensioned PS. For type 1 the soffit form is fixed, whereas for type 2 the soffit form is movable.

19.4.1 Formwork concepts

Pre-tensioned PS need to be produced in an assembly line, since they can only be cast inside the pre-tensioning bed. The above-mentioned two basic formwork concepts used for such assembly lines differ from each other with regard to the soffit form and the movement of the PS within the assembly line. The main elements of these formwork concepts are explained in Figure 19.20.

Basic steps when using the formwork concept type 1, where soffit form is fixed and spans and rebar cages are always moved by portal cranes, are shown in Figure 19.21.

Basic steps when using the formwork concept type II, where soffit form is movable and movement of rebar cage and some of the span movements are done with the soffit form, are given in Figure 19.22.

Comments made about the design and detailing of the pre-tensioning beds in subsection 19.2 ('Production of precast girders') are also applicable for the pre-tensioning beds of PS. Which formwork concept is best for a project depends on many factors. With concept 2, the newly cast span does not need to be lifted off its soffit form immediately after initial curing, as it does with concept 1, but only after an additional cycle.

Table 19.5 Typical cycle for a field segment mould (constant depth, segment weight 700 kN)

Activity	Duration: h	12 am	2 am	4 am	6 am	8 am	10 am	12 pm	2 pm	4 pm	6 pm	8 pm	10 pm
Stripping of deck slab's stop ends; partial stressing of deck slab PT	1				X								
Stripping of outer side forms; stripping of inner form and moving of inner form traveller out of cast segment; breaking bond between new and conjugate segment by tilting of transfer cart	2				X	X							
Moving of conjugate segment with transfer cart into parking position; moving of new segment with transfer cart into conjugate position	1					X							
Lower third set of soffit form with portal/tower crane into casting position; adjust setting-out geometry of conjugate segment; engage soffit form against bulkhead and conjugate segment; close outer side forms; clean form, seal joints and apply form oil	2						X						
Lower rebar cage with portal/ tower crane	0.5							X					
Drive inner form traveller into casting position; close inner form	0.5							X					
Engage PT tendon's duct locators to bulkhead and inflate rubber hoses to stiffen ducts and fix their position against the conjugate segment; fix void formers (for segment lifting holes, diabolos, etc)	2.5							X	X				
Fix geometry control hardware; fine-adjust setting-out position of conjugate segment	0.5								X				
Pour concrete and trowel-off bottom slab's and deck slab's top surface	2									X			
Cure concrete	12	X	X	X							X	X	X

Table 19.6 Risk assessment for the production of precast segments

Activity	Risk	Mitigating measures	Short-line mould	Long-line mould
Work on top of segment in mould and at finishing area (second-stage stressing and grouting of transverse tendons etc.)	Vertical access and work at height: climbing up and down, work performed at height (at >2 m above ground)	Access staircases, working platforms along side forms, working platforms at finishing area, vertical access provisions in storage area (single and double stacking)	X	X
Fixing segment to lifting devices for removal from casting mould (long line) and for storage (single and double stacking) and loading; placing and removal of sandbags (or similar) used as segment support points for double stacking				
Lowering of cage and closing of inner form	Cage does not fit into casting mould	Use of template when preassembling rebar cage	X	X
	Damaged and dislocated concrete spacers	Use of engineered concrete spacers effectively tied to rebars	X	X
Tendon ducts	Crushing of tendons during concreting	Use of inflatable hoses	X	X
	Dislocation at BH	Duct locator	X	X
	Dislocation at conjugate segment	Use of inflatable hoses	X	X
Concreting	Movement of conjugate segment during concreting	Connection between soffit forms and connection to bulkhead	X	
		Removal of segment N – 2 only after casting of segment N; BH fixed to soffit form		X
	Movement of outer side form at bottom	Tie bars to soffit form	X	X
	Movement of inserts (lifting cones etc.)	Use of temporary fixity beams/ frames which can be firmly secured against BH and conjugate segment	X	X
Geometry control	Dislocation of geometry control points	Use of engineered control points with an engineered fixity system to BH and conjugate segment respectively	X	X
	Wrong survey readings	Back-up reading	X	X
Stressing of tendons	Failing strands/wedges with subsequent release of built-up energy in anchorage components and jacks	Exclusion zone behind stressing jacks	X	X
Movement of portal cranes	Interference with people and goods crossing portal crane's rail path	Warning lights and acoustic signals	X	X
Movement of trucks along beds	Collision with people working around casting moulds	Dedicated driveways and dedicated access walkways	X	X

Figure 19.20 The two basic formwork concepts used to produce pre-tensioned PS (VSL International)

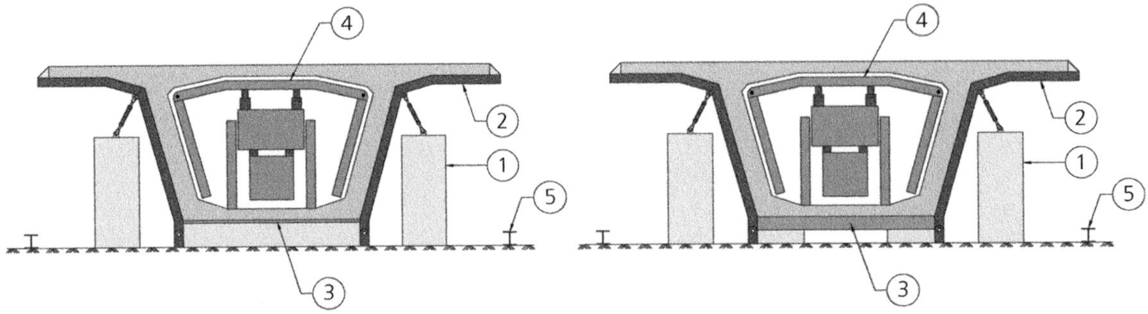

Formwork concept type 1

1. Compression prop of the pre-tensioning bed
2. Outer side form
3. Fixed soffit form
4. Movable inner form
   (self-launching tunnel form with launching noses)

Formwork concept type 2

1. Compression prop of the pre-tensioning bed
2. Outer side form
3. Movable soffit form (by sliding)
4. Movable inner form
   (self-launching tunnel form with launching noses)

Of course, as with precast girders, PS can also be prestressed by post-tensioning tendons or by a combination of pre- and post-tensioning tendons. The latter option is of particular interest. The transfer problem is significantly minimised, because the pre-tensioning, applied at early stage and without profile to simplify the bed, only has to take the self-weight of the span.

19.4.2 Permanent work aspects

As explained in Section III, Chapter 15, the detailing of the box girder's inner geometry should ensure that the inner form can be a self-launching tunnel form.

Comments made about the prestressing in subsection 19.2 ('Production of precast girders') are also applicable for the prestressing of fully precast spans.

19.4.3 Fabrication of reinforcement cages (rebar cages)

Rebar cages for PS are basically always prefabricated to maximise usage of the moulds. Since such cages are heavy and big, decent lifting gear and spreader beams are needed to move such cages from the preassembly area into the moulds.

19.4.5 Placing and curing of concrete

Different methods are used to place the concrete, depending on the cross section and size of the span to be cast.

- For any kind of cross section: pumping of the concrete with stationary placing booms into the spans.
- For open type of cross sections:
  - pouring of the concrete into the span directly from ready-mix trucks driving along the span's mould
  - use of portal cranes to place concrete delivered by ready-mix trucks with buckets.

To reduce the cycle times, especially when spans are produced in pre-tensioning beds, the concrete curing is often accelerated by hot-air or steam curing.

Figure 19.21 Basic steps when using the formwork concept type 1 (VSL International)

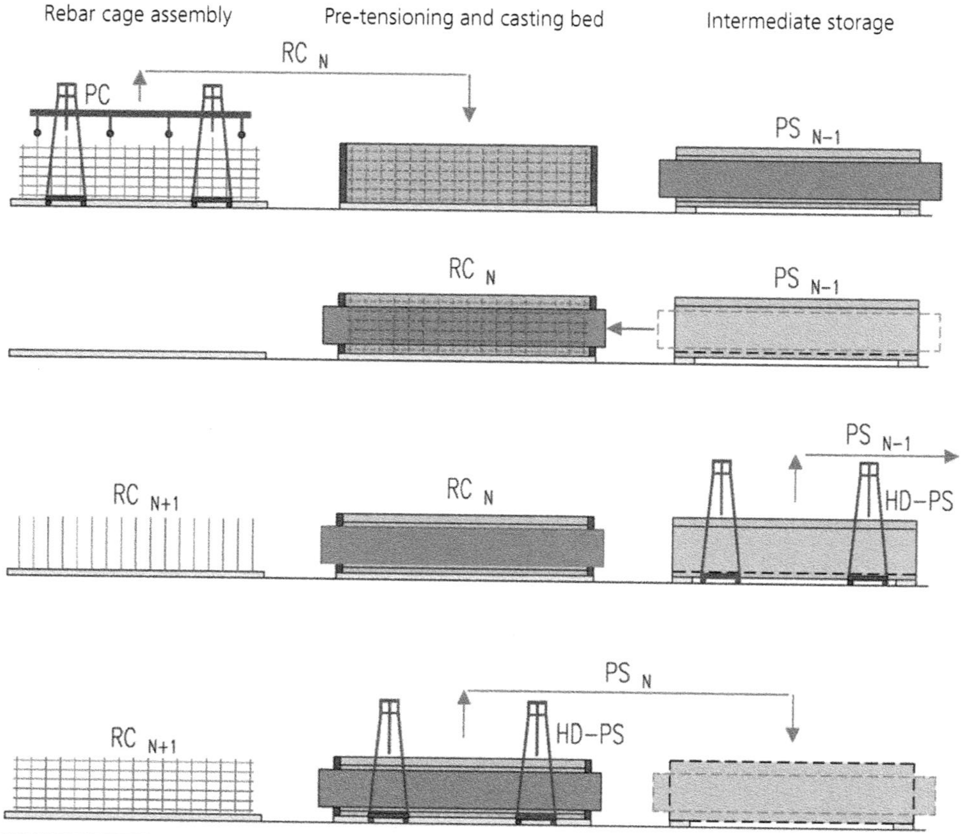

- Placing of rebar cage RC_N by pair of portal cranes (PC) into casting cell

- Launching of inner form out of PS_{N-1} into rebar cage RC_N

- Installation of bulkheads and stressing of pre-tensioning strands (not shown in sketch)
- Casting and curing of PS_N
- Removal of PS_{N-1} from intermediate storage by pair of heavy-duty portal cranes (HD–PS) into storage yard

- Load transfer
- Removal of PS_N together with the inner form from casting bed into the intermediate storage by pair of heavy-duty portal cranes (HD–PS)

Figure 19.22 Basic steps when using the formwork concept type 2 (VSL International)

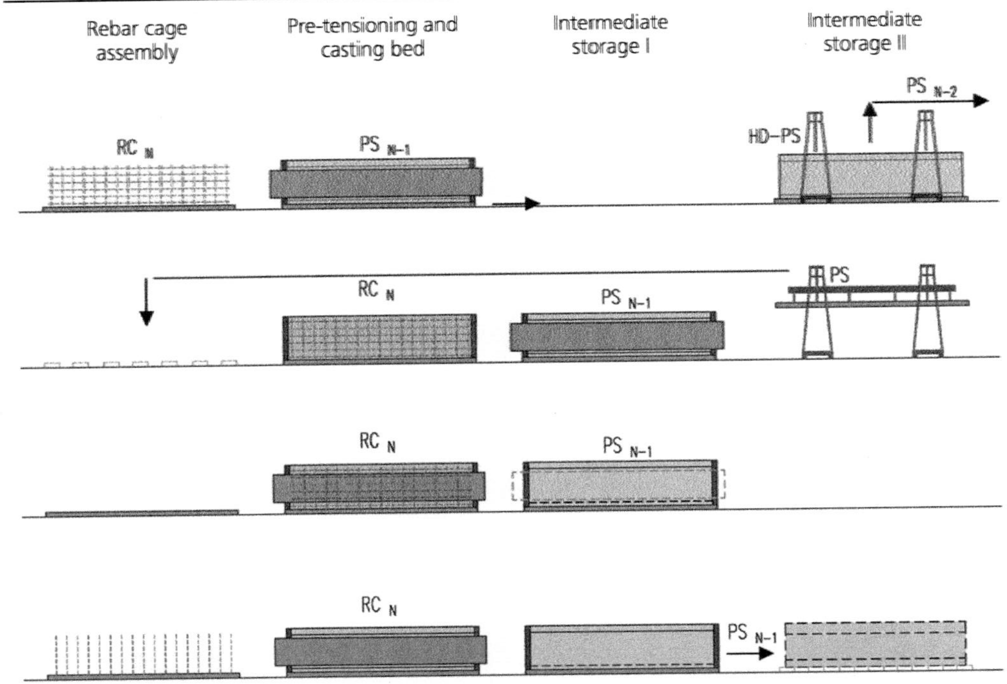

- After load transfer for PS_{N-1}, moving of rebar cage RCn into casting bed and PS_{N-1} into the intermediate storage I by sliding the soffit forms
- Removal of PS_{N-2} by pair of heavy-duty portal cranes (HD–PS) from intermediate storage II to storage yard

- Moving soffit form from intermediate storage II into rebar cage assembly area

- Launching of inner form from PS_{N-1} into rebar cage RC_N

- Installation of bulkheads and stressing of pre-tensioning strands (not shown in sketch)
- Casting and curing of PS_N
- Removal of PS_{N-1} from intermediate storage I by sliding of soffit form into intermediate storage II

19.4.6 Cycle time

A typical casting cycle for production of a fully precast span of 30 m is shown in Table 19.7. The span has a deck width of 12 m and a self-weight of 5600 kN. It is prestressed with a combination of straight pre-tensioning and parabolic post-tensioning strands, with pre-tensioning sufficient for the self-weight

Table 19.7 Cycle of 30 hours for production of a PS of 30 m with a deck width of 12 m and a self-weight of 5600 kN

Activity	Duration: h	Day 1 – 12 am to 12 pm						Day 1 – 12 pm to 12 am						Day 2 – 12 am to 12 pm					
		12	2	4	6	8	10	12	2	4	6	8	10	12	2	4	6	8	10
Closing of side forms; cleaning of form skin; application of form oil	2				X														
Lowering of prefabricated rebar cage together with strands and stop ends	0.5					X													
Launching of inner form from previously cast PS N – 1 into rebar cage of PS N; closing of inner form	1					X													
Installation of downstream bulkhead (which obstructs movement of inner form)	0.5					X													
Coupling of strands and stressing one by one with mono-strand jacks	2						X												
Placing and finishing of concrete	5.5							X	X	X									
Curing	12									X	X	X	X	X	X	X			
Stripping of outer side form; stripping of inner form	1																X		
Load transfer by moving bulkhead with solid ram jacks	1																X		
Cutting of strands and removal of stop ends	2																	X	
Lifting of PS out of casting mould	0.5																		X

of the span. It is produced with a formwork concept type 1. The curing is accelerated by steam curing. It is important to note that the cycle shown is for illustration only and that actual cycles need to be worked out for individual projects, considering project-specific features and constraints.

19.4.7 Risks

The list of risks given in Table 19.8 is by no means comprehensive, but addresses risks which are particular to production of PS. For a specific project, a main contractor needs to carry out a full risk assessment of the operations performed in the casting yard, considering the actual features of the yard, the formwork, the plant and the detailed construction method adopted.

Table 19.8 Risk assessment for the production of fully precast spans

Activity	Risk	Mitigating measures
Work on top of span in mould and at finishing area (second-stage stressing and grouting of transverse tendons if any etc.)	Vertical access and work at height: climbing up and down, work performed at height (at >2 m above ground)	Access staircases, working platforms along side forms, working platforms at finishing area, vertical access provisions in storage area
Fixing span to lifting devices for removal from casting mould and for handling in storage area		
Lowering of cage and closing of inner form	Cage does not fit into casting mould	Use of template when preassembling rebar cage
	Cage distorts excessively during handling	Use of stiff lifting frame with multiple suspension points for placing of deck slab and bottom slab reinforcement
	Concrete spacers get damaged and dislocated	Use of engineered spacers effectively tied to rebar
Stressing of strands	Failing strands/wedges/couplers with subsequent release of built-up energy in bulkheads and jacks	Exclusion zone behind stressing jacks
Inner tunnel form	Squashing workers during operation, operation in confined space during stripping	Acoustic signals during operation and visual controls; ventilation
	Concreting in confined space	Ventilation
Movement of portal cranes	Interference with people and goods crossing portal crane's rail path	Warning lights and acoustic signals
Movement of trucks along beds	Collision with people working around casting moulds	Dedicated driveways and dedicated access walkways

19.5. Production of incrementally launched decks

This subsection covers the precasting of bridge decks moved into final position by incremental launching (Incremental Launching Method/ILM).

19.5.1 Formwork concepts

The two basic types of outer form concepts for precasting bridge decks which are incrementally launched are shown in Figure 19.23.

For the outer form concept type 1, the soffit form has a fixed part underneath the webs and a vertically movable centre part. The outer side forms are rotated about their base support for stripping. When the deck is launched out of the cell, it slides within the casting cell on the fixed part of the soffit form.

For the outer form concept type 2, the entire outer form (soffit and outer side form) is mounted on a support platform, which is vertically movable. When the deck is launched out of the cell, it slides on top of temporary bearings (typically one pair inside the cell and one pair immediately in front of the cell).

Also in Figure 19.23 the two options to cast the cross section are presented. The cross section is either cast in one pour or in two pours. If cast in two pours, the bottom slab and the webs are cast first and the deck slab is cast in a second pour. Both options are possible for any of the two types of outer formwork concepts. Casting a cross section in one pour will reduce the production cycle provided the inner

Figure 19.23 The two basic types of outer formwork concepts for precasting bridge decks which are incrementally launched (VSL International)

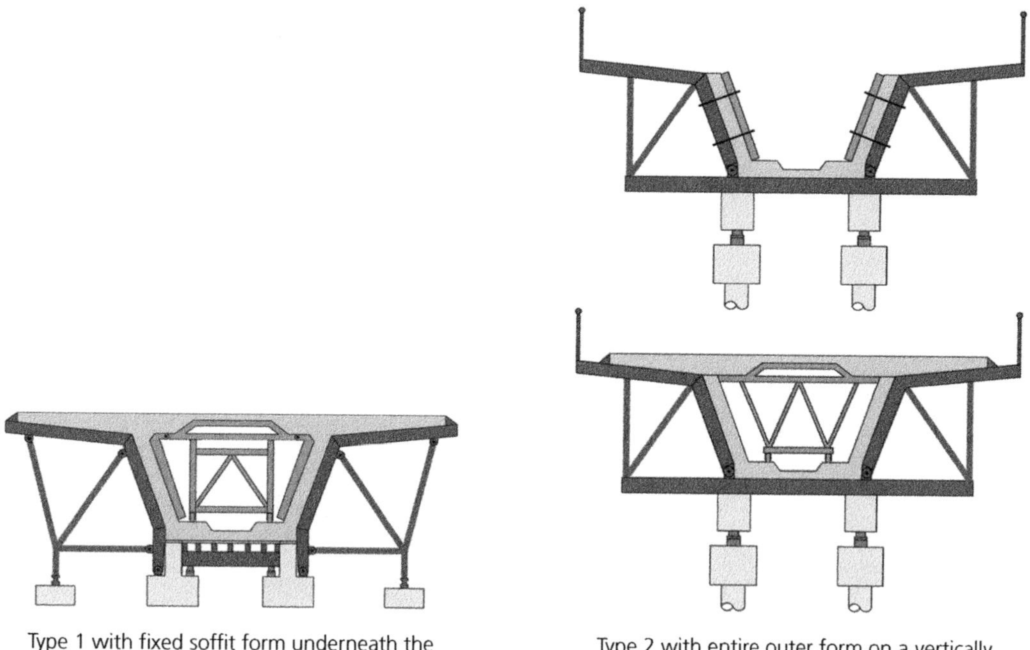

Type 1 with fixed soffit form underneath the webs. In this example the cross section is shown to be cast in one pour

Type 2 with entire outer form on a vertically movable support platform. In this example the cross section is shown to be cast in two stages

Figure 19.24 The main elements of the form concept type 1 with the cross section cast in one pour (VSL International)

1. Centrical PT – tendon ducts in top slab
2. Centrical PT – tendon ducts in bottom slab
3. Base beam and slab
4. Movable part of soffit form noses)
4.1. Support jacks
5. Fixed part of soffit form
6. Outer side form
7. Inner tunnel form

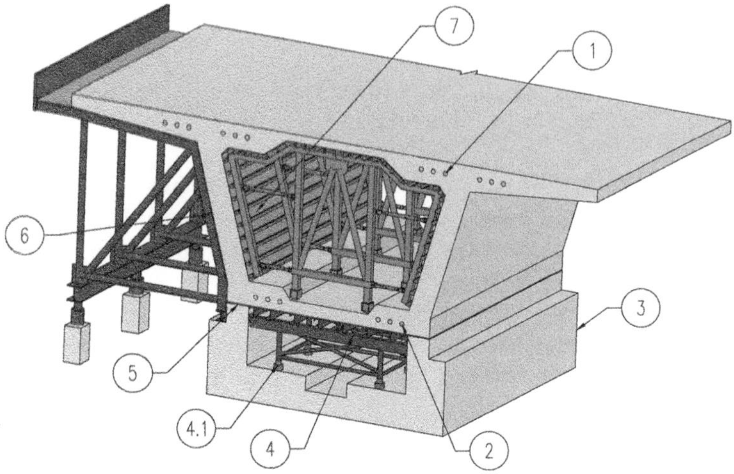

form can be detailed as a tunnel form. To be able to make use of a tunnel form, the segment's inner geometry must be free from oversized blisters and diaphragms must be cast at the front end of a segment. For more information about this topic refer, to Section III, Chapter 15.

The main elements of the form concept type 1 are given in Figure 19.24.

The main elements of the form concept type 2 are given in Figure 19.25.

Figure 19.25 The main elements of the form concept type 2 with the cross section cast in one pour (VSL International)

1. Base slab
2. Support beam in concrete (vertically movable)
2.1. Support jacks
3. Support columns (with temporary bearings)
4. Support platform with outer form
5. Inner tunnel form

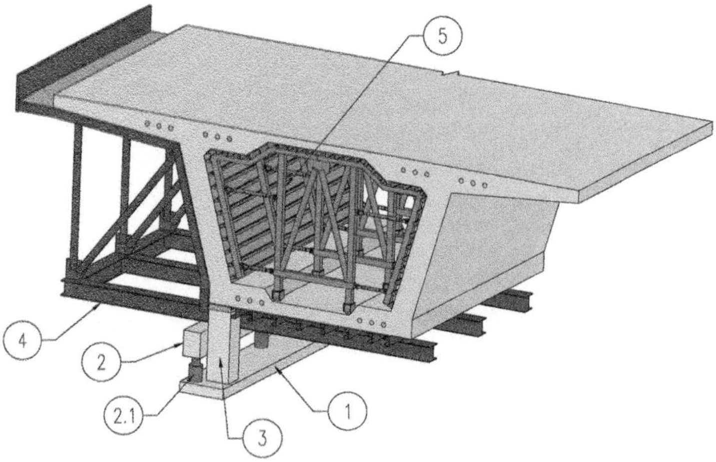

Which type of form concept is the right choice for a given project depends on many factors, but primarily on the way the deck is launched. If the launching is achieved by pushing the deck with solid ram jacks from behind, only the type 1 form concept can be used, since the jacks push against recesses in the beams supporting the fixed part of the soffit form. If the deck is launched with strand lifting units or with lift/push jacks installed at the abutment, both types of form concepts are possible. The type 2 form concept allows easy compensation for settlement of the yard's foundation, but requires hydraulic jacks able to cope with the weight of the entire outer form, the support platform and the support beams.

19.5.2 Permanent work aspects

In addition to comments given in Section II, Chapter 13 and Section III, Chapter 15, it is important to note that a prerequisite for using incremental launching for concrete decks is that the deck has to be cast in segments of similar and constant outer geometry, both in terms of cross section and in terms of horizontal and vertical alignment. This condition can be relaxed with regard to the girder depth and the deck width, but is a must for the soffit and the lower part of the webs.

19.5.3 Fabrication of reinforcement bar cages (rebar cages)

The rebar cages for incrementally launched bridge decks can be produced in situ or can be partially preassembled in elements. A full preassembly is not practical, since the centric PT tendons are generally coupled only at every alternate segment joint. That means that 50% of the centric PT strands protrude from the previously cast segment and need to be inserted into the ducts located in the rebar cage. Whether the rebar cages are produced in situ or are partially preassembled depends mainly on the available cycle time.

19.5.4 Placing and curing of concrete

Different methods are used to place the concrete, depending on the cross section, the size of segment to be cast and whether the production facility is covered or not. These methods are

- use of portal cranes or a crawler crane to place the concrete delivered by ready-mix trucks with buckets (not suitable if the full cross section is cast in one pour)
- pumping of the concrete with stationary placing booms into the segment.

19.5.5 Cycle time

A typical casting cycle for a 25-m-long single-cell box segment of 12 m width, which is cast in one pour with form concept type 1, is shown in Table 19.9. Its centric prestressing consists of straight tendons in the deck and bottom slab, which are anchored at alternate segment joints. The reinforcement is therefore preassembled in three elements: bottom rebars of bottom slab plus web reinforcement (element 1); bottom rebars of deck slab (element 2); top rebars of deck slab (element 3). The deck is launched with strand moving units, which are anchored to the deck by needle beams passing through openings in the deck and bottom slab. The layout of the casting yard is schematically shown in Figure 19.26. It is important to note that the cycle shown is for illustration only and that actual cycles need to be worked out for individual projects, considering project-specific features and constraints.

19.5.6 Risks

The list of risks given in Table 19.10 is by no means comprehensive, but addresses risks which are particular to the production of an incrementally launched deck. For a specific project, a main contractor needs to carry out a full risk assessment of the operations performed, considering the actual features of the yard, the formwork, the plant and the detailed construction method adopted.

Table 19.9 Cycle for production of a 25-m-long ILM segment (single-cell box girder, deck width of 12 m) cast in one pour; rebar cage preassembled in three elements; one 10-hour shift per day

	Duration: h	Day 1		Day 2		Day 3		Day 4		Day 5		Day 6		Day 7		Day 8		Day 9	
Launching of deck by segment length	4 to 6	X																	
Moving inner soffit form up into casting position; closing of outer side form; cleaning of outer form and application of form oil	2		X																
Placing of preassembled rebar cage element 1 and loose rebars at segment joint; rolling-out of strands protruding from bottom slab of previously cast segment and insertion into ducts; fixing and coupling of ducts; installation of tendon couplers; installation of lower part of stop end; fixity of tendon anchorages and coupling of ducts; placing of bottom slab's top rebars	15			X	X	X													
Launching of inner tunnel form parked in previously cast segment N – 1 into mould	5						X												
Resetting of launching needle beam and pulling strands from segment N – 2 to segment N – 1 (going through openings in top and bottom slab)	5							X											
Fixing reinforcement for diaphragm, if applicable, at joint between segment N – 1 and N including formwork towards segment N – 1	20							X	X	X	X								
Placing preassembled rebar cage element 2 and loose reinforcement at segment joint; rolling-out of strands protruding from top slab of previously cast segment and insertion into ducts; fixing and coupling of ducts; installation of tendon couplers; installation of upper part of stop end; fixity of tendon anchorages and coupling of ducts	15							X	X	X									
Placing preassembled rebar cage element 3 and loose rebars at segment joint	15										X	X	X						
Pouring and finishing of concrete (180 m^3)	10													X	X				
Curing	36 (including 2 nights)															X	X		
Removal of stop ends; installation of tendons' anchor blocks; opening of outer side forms, stripping of inner form, lowering of inner soffit form	5																	X	
Stressing of centrical PT tendons	2																		X

Figure 19.26 Schematic casting yard layout (elevation) (VSL International)

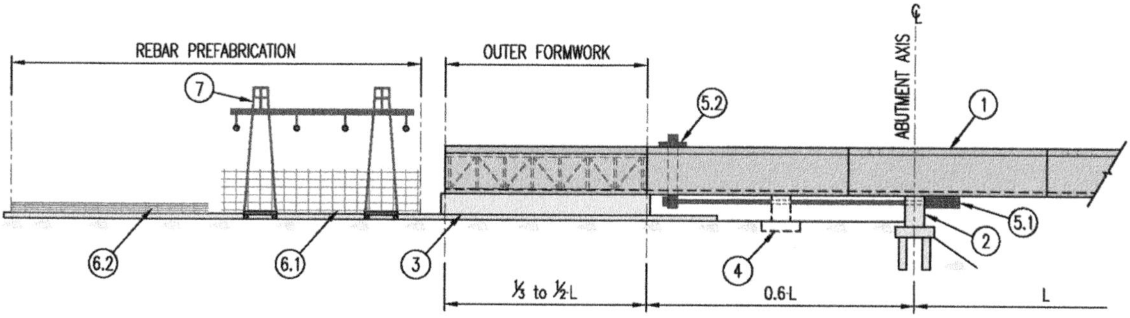

1. Deck
2. Abutment
3. Base slab
4. Temporary support (for support of nose and first launch)

5. Launching set-up
5.1. Pulling jacks
5.2. Needle beams

6. Rebar prefabrication
6.1. Rebar element 1
6.2. Rebar elements 2 and 3

19.6. Handling of precast elements

This subsection covers the handling of precast elements in the precasting yard. As explained above, some formwork systems also include the movement of the precast elements into an intermediate storage location. This is the case for precast segments when precast by the short-line method. The transfer cart moves the segment from the casting cell into the conjugate position and into a parking place. With the formwork concept 2 for the production of fully precast spans (PS), presented in subsection 19.4 ('Production of fully precast spans'), rebar cages and PS are slid with the soffit form along the production line until the completed span reaches an intermediate storage location. For incrementally launched bridges, the deck is longitudinally moved out of the formwork by the deck launching system.

For all other movements and formwork systems, mainly the two plants used to move precast elements in the casting yard are

- portal cranes running on fixed rails
- straddle carriers.

For both types of machines, the hook height is controlled by the need to be able to move an element over stored elements. Precast segments, for example, may be stored in up to three vertical layers.

For details about various ways in which lifting gears can be connected to precast concrete elements refer to chapter III-18.

19.6.1 Portal cranes

If the space available for precasting and storing segments is limited, portal cranes are used to handle the segments in the yard. They run on fixed rails and have gauges generally between 10 and about 30 m and even up to 40 m.

Most portal cranes consist of

- two framed pairs of legs connected at their base to articulated single-line wheel bogies (forming two points of support per framed pair of legs)

Table 19.10 Risk assessment for the production of incrementally launched decks

Activity	Risk	Mitigating measures
Work on top of segment in mould	Vertical access and work at height: climbing up and down, work performed at height (at >2 to 3 m above ground)	Access staircases, working platforms along side forms
Lowering of cages and closing of inner form	Cage does not fit into casting mould	Use of template when preassembling partial rebar cages
	Excessive distortion of cage during handling	Use of rigid lifting frame with multiple suspension points
	Damaged and dislocated concrete spacers	Use of engineered concrete spacers effectively tied to rebars
Inner tunnel form	Squashing workers during operation, operation in confined space during stripping	Acoustic signals during operation and visual controls; ventilation
	Concreting in confined space	Ventilation
Stressing of tendons	Failing strands/wedges/compression fittings with subsequent release of built-up energy in anchorage components and jacks	Exclusion zone behind stressing jacks
	Slippage of strands in tendon couplers	Use of quality PT systems and PT installation work carried out by PT specialist company
Movement of portal cranes	Interference with people and goods crossing portal crane's rail path	Warning lights and acoustic signals
Movement of trucks along beds	Collision with people working around casting moulds	Dedicated driveways and dedicated access walkways

- a pair of main beams (the portal beam)
- the upper cross beam (spanning between the pair of main beams) with the winch and the segment lifting beam
- the power pack with distribution lines
- the access provisions.

To cope smoothly with variations in the rail's gauge, portal cranes are almost always three-pin frames in the transverse direction. That is achieved by designing a flexural hinge connection between the portal beam and one framed pair of legs and a fixed connection to the other framed pair of legs.

In temporary storage yards power supply is by a combustion-engine-driven generator either attached to the bottom of one leg if space permits or on top of the portal beam. For an example of a typical portal crane refer to Figure 19.27.

Figure 19.27 Typical portal crane (DEAL)

19.6.2 Straddle carriers

If there is enough space available for precasting and storing segments, portal cranes with rubber wheels can be used to handle the precast elements in the yard. These kind of portal cranes are called straddle carriers. The yard area required is bigger than when using portal cranes. No foundation for rails has to be built. But the areas where the straddle carriers are operating needs to be compacted sufficiently.

The wheels (or pair of wheels) of straddle carriers are articulated about their vertical axis. Straddle carriers have therefore extremely small turning radii and the wheels can operate in any direction against the upper part of the carrier.

Straddle carriers are commonly used in ports for handling containers. Unlike straddle carriers used in ports, which are equipped with two winches and four suspension points, straddle carriers built for casting yards to handle precast segments have a single winch with a single lifting point. Such a straddle carrier is shown in Figure 19.28. It consists of

- rubber wheel bogies
- a rigid portal consisting of two A-type framed legs and a single or a pair of main beams (the portal beam)
- the upper cross beam/trolley with the winch and the segment lifting beam
- the power pack with distribution lines
- the access provisions.

Precast beams or fully precast spans (PS) are moved around in casting yards with a pair of straddle carriers. In the case of heavy PS, such straddle carriers are basically portal beams fixed onto two A-type framed legs, each of which is sitting on top of an SPMT (Self-Propelled Modular Transporter). For more details of SPMTs, refer to Section II, Chapter 14. For illustration refer to Figure 19.29.

Figure 19.28 Typical purpose-built straddle carrier for handling of precast segments in casting yards (DEAL)

Figure 19.29 Use of a pair of straddle carriers to handle precast beams and fully Precast Spans (PS) (DEAL)

Handling of U-beams

Handling of fully Precast Span (PS)

REFERENCES

Figure 19.1: Meyer M (2009) 'State of the Art Bridge Construction – The Challenge to choose the right method', IABSE Symposium BKK.

Figures 19.2 to 19.11, 19.13, 19.14, 19.18 to 19.26: VSL International

Figures 19.12, 19.15 to 19.17: adapted from Florida Department of Transportation's (FDOT's) Segmental Manual, 1989

Figures 19.27 to 19.29: DEAL

Section IV

Recommendations for the Industry

Bridge Deck Erection Equipment
ISBN 978-0-7277-6193-4

https://doi.org/10.1680/bdee.61934.439

Chapter 20
Recommendations for the industry

20.1. Introduction

The members of the IABSE Working Group 6 believe the bridge construction community may find a list of recommendations useful in moving BDE selection, design, fabrication and operation tasks forward positively. The expectation is that a better understanding of the ways in which BDE work will make bridge-building safer, faster and cheaper. This section lists the group's recommendations.

It is considered important that all the stakeholders within a project have involvement as a team from the very first stages of the project to get the best outcome for the project. The interaction between the permanent bridge design and the BDE is crucial. The bridge designer's input is therefore of great importance. The form of contract is not of vital significance to the successful use of the BDE on the project, as has been demonstrated in many projects that have used a variety of contract forms in the past. However, we believe that the more recent forms of contract that encourage a 'team' approach to the construction operation help to create an atmosphere of positive cooperation between the stakeholders. This is not always the case with the more traditional forms of contract where the owner's team are more remote from the operations on the site.

BDE design, peer review, checking, fabrication and operation are all highly specialised techniques and the success of a project in which BDE plays a part will be highly dependent on getting suitably experienced and competent companies involved. BDE work is certainly not a situation in which 'cheapest is best' is wisely applied. Consideration has been given to 'approving' or 'accrediting' suitably experienced individuals that work in the BDE field. The authors recommend that an Accreditation System is created whereby suitably experienced designers, peer reviewers and supervising operatives are registered. Such a scheme will make it easier for clients and project management teams to select key people to take charge of BDE, knowing they have the experience necessary to ensure their BDE works in the way their project demands.

There are various key actions which should be required by the owner or the Main Contractor (MC) that will help to achieve safe and efficient operation of a BDE.

- Use only competent and experienced designers and specialised subcontractors, equipment suppliers and operators on the site.
- Keep the BDE operating systems as simple as possible.
- Make a risk assessment of all operations from assembly to dismantling and review the results on a regular basis (an example of a typical risk assessment sheet is given in Section I, Chapter 4).
- Make procedures as repetitive as possible.
- Use checklists (an example of a checklist is given in Section I, Chapter 4).
- Respect structural engineering fundamentals: lateral and longitudinal restraint (horizontal fixity), equilibrium and stability.

Table 20.1 Delivery process, stakeholders and key documents for a Design–Bid–Build type of delivery model

Process Step	Key Document	The Stakeholders								
		Owner (RE)	Consultant permanent work	Main Contractor	Construction Engineering	BDE				
						Supply	Operation	3rd Party Inspection	Independent Design Check	Peer Review
Means & Method										
■ Evaluation of Different Options		I	L							
■ Selection of Best Option		I	L							
■ Confirmation of Need for BDE		I	L							
Selection of BDE										
■ Performance Criteria	Preliminary	R	R	L	I					
■ Call for Tender	Basic Methodology			L	I	I				
■ Evaluation, Clarifications, LOI				L	I	I				
■ Confirmation of Basic Design						L				
■ Finalisation of Basic Methodology	Equipment Description Report (including design basis)	R	R	R	I	L				
Award of Supply Contract				L						

Procurement										
■ Design & Detailing	Calculations, GA, Fabrication Dwgs, Operation Manuals	R	I	I		L			R	R
■ Fabrication	QC Records					L		R		
■ Trial Assemblies	Trial Assembly Report			I		L	I	R		
■ Delivery & Assembly on site				I		L	I			
Operation										
■ Preparation of Detailed Method statements and Check Lists	Method Statements, Check Lists			I	L		I			
■ Commissioning	Commissioning Report			R		L	I			
■ Load and Deflection Tests	Load and Deflection Test Reports					L	I	I	I	
■ Training – Initial Training – Subsequent Training/Retraining	Training Certificates					L	I L			
■ Operation of BDE							L			
■ Regular Maintenance							L			
■ Regular Technical Audits				L			I			
Post Operation										
■ Dismantling	Method Statement			L			I			
■ Storage	Storage Plan			L			(I)			
■ Sale	Technical Documentation (as built)			L						

- Establish a culture of checking and reviewing all activities from start to finish.
- Do not try to improve the BDE system without full reference back to the BDE's designer.

20.2. Project delivery

Tables 20.1 and 20.2 show the delivery process for a BDE on a project for a Design–Bid–Build type of delivery model (Table 20.1) and for a Design–Build type of delivery model (Table 20.2). The tables indicate the stakeholders' involvement during the different process steps and the key documents which have to be produced. A stakeholder is either leading (L), reviewing (R) or kept informed (I).

20.3. Preparation

20.3.1 Definition of general erection means and methods

After analysis of all the project-specific criteria and constraints and based on the identified erection method, the main contractor develops **the general construction means and methods**. These methods and sequences have to be in line with the erection sequences as assumed by the bridge designer. The **construction schedule** and an **initial risk assessment** are also developed at this stage.

20.3.2 Selection of BDE

Subsequently, if the decision is that a BDE shall be used, the **type of BDE** is selected and its **kinematics** can be defined. The **number of BDEs** needed for the project is also established.

The **BDE's performance specifications** are defined, based on studies carried out when working out the general erection means and methods. However, the performance specifications always have project-specific aspects and should therefore be tailor-made for the project.

The BDE's performance specifications refer to the construction means and method and sequences selected. They address:

- safety
- the permanent work interfaces
- other project-specific requirements, if any
- the target cycle time
- the crew size
- the degree of mechanisation
- the access requirements
- the control systems
- the assembly, commissioning and relocation durations.

The contractor clearly identifies the 'base model' functionality as well as the desired 'upgraded options'.

All this information is summarised in a report (**Basic Methodology Report**).

20.3.3 Call for tender and award

Unless the MC owns the BDE, a tender for supply of BDE is called. The **Basic Methodology Report** is the technical part of the tender documentation sent out to the bidders. The bids are evaluated to assess whether proposed BDE complies with the defined performance criteria.

After technical and contractual clarifications between the MC and the bidders, a supplier is shortlisted and a letter of intent is issued. The selected supplier now has to be given enough time to develop the basic design of the BDE in order to confirm its basic methodology and operational characteristics. The supplier is requested to summarise the results of the study in a draft version of the **Equipment Description Report**. The issues that need to be addressed in this report are listed fully in Section I, Chapter 3 and reference must be made to that chapter, but the key elements are:

- description of the equipment
- **design basis** reflecting specific characteristics of the BDE (loading, load cases, load factors, dynamic factors, friction coefficients, geometrical imperfection limits, etc.) and the design, quality and safety codes proposed to be used
- dimensions and weights of the BDE and its elements
- performance characteristics
- components' operational characteristics (capacities, speed, etc.)
- detailed launching sequences/kinematics (covering all controlling cases), including reactions on the structure (location, contract area, magnitude)
- load and deflection test
- access provisions on BDE (shown on drawings)
- anticipated details of cycles and crew sizes for different types of typical operations
- deliverables
- fabrication information
- delivery.

It is critical that the Equipment Description Report includes a clause requiring the confirmation of the equipment weight, reactions on the structure and anticipated structural behaviour of the BDE. These verifications can be performed during the load and deflection test and commissioning phases of the project. Any discrepancies with the design documents have to be addressed and resolved prior to the commencement of the erection operations.

Detailed performance specifications are a key factor for the successful procurement of the BDE that will meet the project requirements and the stakeholders' expectations. Further information is needed in addition to that listed above, relating to commercial matters, and this is listed fully in Section I, Chapter 3.

Enough time has to be allocated for all the stakeholders to properly review, comment on and revise the Equipment Description Report. Of particular importance is the verification that the permanent work is able to cope with the loading imposed by the BDE during its operation. An interactive process should follow to optimise the BDE concept and fine-tune the details. At the conclusion of this period, the BDE Equipment Description Report is accepted, the details are frozen, and the scope of works and supply are defined. A final version of the Equipment Description Report is issued before the supplier can start with the final design and detailing of the BDE. It goes without saying that the more detailed the Basic Methodology Report is, the faster it will be to finalise the BDE Equipment Description Report.

20.4. Procurement

20.4.1 Design and Design Checks

There are seven distinctly different design activities on a project in which BDE is involved. These are:

1. permanent bridge design
2. permanent bridge design independent checking
3. BDE design

Table 20.2 Delivery process, stakeholders and key documents for a Design–Build type of delivery model

Process Step	Key Document	The Stakeholders								
		Owner (RE)	Consultant permanent work	Main Contractor	Construction Engineering	BDE				
						Supply	Operation	3rd Party Inspection	Independent Design Check	Peer Review
Means & Method										
■ Evaluation of Different Options			I	L	L	(I)				
■ Selection of Best Option			I	L	I	(I)				
■ Confirmation of Need for BDE			I	L	I	(I)				
Selection of BDE										
■ Performance Criteria	Preliminary		R	L	I					
■ Call for Tender	Basic Methodology			L	I	I				
■ Evaluation, Clarifications, LOI				L	I	I				
■ Confirmation of Basic Design						L				
■ Finalisation of Basic Methodology	Equipment Description Report (including design basis)	R	R	R	I	L				
Award of Supply Contract				L						

Procurement										
■ Design & Detailing	Calculations, GA, Fabrication Dwgs, Operation Manuals		I	I		L			R	R
■ Fabrication	QC Records					L		R		
■ Trial Assemblies	Trial Assembly Report			I		L	I	R		
■ Delivery & Assembly on site				I		L	I			
Operation										
■ Preparation of Detailed Method statements and Check Lists	Method Statements, Check Lists			I	L		I			
■ Commissioning	Commissioning Report			R		L	I			
■ Load and Deflection Tests	Load and Deflection Test Reports					L	I		I	
■ Training – Initial Training – Subsequent Training/ Retraining	Training Certificates					L	I L			
■ Operation of BDE							L			
■ Regular Maintenance							L			
■ Regular Technical Audits				L			I		I	
Post Operation										
■ Dismantling	Method Statement			L			I			
■ Storage	Storage Plan			L			(I)			
■ Sale	Technical Documentation (as built)			L						

4 check of the effect that BDE has on the permanent works and the stage-by-stage analysis/design verification
5 BDE independent peer review *('sanity check')*
6 BDE independent detailed design check
7 'Professional Engineer (PE) certification' to comply with local regulations.

Upon completion of the fabrication drawings and the Operation Manual, the design and detailing of the BDE should always be subjected to an independent detailed design check and a peer review by an expert. Such checks and reviews can be under the supplier's scope of work, as long as they are performed by designers and experts not directly involved in the design and detailing of the BDE.

The design check must be made from base drawings and the Operation Manual. A check that only covers the arithmetic of the design calculations is inadequate.

Detailed descriptions of the process and scope of peer review and the independent design check are to be found in Section III, Chapter 16.

In addition, it is important to have a process to ensure that the BDE was fabricated and assembled in accordance with the specifications. This requires an independent inspection of the fabrication and overseeing the trial assembly in the fabrication shop. It is best done by the independent design checker.

While in many cases design activities 1 and 2 may take place well in advance of the other design activities, it is considered beneficial to the project to have all the other designers involved from as early as possible. Matters relating to the programme for the various design activities, and the local 'PE' certification, are dealt with in greater detail in Section III, Chapter 16.

20.4.2 Fabrication, trial assembly and delivery

The fabrication specification for a BDE should be at least as high as for permanent works. The loads for which it will be designed are real and stresses within the structure are deliberately kept high so as to achieve the power-to-weight ratio demanded by good BDE design. It is possible that the BDE may not be subject to fatigue effects. That is a matter for the designer to check. The BDE drawings must specify the required fabrication quality that is appropriate for the product and the fabricator must comply with those requirements.

It may be possible to accept a lower degree of paint treatment for an item of BDE than would be demanded for permanent works. However, experience shows that the cost of reasonable paint treatment is worth spending for several reasons. More detailed discussion on this is in Section I, Chapter 3.

Objectives of the trial assembly are:

- verification that mating components match
- verification that equipment components are physically fit and function.

It is not essential to trial-assemble the whole BDE, although, if reasonably possible, it is an excellent check and gives great confidence to all the stakeholders. Experience shows that the involvement of operating staff during trial assembly or even earlier creates a climate of ownership among the site staff and increases the level of knowledge about the BDE.

20.5. Operation

The equipment supplier has to provide the contractor with a comprehensive Operation Manual tailor-made for the equipment, also addressing in detail the risks connected with the equipment's use. Refer to Section I, Chapter 4 and the risk sections in each chapter of Section II.

The party in charge of the operation of the BDE – the main contractor or specialist subcontractor – needs to have a qualified team familiar with detailed planning of the erection operations and capable of producing all the work package documentation needed for the safe and efficient operation of the BDE in the project. For details about preparation of detailed method statements and checklists refer to Section I, Chapter 3.

20.5.1 Commissioning

Commissioning of the BDE can be a complex and expensive stage of the project and should not be underestimated in any of its aspects. An exact erection sequence for all the assemblies and sub-assemblies has to be developed by the supplier, while the actual assembly plan has to be developed by the site personnel, considering in detail the actual local constraints, the available lifting devices and environmental conditions.

If, due to any reason, field modifications are required, these modifications can only be performed after they are reviewed and approved by all the relevant stakeholders, including the BDE designer.

Prior to the start of the commissioning activities, the BDE needs to be checked and certified to ensure that it was properly assembled in accordance with assembly documents.

Before the start of any construction operation, the equipment needs to be load and deflection tested and certified according to approved procedures. For details of recommended load and deflection tests refer to Chapter 3. If a BDE is dismantled and reassembled on the same project, load and deflection tests have to be repeated.

20.5.2 Operation of a BDE and training of operating crew

The BDE should be operated only by trained and certified crews.

From the planning of site work, it is important to implement a project-specific training and certification programme that involves all the team members who will be associated with the operations of the BDE. The training plan must be maintained throughout the duration of the project.

It is beneficial to perform occasional independent audits by qualified individuals who are familiar with the BDE operations to ensure proper adherence to operational procedures and documentation of the BDE operations.

20.5.3 Maintenance and spare parts

A BDE is a complex machine that requires regular inspections and maintenance. This work needs to be performed as recommended in the Operation and Maintenance Manuals by qualified and trained mechanics and electricians. The qualification of personnel for this role is based on training and expertise in the specific field. Getting these personnel involved during the trial assembly helps to build up their level of understanding of the specific BDE.

Regular inspections and maintenance will drastically reduce the severity and duration of work stoppage due to equipment breakdowns and accidents. These activities should be defined and planned in accordance with the Operation and Maintenance Manuals of the BDE.

An appropriate set of spare parts and consumables should always be available on site. The supplier of the equipment should define the minimum requirements and the management should arrange to have it available on site from the commissioning stage.

20.6. Post operation

At the completion of the project, the equipment should be dismantled as per the approved work plan and stored properly. Care should be taken to protect the electrical, mechanical and hydraulic components from exposure to the elements. Electrical and electronic components can suffer severe deterioration and damage if not stored in climate-controlled facilities. All the available documentation, including fabrication records, QA/QC paperwork, load and deflection testing, shop drawings, calculations, checks, certifications, modification, upgrades, accident records, inspections, maintenance, etc. should be kept in a safe place.

Bridge Deck Erection Equipment
ISBN 978-0-7277-6193-4

https://doi.org/10.1680/bdee.61934.449

Index

Page numbers in *italics* refer to figures and photographs. Page numbers followed by a t refer to tables.

Printed and bound by CPI Group (UK) Ltd, Croydon, CR0 4YY
15/09/2026
14962492-0001